Controlled/Living
Radical Polymerization

ACS SYMPOSIUM SERIES **768**

Controlled/Living Radical Polymerization

Progress in ATRP, NMP, and RAFT

Krzysztof Matyjaszewski, EDITOR

Carnegie Mellon University

American Chemical Society, Washington, DC

Library of Congress Cataloging-in-Publication Data

Controlled/living radical polymerization : progress in ATRP, NMP, and RAFT / Krzysztof Matyjaszewski, editor.

    p. cm.—(ACS symposium series, ISSN 0097–6156 ; 768)

  Includes bibliographical references and index.

  ISBN 0–8412–3707–7

  1. Polymerization—Congresses. 2. Free radical reactions—Congresses.

  I. Matyjaszewski, K. (Krzysztof), 1950– . II. Series.

QD281..P656   2000
547´.28—dc21   00–32254

The paper used in this publication meets the minimum requirements of American National Standard for Information Sciences—Permanence of Paper for Printed Library Materials, ANSI Z39.48–1984.

PRINTED IN THE UNITED STATES OF AMERICA

Foreword

The ACS Symposium Series was first published in 1974 to provide a mechanism for publishing symposia quickly in book form. The purpose of the series is to publish timely, comprehensive books developed from ACS sponsored symposia based on current scientific research. Occasionally, books are developed from symposia sponsored by other organizations when the topic is of keen interest to the chemistry audience.

Before agreeing to publish a book, the proposed table of contents is reviewed for appropriate and comprehensive coverage and for interest to the audience. Some papers may be excluded in order to better focus the book; others may be added to provide comprehensiveness. When appropriate, overview or introductory chapters are added. Drafts of chapters are peer-reviewed prior to final acceptance or rejection, and manuscripts are prepared in camera-ready format.

As a rule, only original research papers and original review papers are included in the volumes. Verbatim reproductions of previously published papers are not accepted.

ACS BOOKS DEPARTMENT

Contents

ADVANCES IN CONVENTIONAL AND CONTROLLED
RADICAL POLYMERIZATION

NITROXIDE-MEDIATED POLYMERIZATION

ATOM TRANSFER RADICAL POLYMERIZATION

ix

INDEXES

Preface

Radical polymerization has again become one of the hottest areas of research. This originates with the old advantages of a radical process including a large number of monomers that can be copolymerized to form a nearly infinite number of different polymers under relatively undemanding conditions but also in the new opportunities provided by controlled/living systems. Although termination reactions cannot be fully suppressed, their proportion is strongly limited, enabling preparation of polymers with precisely controlled molecular weights, low polydispersities, novel architectures, compositions, and functionalities.

This book comprises both topical reviews and specialists' contributions presented at the American Chemical Society (ACS) Symposium on "Controlled Radical Polymerization" held in New Orleans, August 22–24, 1999. It was a sequel to the previous ACS Symposium held in San Francisco in 1997, which was summarized in the ACS Symposium Volume 685, *Controlled Radical Polymerization*. The first seven chapters of the present volume provide a general introduction to conventional and controlled radical polymerizations and describe approaches toward stereocontrolled systems. The next four chapters are focused on recent advances in nitroxide-mediated polymerization (NMP). Atom transfer radical polymerization (ATRP) is described in the following eight chapters. The next three chapters refer to the application of reversible (degenerative) transfer systems, with the most notable examples being reversible addition fragmentation chain transfer (RAFT) and catalytic chain transfer (CCC) systems. All of these techniques provide access to new materials that may lead to new commercial applications, as discussed in the remaining eight chapters.

Papers published in this book demonstrate that radical polymerization is becoming a preferred method for making well-defined polymers, both in academic and industrial laboratories. Many new controlled radical systems have been discovered; substantial progress has been achieved in the understanding of these reactions and in quantitative measurements of the rates, equilibria, and concentrations of the involved species. Comprehensive studies of controlled/living radical polymerization involve a wide range of scientific disciplines, including theoretical–computational chemistry, kinetics, physical organic chemistry, organic synthesis, and organometallic–coordination chemistry, as well as polymer synthesis and characterization, materials science, and so on. The facile synthesis of many new polymeric materials calls for their precise characterization. This should lead to the comprehensive structure–property relationship which may lead to a foundation for new materials for specific applications.

This book is addressed to chemists interested in radical processes and especially in controlled/living radical polymerization. It provides an introduction to the field and summarizes the most recent progress in ATRP, NMP, and RAFT systems.

The financial support for the symposium from the following organizations is gratefully acknowledged: ACS Division of Polymer Chemistry, Inc., ACS Petroleum Research Foundation, Aristech, BASF, Bayer, BF Goodrich, DuPont, Eastman Chemicals, W. R. Gore, ICI, Lubrizol, Mitsui Chemicals, National Starch, Polaroid, PPG Industries, Procter & Gamble, Rohm and Haas, and Solvay.

KRZYSZTOF MATYJASZEWSKI
Department of Chemistry
Carnegie Mellon University
4400 Fifth Avenue
Pittsburgh, PA 15213

ADVANCES IN CONVENTIONAL AND CONTROLLED RADICAL POLYMERIZATION

Chapter 1

Comparison and Classification
of Controlled/Living Radical Polymerizations

Krzysztof Matyjaszewski

Department of Chemistry, Carnegie Mellon University,
4400 Fifth Avenue, Pittsburgh, PA 15213

Controlled/living radical polymerization employs the principle of equilibration between growing free radicals and various types of dormant species. There are several approaches to achieve good control over molecular weights, polydispersities and end functionalities in these systems. They can be classified depending on the mechanism and chemistry of the equilibration/exchange process, as well as on the structure of the dormant species. Some of them are catalyzed and some are not, some of them exhibit the persistent radical effect and some do not. Currently three methods appear to be most efficient and may lead to commercial applications: nitroxide mediated polymerization, atom transfer radical polymerization and degenerative transfer processes. Their relative advantages and limitations are compared herein.

INTRODUCTION

There are several approaches to controlling free radical polymerization by suppressing the contribution of chain breaking reactions and assuring quantitative initiation.*(1)* All of these approaches employ dynamic equilibration between growing free radicals and various types of dormant species. These reactions are described as controlled radical polymerizations (CRP) or controlled/living radical polymerizations rather than as true living radical polymerizations, due to the presence of unavoidable termination, which is intrinsically incompatible with the concept of living polymerizations.*(2)* It is feasible to imagine that in the future a real living radical polymerization will be invented in which termination will be absent or at least the termination rate coefficients will be strongly reduced in comparison with conventional free radical polymerization. This could be accomplished by selective complexation of the free radical or by carrying out the reaction in a confined space.

The exchange process is at the very core of the CRP methods and can be approached in several ways depending on the structure of the dormant and deactivating species, the presence of the catalyst and the particular chemistry and mechanism of the exchange. Currently, three methods appear to be most efficient and can be successfully applied to a large number of monomers: stable free radical polymerization (SFRP), best represented by nitroxide mediated polymerization (NMP), metal catalyzed atom transfer radical polymerization (ATRP), and reversible addition-fragmentation chain transfer (RAFT) along with other degenerative transfer processes. Their relative advantages and limitations will also be discussed.

CLASSIFICATION OF CRPS

Operationally, all CRPs employ the principle of dynamic equilibration between dormant and active species. The position and dynamics of this equilibrium define the observed rates as well as affect molecular weights and polydispersities of the formed polymers. It is possible to group CRPs into several categories, depending on the chemistry of exchange and structure of the dormant species. Although it may be simpler to divide CRPs based on the structure of the dormant species, the mechanistic classification may be more appropriate, since it enables better correlation of the rates, molecular weights and polydispersities of the obtained polymers with the concentration of the involved reagents. Thus, mechanistically, CRPs can be classified into four different cases:

$$\text{\textasciitilde}P_n\text{-}X \quad \underset{k_d}{\overset{k_a}{\rightleftharpoons}} \quad \text{\textasciitilde}P_n^\circ + X^\circ \tag{1}$$

$$\text{\textasciitilde}P_n\text{-}X + Y \quad \underset{k_d}{\overset{k_a}{\rightleftharpoons}} \quad \text{\textasciitilde}P_n^\circ + X\text{-}Y^\circ \tag{2}$$

$$\{\text{\textasciitilde}P_n\text{-}Z\}^\circ \quad \underset{k_d}{\overset{k_a}{\rightleftharpoons}} \quad \text{\textasciitilde}P_n^\circ + Z \tag{3}$$

$$\text{\textasciitilde}P_n\text{-}X + P_m^\circ\text{\textasciitilde} \quad \underset{k_d}{\overset{k_a}{\rightleftharpoons}} \quad \text{\textasciitilde}P_n^\circ + X\text{-}P_m\text{\textasciitilde} \tag{4}$$

In all of these cases dormant, non-propagating species are reversibly activated with the rate constant of activation (k_a) to form the active species, P_n^*, which react with monomers, M, with the propagation rate constant, k_p. The propagating radicals can be deactivated with the rate constant of deactivation, k_d, or can terminate with other growing radicals with the termination rate constant, k_t. Because in all CRPs the concentration of radicals is kept very low, termination can sometimes be neglected since it does not significantly affect polymer properties. However, since in cases 1-3, each act of termination generates radical traps (X*, XY* or Z, respectively) irreversibly, termination may have some effect on kinetics.

It must be stressed that some of these methods have been successfully applied also in the organic synthesis with atom transfer radical addition, and degenerative transfer with alkyl iodides and xanthates being perhaps the best examples.*(3,4)*

Case 1 is best exemplified by nitroxide mediated polymerization in the presence of TEMPO,*(5,6)* and bulky acyclic nitroxides, *(7,8)* triazolinyl radicals,*(9)* some bulky organic radicals, e.g. trityl*(10,11)*, photolabile C-S bonds *(12)* and organometallic species.*(13,14)*

Case 2 is a subset of Case 1, since it is based on the catalyzed, reversible cleavage of the covalent bond in the dormant species via a redox process. Because the key step in controlling the polymerization is atom (or group) transfer between growing chains and a catalyst, this process was named atom transfer radical polymerization (ATRP) and is catalyzed by various Ru, Cu, Fe and other transition metal derivatives. *(15-19)*

Case 3 has not yet been as successful as the above two systems. This process involves the reversible formation of persistent radicals (PR) by reaction of the growing radicals with a species containing an even number of electrons. PRs do not react with each other or with monomer. Here, the role of a reversible radical trap may be played by phosphites,*(20a)* some reactive, but non-polymerizable alkenes, such as tetrathiofulvalenes*(21)* or stilbene *(22)* and also metal compounds with an even number of electrons.*(23)* It seems that some reversible addition fragmentation chain transfer (RAFT) systems may also behave in a similar way, especially when a strong decrease in the polymerization rate is observed. (cf. also case 4).

Case 4 is based on a thermodynamically neutral exchange process between a growing radical, present at very low concentrations, and dormant species, present at much higher concentrations (at least three or four orders of magnitude). This degenerative transfer process can employ alkyl iodides, *(24)* unsaturated methacrylate esters,*(25)* or thioesters.*(26)* The latter two processes operate via addition-fragmentation chemistry with the redundant or degenerate loops which force radicals in the desired direction. A recent paper on the application of xanthates in organic synthesis "Riding the tiger:

Using degeneracy to tame wild radical processes", describes well the essence of these reactions.*(4)*

In the first three cases the equilibrium is strongly shifted towards dormant non-propagating species and rates depend on the concentration of the persistent radical species, such as X^*.

The only difference between the first two cases is the bimolecular activation and catalyzed nature of the second system. The $X-Y^*$ species plays the role of PR. Case 3 is nearly identical to case 1, the only difference is that in case 3, the dormant species has an odd number of electrons and the radical trap is not a radical. Thus, formally the role of persistent radical is played by a non-radical species. Currently this is the least explored chemistry. Case 4 is very different from the other three. There is no PR in it, the equilibrium constant equals $K=1$ ($k_a=k_d=k_{exch}$) and rates should conform closely to conventional radical systems with 1/2 external order in radical initiator. However, some rate reduction may be expected if the transition state in which atom or group is being transferred from one to another chain is strongly stabilized and becomes an intermediate product.

Thus, each system has a specific dependence of the polymerization rates, molecular weights and polydispersities on conversion and concentrations of the involved reagents. They are summarized in Table 1 below.

Table 1. Typical Kinetic Laws and Dependence of Degree of Polymerization and Polydispersity on Conversion for Controlled Radical Polymerization Systems

No	Example	Kinetic Law	DP_n	Polydispersity
1	NMP/TEMPO	$R_p=k_pK_{eq}[I]_o/[X^*]$	$DP_n=\Delta[M]/[I]_o$	$M_w/M_n=1+(2/p-1)(k_p[I]_o)/(k_d[X^*])$
2	ATRP	$R_p=k_pK_{eq}[I]_o[Y]/[XY^*]$	$DP_n=\Delta[M]/[I]_o$	$M_w/M_n=1+(2/p-1)(k_p[I]_o)/(k_d[XY^*])$
3	?	$R_p=k_pK_{eq}[I]_o/[Z]$	$DP_n=\Delta[M]/[I]_o$	$M_w/M_n=1+(2/p-1)(k_p[I]_o)/(k_d[Z])$
4	Deg. Tr./RAFT	$R_p=k_pfk_d(k_t)^{-1/2}([I]_o)^{1/2}$	$DP_n=\Delta[M]/([TA]_o+\Delta[[I])$	$M_w/M_n=1+(2/p-1)(k_p/k_d)$

Polymerization rates in the first three systems depend on the free radical propagation rate constant, k_p, on the equilibrium constant, $K_{eq}=k_a/k_d$, on the concentration of initiator, $[I]_o$, and inversely on the concentration of a radical trap ($[X^*]$, $[XY^*]$, or $[Z]$). It has to be realized that the concentration of the trap, i.e. PR, changes during the polymerization, which may lead to some peculiar kinetic behavior. In case 2, rates increase with the concentration of activator, $[Y]$. For degenerative transfer, the rate law should be similar to that for the conventional radical process, but some retardation may also be observed.

Degrees of polymerization (DP) increase linearly with conversion and depend inversely on the initiator concentration, provided that the initiation is fast. DP may be higher than predicted, if primary radicals terminate too fast before addition to monomer, resulting in lower initiator efficiency. In degenerative transfer, the number of chains is the sum of the used transfer agent and consumed initiator.

Polydispersities, M_w/M_n, decrease with conversion, p, and chain length, meaning that they depend inversely on the concentration of the used initiator. Polydispersities also depend on the ratio of rates of propagation and deactivation, the latter being a product of the rate constant of deactivation and the concentration of the radical trap. However, in case 4, the exchange (deactivation) proceeds by the reaction with the transfer agent which also (predominantly) defines chain length. Therefore, in this unique system, polydispersities should not depend on the chain length provided that the proportion of the used initiator is small in comparison with that of the transfer agent. *(27)*

It has to be noted, that if the equilibrium constants are very low and/or concentrations of the reagents are small enough, then the overall rate may be defined by the rate of the thermal process and give apparent zero orders with respect to the involved reagents. Also, when the PRs are formed spontaneously, the deviations from both internal and external first orders with respect to monomer, initiator and catalysts may be observed.

It is possible to further mechanistically subdivide the four systems into several additional categories, depending on the molecularity of exchange, catalytic nature of the process, occurrence of PRE (persistent radical effect) and the particular chemistry involved in the exchange.

The molecularity of the exchange. The dormant species can be converted to active species either via an unimolecular (cases 1,3) or a bimolecular process (cases 2,4).

The catalytic nature of activation of dormant species. In most CRPs, the activation process is spontaneous and can be accelerated only by increasing the reaction temperatures. The only example of a catalytic process is ATRP, in which rates depend on the concentration of transition metals in their lower oxidation state, e.g. case 2. The rates in degenerative transfer systems, including RAFT, depend on the concentration of the radical initiator. Thus, although it should not be considered as a catalyst, since it is irreversibly consumed, it plays a role similar to catalyst in affecting the polymerization rate.

Persistent radical effect. (PRE). *(28)* Most CRP systems conform to the PRE model: (cases 1,2,3). Systems based on degenerative transfer, including addition-fragmentation chemistry do NOT conform to PRE model. Hypervalent iodine-based radicals and sterically hindered tertiary radicals are considered only as short-lived intermediates (case 4). However, under some conditions they may be present at higher concentrations and retardation is plausible. These systems may behave partially like those in case 3.

Mechanism of equilibration. There are three general mechanisms for the reversible formation of propagating radicals:

4.A. Bond scission-recombination. This most common mechanism includes all unimolecular activation systems (cases 1,3). One of the most common cases, TEMPO mediated polymerization of styrene, is shown below:

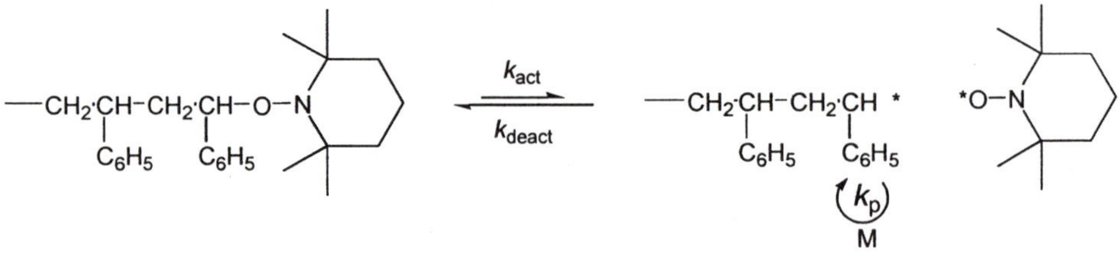

(5)

4.B. Reversible atom or group transfer. This describes both iodine mediated degenerative transfer systems and ATRP. (cases 4 and 2)

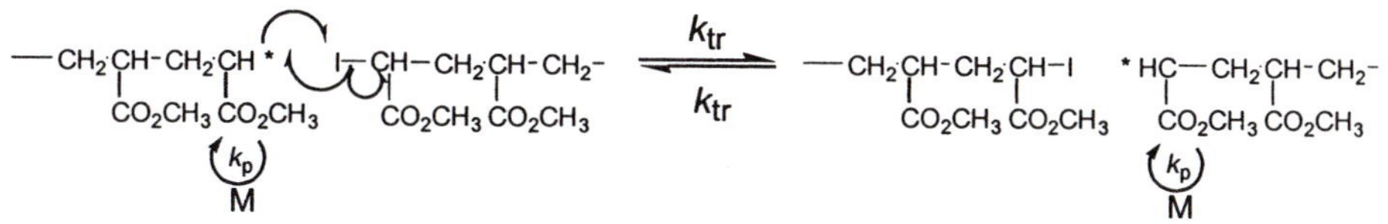

(6)

ATRP is schematically shown below together with the relevant rate constants for the styrene polymerization in bulk at 110 °C catalyzed by CuBr complexed by 4,4'-di (5-nonyl)-2,2'-bipyridine (dNbpy):

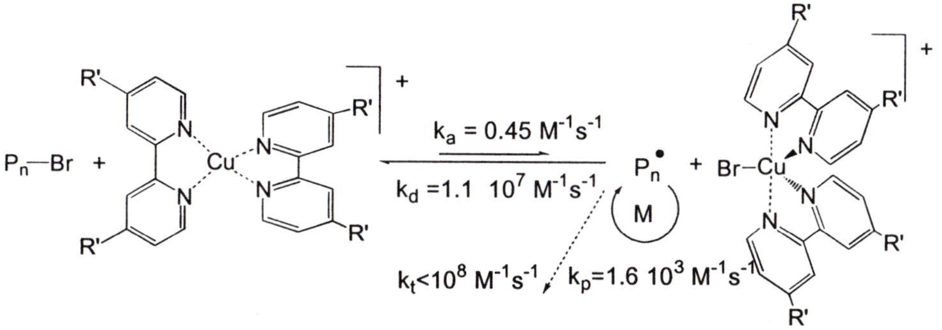

(7)

4.C. Addition-fragmentation. This type of bimolecular exchange process employs reversible addition of the radicals to compounds with non-polymerizable multiple bonds (case 4 and potentially case 3). It may include methacrylate oligomers prepared by catalytic chain transfer:

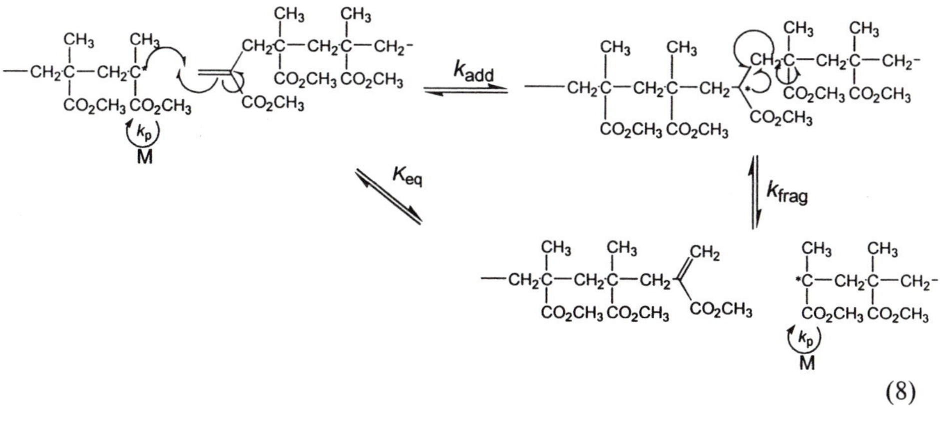

(8)

but also dithioesters in polymerization of styrenes and various (meth)acrylates

$$(9)$$

In principle, addition-fragmentation chemistry could be also described as a reversible group transfer process.

Structure of dormant species. Nevertheless, perhaps the most transparent criterion for classification of various CRP is one based on the structure of the dormant species and the type of bond which is reversibly broken. However, mechanistically, CRP will always follow one of the four examples shown as cases 1-4 and summarized in Table 1 above.

5.A. C-C bonds. The triphenylmethyl radical was used first as a reversible trap in methyl methacrylate polymerization (case 1). *(10,11)* Polymerization of methacrylate macromonomers mediated by the addition-fragmentation chemistry also proceeds by C-C bond cleavage (case 4, eq. 8) *(25)*

5.B. C-N bonds. To this category belong triazolinyl *(9)* and verdazyl derivatives.*(29)* (case 1). The most efficient appears triazolinyl radical with all phenyl substituents as shown in eq. 10; apparently a derivative with spiroindanyl substituent at 3,3'-position is less efficient due to its higher thermal stability.

$$(10)$$

5.C. C-O bonds. The classic example here is alkoxyamine, with TEMPO being the most common and most thoroughly studied nitroxide. (case 1). However, some new more bulky nitroxides provide better control and can be extended to a larger number of monomers *(7,20b,8)*:

$$(11)$$

5.D. C-S bonds. Iniferters based on dithiocarbamates were the first species with photochemically labile C-S bonds (predominantly case 1) *(12)*. Dithioesters act as very efficient degenerative transfer reagents in the RAFT process with transfer coefficients more than 100 times higher than the dithiocarbamates. (case 4, eq. 9) *(26,30)*

5.E. C-Halogen. ATRP. (case 2). Degenerative transfer with alkyl iodides has been successfully used for styrene, acrylates and vinyl acetate. (case 4, eqs. 6 and 7)

5.F. C-Metal bonds. Homolytic cleavage of organometallic bonds has been successfully applied in polymerization of acrylates and styrene using cobalt derivatives, especially complexed by sterically demanding porphyrins. (case 1) It is also feasible that some transition metals with even numbers of electrons (high spin?) may reversibly react with organic radicals to form paramagnetic organometallic species (case 3)

5.G. Other systems. In principle, other systems can be simply formed by extrapolating group 5A to P, As, and Sb, group 6A to Se and Te. One can also use heavier halogens and other transition metals. In addition, there have been some reports on polymerization in the presence of B- and Al- species, however, more detailed studies in the latter systems excluded the presence of hypervalent aluminum radicals.*(31)*

BASIC REQUIREMENTS FOR CONTROLLED/LIVING RADICAL POLYMERIZATION

Perhaps the most important difference between conventional and controlled/living radical polymerizations is the lifetime of an average chain. In conventional systems chain is born, grows and dies within approximately 1s; during this time it is not possible to perform any synthetic manipulations such as chain extension, end functionalization, variation of monomer feed, etc. On the other hand, under controlled/living conditions, chains grow during several hours enabling precise macromolecular engineering. Long lifetime of chains requires sufficiently low concentrations of macroradicals but also sufficiently high concentration of propagating chains. This can be accomplished via equilibration between active free radicals and various dormant chains which should exchange rapidly. In many CRP systems concentration of radicals is similar to those in conventional systems, meaning that overall polymerization rates are similar and the absolute concentration of terminated chains are similar. However, the proportion of the terminated chains in CRP systems is much lower (usually <10%). Finally, because most chain are in the dormant state and radicals remain at low concentration, it is possible to initiate all chains at approximately same time and achieve growth in several directions making well-defined stars, bottle brush structures, but also end functional polymers and block copolymers.

Thus, there are three general requirements for CRP*(1)*. They include:

fast exchange between dormant species and growing radicals

a small proportion of chains involved in chain breaking reactions

fast and quantitative initiation

Exchange. It is perhaps more instructive to first analyze the dynamics of exchange since all CRPs employ this concept in one or the other way. Here, one of the most important parameters is the number of monomer molecules added to the growing chain during one activation step. If this number is very low, preferentially less than unity, then polymers with low polydispersities and good control of molecular weights are obtained. If this number is very large, exceeding the targeted degree of polymerization, then poor control is obtained and the polymerization may resemble a conventional, thermally or redox initiated system. As discussed before, the evolution of polydispersities in systems based on degenerative transfer is different from those in other CRP systems which obey the persistent radical effect (PRE). In the PRE systems, represented by nitroxide mediated polymerization and atom transfer radical polymerization, polydispersities depend on the ratio of rate constants of propagation to deactivation (k_p/k_d), the concentration of deactivating persistent radicals ($[D]$), the chain length, which is reciprocally correlated with the concentration of the initiator ($[RX]_o$), and conversion, p.*(32)* For sufficiently high polymerization degrees, the following equation operates:

$$M_w/M_n = 1 + ([RX]_o\, k_p)/([D]\, k_d)\, (2/p - 1)$$

(12)

It is also possible to express the polydispersities as a function of the reaction time and rate constant of activation, k_a (for a bimolecular activation like in ATRP, a product of the concentration of activator and k_a must be used)*(33)*:

$$M_w/M_n = 1 + 2/(k_a\, t)$$

(13)

Both equations can be applied leading to similar results, since $([RX]_o\, k_p)/([D]\, k_d) = [P^*]k_p/k_a = R_p/[M]\, k_a$, especially under pseudo-stationary conditions, when the numbers of activation and deactivation steps are equal. However, kinetic data (as a function of time) may sometimes be affected by adventitious impurities, the mixing process, etc. Usually the evolution of molecular weights and polydispersities with conversion is more reliable and easier to reproduce. However, the first method requires either direct measurements of the concentration of the deactivator, by e.g. EPR, or application of a known excess of deactivator.

Figures 1 and 2 illustrate the evolution of polydispersities for several simulated ATRP systems.

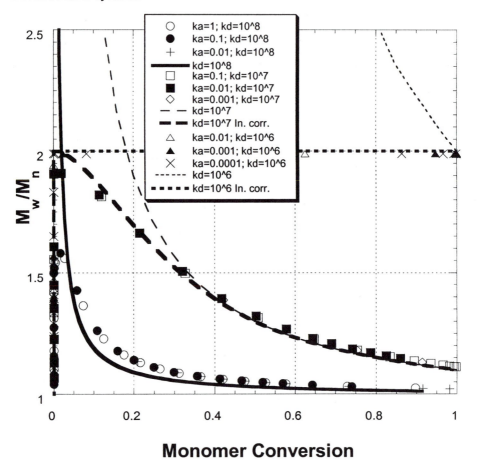

Monomer Conversion

Fig. 1. Evolution of polydispersities with conversion. Simulations were performed for the systems resembling polymerization of acrylates initiated by alkyl bromides and catalyzed by CuBr/2dNbpy at 80 °C using the following concentrations: $[M]_o=10$ M, $[RX]_o=0.1$ M, $[Cu(I)]_o=0.1$ M, $[Cu(II)]_o= 0.05$ M and rate constants: $k_p=5 \ 10^4 \ M^{-1} \ s^{-1}$, $k_t= 10^7 \ M^{-1} \ s^{-1}$, $k_t^o=5 \ 10^9 \ M^{-1} \ s^{-1}$, and the following values of the exponents of the rate constants of activation (k_a) and deactivation (k_d) =1,8; -1,8; -2,8; .1,7; -2,7; -3,7; -2,6; -3,6, -4,6. In addition, the theoretically predicted evolution of polydispersities is plotted according to eq. 12 and 14 (thicker lines).

A correlation of simulated results (symbols refer to different values of k_a and k_d) with dependence predicted by Eq. 12 is better than with Eq. 13 (cf. Figures 1 and 2, respectively).

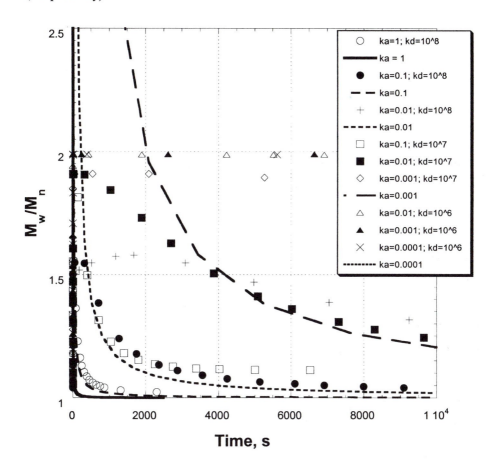

Fig. 2. Evolution of polydispersities with time. Simulations were performed for the systems identical to those in Fig.1. Lines represent theoretically predicted evolution of polydispersities according to eq. 13.

It can be also noticed in Fig. 1, that polydispersities simulated by Predici software are generally higher than those predicted by eq. 12. This is due to termination reactions which broaden MWD. Also, the fit is poor when deactivation becomes slow. This is due to incomplete initiation and requires using a more complex equation to account for the unreacted initiator *(27)* as shown by eq. 14 and the thick lines in Fig. 1, which fit the results better than the thin lines derived from eq. 12:

$$M_w/M_n = \{1 + ([RX]_o k_p)/([D] k_d) (2/p - 1)\}\{1-(1-p)^{[([D] k_d)/([RX]_o k_p)]}\}$$

$$(14)$$

Nevertheless at higher conversion there is relatively good agreement with values calculated based on the simpler eq. 12. It is worth noting that polydispersities depend only on values of k_d and, for the same k_d, are not affected by k_a (Fig. 1)

Although there is a rough correlation between evolution of MWD with time and values of k_a, small differences remain, due to different values of k_d. (Fig. 2) It is difficult to plot all data and compare them, because of the dramatic differences in the polymerization rates (approaching five orders of magnitude).

Termination. In all controlled and conventional radical polymerization systems termination does occur. In the first approximation, the concentration of terminated chains depends only on polymerization rate (i.e. concentration of propagating radicals). However, the ratio of the concentrations of terminated chains to the sum of the active and dormant chain concentrations can be dramatically different in both systems. In conventional radical polymerization, nearly all chains are terminated (except a tiny fraction, ppm, of actually growing chains). In controlled systems, the proportion of terminated chains can be very small, usually less then 10%. It is important to know precisely the proportion of these chains since they can not be functionalize or extended in the form of block copolymers, etc. It has to be remembered that the overall polydispersity may not always indicate a low proportion of terminated chains.

Rate coefficients of radical termination are very large, often diffusion controlled ($k_t > 10^7$ mol^{-1} L s^{-1}). They decrease with the chain length and also with the viscosity of the reaction medium.*(34)* Up to now, attempts to further reduce these rate coefficients have not been fully successful. However, a nearly terminationless radical polymerization of methyl methacrylate (MMA) in the presence of large amounts of phosphoric acids has been reported, presumably due to the complexation of PMMA radical by the acid and reduction of termination rate coefficients due to steric or electronic reasons. *(35)* Termination between two growing radical chains may also be absent in polymerizations carried out in a confined space such as in cavities of zeolites or other three dimensional compounds.*(36,37)* Such approaches, based on the physical or chemical stabilization of growing radicals, may lead to a living radical polymerization in which termination would be absent. However, in these early systems, polymers with uncontrolled molecular weights and high polydispersities were formed due to slow initiation and/or variation of the propagation rate with the steric/electronic effects in the surrounding of growing chains. It would be interesting to combine both approaches which would rely on the dynamic equilibration between active and dormant chains and quantitative initiation together with the physical or chemical stabilization of the growing radicals. The latter systems may potentially have chemoselectivities and stereoselectivities which are different from those of conventional systems and of currently available controlled/living free radical polymerizations, possibly leading to new not yet accessible polymers.

When the proportion of terminated chains is relatively low, it depends on the ratio k_t/k_p^2, increases with the polymerization rate ($d\ln[M]/dt$), with conversion ($\Delta\ln[M]$) and with the chain length (reciprocally correlated with the concentration of initiator, $1/[RX]_o$):

$$\%_t = (k_t/k_p^2)(d\ln[M]/dt)\,\Delta\ln[M]/[RX]_o$$

(15)

Figure 3 illustrates more precise evolution of the proportion of terminated chains with the progress of the reaction in polymerization of MMA in diphenyl ether at 80 °C, using a 100:1:1 ratio of $[M]_o:[RBr]_o:[CuBr/2dNbpy]_o$, whereas Figure 4 shows the final simulated MWD of the obtained polymers.

The simulations were carried out using Predici software. *(38)* A simplified scheme was used, in which the rate constants of activation and deactivation were the same for the low and high molar mass species; addition of initiating radical to monomer was 10 times faster than propagation; rate coefficients of termination of initiating radicals were diffusion controlled ($k_t^o= 10^9$ $M^{-1}s^{-1}$), but those for the polymeric species did not vary with the chain length and were identical for recombination and disproportionation.

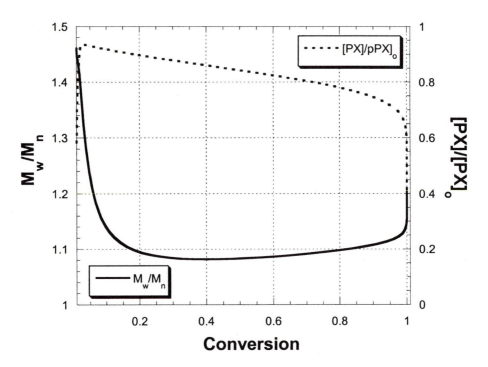

Fig. 3. Predici simulations of the evolution of polydispersities and the proportion of preserved end functionalities in polymerization of $[MMA]_o= 5$ M , initiated by [ethyl 2-bromoisobutyrate]$_o$=0.05 M and catalyzed by $[CuBr/2dNbpy]_o$=0.05 M at 80 ^oC in diphenyl ether. Simulations were carried out using the following rate constants: $k_p=10^3$ $M^{-1}s^{-1}$, $k_t=10^7$ $M^{-1}s^{-1}$, $k_d=10^7$ $M^{-1}s^{-1}$, $k_a=10$ $M^{-1}s^{-1}$.

Therefore, when approximately 60 % of the chains were deactivated, half of the dead chains formed a polymer with double molecular weight, resulting in an increase of the overall polydispersities but only to the value of M_w/M_n=1.2. It has to be recognized that such polymers with relatively low polydispersities will be very inefficient precursors for block copolymerization due to significant loss of functionalities. Thus, lower temperatures and/or lower concentrations of catalyst are needed to reduce the concentration of free radicals in similar systems. Another possibility would be to limit polymerization to lower conversion or target lower molecular weight polymer.

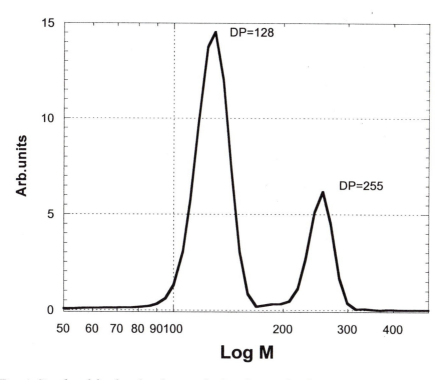

Fig. 4 Simulated final molecular weight distribution for the system described in Fig. 3.

Initiation. Some attributes of controlled/living polymerizations include the linear evolution of molecular weights with conversion, and the formation of polymers with low polydispersities and high degrees of end functionality as well as efficient block copolymerization. All of these features require quantitative initiation. In CRP systems this is translated into a constant or nearly constant number of growing chains. In NMP and ATRP, it is necessary to select an initiator which resembles the structure of dormant chains or has somewhat higher intrinsic reactivity. If the initiator generates too many radicals, termination may reduce its efficiency. Here, a notable exception are sulfonyl halides which do not dimerize fast enough and provide fast initiation especially for methacrylate monomers.*(39)*

In most controlled/living systems the highest polydispersities for slow initiation systems are defined by the Gold distribution ($M_w/M_n<1.35$). However, in CRP, slow initiation is also accompanied by a lower concentration of spontaneously formed PR and therefore slower exchange. As a result polymers with much higher polydispersities can be obtained.

Systems based on degenerative transfer, including RAFT, belong to a special case where initiation is very slow and, in fact, never completed. However, a large majority of the chains are formed early on by an efficient transfer agent which is usually present in a large excess over the initiator. Therefore, a small amount of continuously generated chains does not contribute significantly to the overall number of chains and good control over molecular weights and polydispersities can be achieved.

COMPARISON OF NMP, ATRP AND RAFT

Currently, the three most efficient methods of controlling radical polymerization with a commercial promise include NMP, ATRP and RAFT. Each of these methods has advantages as well as limitations. The relative advantages and limitations of each method can be discussed by comparing four typical features. They include the range of polymerizable monomers, typical reaction conditions (temperature, time, sensitivity to impurities,etc.), the nature of transferable end groups/atoms and various additives such as catalysts, accelerators, etc.

NMP was originally best represented by TEMPO. TEMPO could be successfully applied only to styrene and copolymers due to its relatively small equilibrium constant. For acrylates and methacrylates it yields either unsaturated oligomers/polymers or poorly controlled polymers. Some improvement was reported for polymerization of acrylates when nitroxides with a lower thermal stability (4-oxy TEMPO) were used. Most recently, much better results were obtained with more sterically bulky nitroxides, such as phosphonate ester derivatives. Due to a higher equilibrium constant, polymerization of acrylates and acrylamides was successful. When TEMPO is used for NMP, reactions are carried out in bulk and at high temperatures (>120 °C) because they are inherently slow, although some acceleration was described in the presence of sugars, acyl compounds and acids. Polymerizations in solution, dispersion and emulsion have been also reported. As initiators, either a combination of conventional initiator (BPO, AIBN) and free nitroxide (1.3:1 ratio is apparently the best) or preformed alkoxyamines can be used. End groups in the dormant species are alkoxyamines and some unsaturated species formed by abstraction of β-H atoms or other inactive groups formed by side reactions, e.g., termination. Conversion of alkoxyamines to other useful functionalities generally requires radical reactions. For example, via reaction with $FeCl_3$, halogen terminal groups have been successfully incorporated. Alkoxyamines are relatively expensive, generally difficult to remove from the chain end, not yet commercially available and, hence, need to be synthesized. However, the process typically does not require a catalyst and is carried out at elevated temperatures. The most promising new nitroxides provide easier thermal cleavage of alkoxyamines and can be applied to other monomers such as acrylates, but not yet methacrylates. The synthesis and evaluation of new nitroxides is probably the most interesting and important area of research for the future of NMP.

ATRP has been used successfully for the largest range of monomers, although the homopolymerization of vinyl acetate has not yet been successful. The careful selection of pH and initiating systems led recently to the controlled polymerization of methacrylic acid. ATRP has been carried out in bulk, solution, suspension, dispersion and emulsion at temperatures ranging from –20 °C to 130 °C. Some tolerance to oxygen and inhibitors has been reported in the presence of zero-valent metals, indicating that ATRP may be the most robust CRP method.*(40)* The catalyst must be available for the reaction and it should be sufficiently accessible in the reaction medium. The catalyst is based on a transition metal which regulates the polymerization rate and polydispersities. It can facilitate cross-propagation for the synthesis of difficult block copolymers(e.g. polyacrylates to polymethacrylates) and can scavenge some oxygen, but it should be removed or recycled from the final polymerization product. Some interesting results were obtained by using an immobilized catalyst system.*(41,42)* Perhaps the biggest advantage of ATRP is the inexpensive end group consisting of simple halogens. This is especially important for short chains due to the high proportion of the end groups. The terminal halogen can be easily displaced with other useful functionalities using S_N2, S_N1, radical processes and other chemistries.*(43)* There is a multitude of commercially available initiators and macroinitiators for ATRP. They include alkyl halides with α-phenyl, vinyl, carbonyl, cyano and multiple halogen atoms as well as any compound with a weak halogen – heteroatom bond, such as sulfonyl halides. The halogen from the chain end can be exchanged with that bound to the catalyst. Such an exchange process improves efficiency of crosspropagation from polyacrylates to e.g. polymethacrylates which can not be achieved by other CRP methods.*(44)* The future research directions in ATRP should address better removal/recycling (immobilization?) of the catalyst, the development of new catalytic systems to expand the range of monomers to less reactive ones (e.g. olefins) and elucidation of a comprehensive structure-reactivity relationship for alkyl halides and transition metal complexes.

Degenerative transfer systems, and RAFT in particular, can be potentially used for any radically polymerizable monomer. However, reactions of vinyl esters are apparently more difficult and RAFT of vinyl esters requires either very high temperatures (T>140 °C) or replacement of dithioesters by xanthates. It may be difficult to assure an efficient crosspropagation for some systems. In principle, all classic radical systems can be converted to RAFT or to another degenerative transfer systems in the presence of efficient transfer reagents. The end groups are either alkyl iodides, methacrylates or thioesters (RAFT). The latter are colored and can give some odor for low molar mass species and may require radical chemistry for removal and displacement. Dithioesters are not commercially available. Methacrylate oligomers formed by the catalytic chain transfer are efficient only in the polymerization of methacrylates. No catalyst is needed for degenerative transfer but, in fact, the role of the catalyst is played by the radical initiator. This also means that the initiator may incorporate some undesired end groups and the amount of termination is governed by the amount of the decomposed initiator. A potential disadvantage of degenerative transfer is that there is always a low molecular weight radical available for termination. By contrast, in the ATRP and NMP systems, at sufficient conversions only long chains exist, and, therefore, they terminate more slowly. The future research in DT and RAFT should address better (cheaper, less toxic, with less color and odor) transferable groups and methods for crosspropagation.

Table 2. Comparison of NMP, ATRP and Degenerative Transfer Systems

Feature	Systems		
	NMP	**ATRP**	**Deg. Transfer (RAFT)**
Monomers	-styrenes for TEMPO -also acrylates & acrylamides for new nitroxides -NO methacrylates	-nearly all monomers with activated double bonds -NO vinyl acetate	-nearly all monomers
Conditions	-elevated temp (>120 ^{o}C for TEMPO) -water-borne systems OK -sensitive to oxygen	-large range of temperatures (-30 to 150 ^{o}C) -water-borne systems OK -tolerance to O_2 and inhibitor with Mt^0	-elevated temperatures for less reactive monomers -water-borne systems OK -sensitive to oxygen
End Groups	alkoxyamines -requires radical chemistry for transformations -relatively expensive -thermally unstable	alkyl (pseudo)halides -either S_N, E or radical chemistry for transformations -inexpensive & available -thermally and photostable -halogen exchange for enhanced cross-propagation	dithioesters, iodides & methacrylates -radical chemistry for transformations (S_N for RI) -relatively expensive -thermally and photo less stable -color/odor
Additives	none -NMP may be accelerated with acyl compounds	transition metal catalyst -should be removed / recycled	conventional radical initiator -may decrease end functionality -may produce too many new chains

Table 2 above summarizes the main features of the three discussed systems. Thus, the main advantage of the nitroxide mediated system is the absence of any metal. TEMPO is applicable for styrene but new nitroxides are useful for other monomers. ATRP may be especially well suited for low molar mass functional polymers due to the cost of end groups and easier catalyst removal. It may be also very successful for the synthesis of "difficult" block copolymers and some special hybrids with end functionalities. It does require catalyst recycling, however, it is able to tolerate small amounts of oxygen, inhibitor and impurities. DT and especially RAFT should be successful for the polymerization of many less reactive monomers and for the preparation of high molecular weight polymers. Due to some limitations of the sulfur containing compounds, the search for new efficient transferable groups should be continued.

GENERAL FEATURES AND FUTURE PERSPECTIVE OF CRP

It appears that the above discussed CRP systems resemble in many cases conventional radical processes. They have chemoselectivities (reactivity ratios), regioselectivities (head-to-tail connectivity) and stereoselectivities (tacticities) and termination rate coefficients very similar to those of conventional systems.(45) Though extremely important, the only difference is that it is possible, by assuring fast initiation and by extending the time of building up the polymer chains from a fraction of a second to hours, to perform many synthetic manipulations, i.e. employ macromolecular engineering for each chain. It is possible to control chain topologies, functionalities and composition in a way similar to living ionic polymerizations.

However, it has to be recognized that the proportion of terminated chains and loss of functionalities increases with the chain length and concentration of growing radicals. Thus, to grow long chains, especially using monomers which propagate relatively slowly such as styrenes and methacrylates, will require long reaction times during which radical termination may be suppressed but other chain breaking reactions may become more prevalent.

Therefore, it should be recognized that at present CRPs could be most efficiently applied to the synthesis of polymers with low and moderate molecular weights, unless propagation rate constants are very high (e.g. acrylates). At the same time, there are so many interesting structures to be prepared by CRP that this technique should find commercial use and many new industrial applications in the very near future. It is possible to design novel surfactants, dispersants, lubricants, adhesives, gels, coatings, and many other materials which can only be prepared by controlled/living radical polymerization.

Acknowledgments. The financial support from NSF, EPA and ATRP Consortium Members is gratefully acknowledged.

References

1. Matyjaszewski, K. *Controlled Radical Polymerization*; Matyjaszewski, K., Ed.; ACS: Washington, D.C., 1998; Vol. 685.
2. Matyjaszewski, K.; Müller, A. H. E. *Polym. Prepr. (Am. Chem. Soc., Div. Polym. Chem.)* **1997**, *38(1)*, 6.
3. Curran, D. P. *Synthesis* **1988**, 489.
4. Quiclet-Sire, B.; Zard, S. Z. *Pure & Appl. Chem.* **1997**, *69*, 645.
5. Solomon, D. H.; Rizzardo, E.; Cacioli, P. , **1986**, U. S. Pat. 4, 581, 429.
6. Georges, M. K.; Veregin, R. P. N.; Kazmaier, P. M.; Hamer, G. K. *Macromolecules* **1993**, *26*, 2987.
7. Benoit, D.; Grimaldi, S.; Finet, J. P.; Tordo, P.; Fontanille, M.; Gnanou, Y. *ACS Symp. Ser.* **1998**, *685*, 225.
8. Benoit, D.; Chaplinski, V.; Braslau, R.; Hawker, C. J. *J. Am. Chem. Soc.* **1999**, *121*, 3904.
9. Steenbock, M.; Klapper, M.; Muellen, K.; Bauer, C.; Hubrich, M. *Macromolecules* **1998**, *31*, 5223.
10. Borsig, E.; Lazar, M.; Capla, M.; Florian, S. *Angew. Makromol. Chem.* **1969**, *9*, 89.
11. Braun, D. *Macromol. Symp.* **1996**, *111*, 63.
12. Otsu, T.; Yoshida, M. *Makromol. Chem. Rapid Commun.* **1982**, *3*, 127.
13. Wayland, B. B.; Poszmik, G.; Mukerjee, S. L.; Fryd, M. *J. Am. Chem. Soc.* **1994**, *116*, 7943.
14. Harwood, H. J.; Arvanitopoulos, L. D.; Greuel, M. P. *ACS Polymer Preprints* **1994**, *35(2)*, 549.

15. Kato, M.; Kamigaito, M.; Sawamoto, M.; Higashimura, T. *Macromolecules* **1995**, *28*, 1721.
16. Wang, J. S.; Matyjaszewski, K. *J. Am. Chem. Soc.* **1995**, *117*, 5614.
17. Matyjaszewski, K. *Chem. Eur. J.* **1999**, *5*, 3095 .
18. Patten, T. E.; Matyjaszewski, K. *Acc. Chem. Res.* **1999**, *32*, 895.
19. Patten, T. E.; Matyjaszewski, K. *Adv. Mater.* **1998**, *10*, 901.
20. a)Greszta, D.; D. Mardare; Matyjaszewski, K. *ACS Polym. Preprints* **1994**, *35(1)*, 466; b) Matyjaszewski, K. et al. *J. Phys. Org. Chem.* **1994**, *8*, 306
21. Steenbock, M.; Klapper, M.; Muellen, K.; Pinhal, N.; Hubrich, M. *Acta Polym.* **1996**, *47*, 276.
22. Harwood, H. J.; Christov, L.; Guo, M.; Holland, T. V.; Huckstep, A. Y.; Jones, D. H.; Medsker, R. E.; Rinaldi, P. L.; Soito, T.; Tung, D. S. *Macromol. Symp.* **1996**, *111*, 25.
23. Espenson, J. H. *Acc. Chem. Res.* **1992**, *25*, 222.
24. Matyjaszewski, K.; Gaynor, S.; Wang, J. S. *Macromolecules* **1995**, *28*, 2093.
25. Moad, C. L.; Moad, G.; Rizzardo, E.; Tang, S. H. *Macromolecules* **1996**, *29*, 7717.
26. Chiefari, J.; Chong, Y. K. B.; Ercole, F.; Krstina, J.; Jeffery, J.; Le, T. P. T.; Mayadunne, R. T. A.; Meijs, G. F.; Moad, C. L.; Moad, G.; Rizzardo, E.; Thang, S. H. *Macromolecules* **1998**, *31*, 5559.
27. Litvinienko, G.; Müller, A. H. E. *Macromolecules* **1997**, *30*, 1253.
28. Fischer, H. *J. Polym. Sci., Part A: Polym. Chem.* **1999**, *37*, 1885.
29. Kinoshita, M.; Miura, Y. *Makromol. Chem* **1969**, *124*, 211.
30. Chong, Y. K.; Le, T. P. T.; Moad, G.; Rizzardo, E.; Thang, S. H. *Macromolecules* **1999**, *32*, 2071.
31. Granel, C.; Jerome, R.; Teyssie, P.; Jasieczek, C. B.; Shooter, A. J.; Haddleton, D. M.; Hastings, J. J.; Gigmes, D.; Grimaldi, S.; Tordo, P.; Greszta, D.; Matyjaszewski, K. *Macromolecules* **1998**, *31*, 7133.
32. Matyjaszewski, K. *ACS Symp. Ser.* **1998**, *685*, 258.
33. Fukuda, T.; Goto, A.; Ohno, K.; Tsujii, Y. *ACS Symp. Ser.* **1998**, *685*, 180.
34. Shipp, D. A.; Matyjaszewski, K. *Macromolecules* **1999**, *32*, 2948.
35. Kabanov, V. A. *J. Polym. Sci., Polym. Symp.* **1975**, *50*, 71.
36. Farina, M.; Silvestro, G. D. *Chem. Comm.* **1976**, 842.
37. Ng, S. M.; Ogino, S.; Aida, T.; Koyano, K. A.; Tatsumi, T. *Macromol. Rapid Commun.* **1997**, *18*, 991.
38. Wulkow, M. *Macromol. Theory Simul.* **1996**, *5*, 393.
39. Percec, V.; Barboiu, B.; Kim, H. J. *J. Am. Chem. Soc.* **1998**, *120*, 305.
40. Matyjaszewski, K.; Coca, S.; Gaynor, S. G.; Wei, M.; Woodworth, B. E. *Macromolecules* **1998**, *31*, 5967.
41. Haddleton, D. M.; Kukulj, D.; Radigue, A. P. *Chem. Commun.* **1999**, 99.
42. Kickelbick, G.; Paik, H.-j.; Matyjaszewski, K. *Macromolecules* **1999**, *32*, 2941.
43. Matyjaszewski, K.; Coessens, V.; Nakagawa, Y.; Xia, J.; Qiu, J.; Gaynor, S.; Coca, S.; Jasieczek, C. *ACS Symp. Ser.* **1998**, *704*, 16.
44. Shipp, D. A.; Wang, J.-L.; Matyjaszewski, K. *Macromolecules* **1998**, *31*, 8005.
45. Matyjaszewski, K. *Macromolecules* **1998**, *31*, 4710.

Chapter 2

Kinetics of Living Radical Polymerization

Takeshi Fukuda and Atsushi Goto

Institute for Chemical Research, Kyoto University, Uji, Kyoto 611–0011, Japan

Activation processes and polymerization rates of several variants of living radical polymerization (LRP) are discussed. Despite the presence of side reactions such as termination and initiation, the products from LRP can have a low polydispersity, provided that the number of terminated chains is small compared to the number of potentially active chains. A large rate constant of activation, k_{act}, is another fundamental requisite for low polydispersities. Experimental studies on k_{act} have clarified the exact mechanisms of activation in several LRP systems. The magnitudes of k_{act} have been found to largely differ from system to system. Because of bimolecular termination, which is inevitable in LRP as well as in conventional radical polymerization, the time-conversion curves of LRP have several characteristic features depending on experimental conditions, such as the presence or absence of conventional initiation and/or an extra amount of stable free radicals or the like. A new analytical rate equation applicable to such general cases is presented.

Living (or controlled/"living") radical polymerization (LRP) has opened up a simple and versatile route to the synthesis of well-defined, low-polydispersity polymers with various architectures (*1-11*). The basic mechanism common to all the variants of LRP is the alternating activation-deactivation process depicted in Scheme 1: wherein a potentially active (dormant) species P-X is supposed to be activated to the polymer radical P° by thermal, photochemical, and/or chemical stimuli. In the presence of monomer M, P° will undergo propagation until it is deactivated to the dormant species P-X. This cycle is supposed to be repeated enough times to give every "living" chain an almost equal chance to grow. Here we define a "living" chain as either an active or a dormant chain with the quotation specifying the presence of the two states. In a practically useful system, it usually holds that $[P°]/[P-X] \leq 10^{-5}$, which means that a "living" chain spends most of its polymerization time in the dormant state. Examples of blocking agents X include sulfur compounds (*12-14*), stable nitroxides (*15-33*), iodine (*34-36*), halogens with a transition metal catalysis (*37-48*), transition metal complexes (*49-51*), and others (*52-57*).

It is instructive to consider "ideal LRP" in which reactions other than activation, deactivation, and propagation are absent and the P* concentration is constant. The polydispersity factor $Y (= M_w/M_n - 1)$ of this system is given by (*58*)

$$Y = w_A^2 Y_A + w_B^2 Y_B \qquad (1)$$
$$Y_B = F(C)y_n^{-1} + z_{n,B}^{-1} \qquad (2)$$

where Y_A is the Y value of the initiating adduct P_0-X, and Y_B, $z_{n,B}$, and w_B are the Y value, the number-average degree of polymerization, and the weight fraction of the propagated portion of the chains, respectively, $(w_A + w_B = 1)$; $F(C) = 2$ when the monomer concentration is constant, and $F(C) = (1-2C^{-1})\ln(1-C)$ for a batch system (C = conversion) (*59*); y_n is the average number of activation-deactivation cycles that a chain experiences during polymerization time t, and is related to the activation rate constant by

$$y_n = k_{act}t \qquad (3)$$

Any deviations from equations 1-3 are ascribed to the non-stationarity of [P*] and/or "side reactions" such as termination, *conventional* initiation, irreversible chain transfer and the degradation of the active chain-ends. Since these side reactions (excepting chain-end degradation) should be common to those in conventional radical polymerization, the activation rate constant k_{act} and polymerization rate R_p, which is proportional to [P*], are particularly important parameter/quantity to characterize a given LRP system. (The deactivation rate constant k_{deact}, another fundamental parameter, can usually be deduced by knowing k_{act} and R_p in the stationary state.)

In this paper, we summarize our studies on k_{act} and R_p. It includes some new data on k_{act} and a new analytical equation for R_p that were obtained after the submission of our recent review article for publication (*11*).

Rate constants of Activation and Activation Mechanisms

The k_{act} of a LRP system can be determined by the GPC methods in which the concentration of the initiating polymer adduct P_0-X or the polydispersity of the polymer components produced at an early stage of polymerization is followed by GPC (*58,60*).

There are three main mechanisms of reversible activation, which are (i) dissociation-combination, (ii) degenerative (exchanging) transfer and (iii) atom transfer (Scheme 1). The activation rate constant for each mechanism will take the form

$$k_{act} = k_d \qquad \text{(dissociation-combination)} \qquad (4)$$
$$k_{act} = k_{ex}[\text{P}^\bullet] = C_{ex}R_p/[\text{M}] \qquad \text{(degenerative transfer; } C_{ex} = k_{ex}/k_p) \qquad (5)$$
$$k_{act} = k_A[\text{A}] \qquad \text{(atom transfer)} \qquad (6)$$

where the rate constants k_d, k_{ex}, and k_A are defined in Scheme 1, $C_{ex} = k_{ex}/k_p$, R_p is the polymerization rate, and A is an activator like a transition metal complex. A LRP system is not necessarily driven by a single mechanism. Thus measurements of k_{act} as a function of [P˙], [A] or R_p help us understand mechanistic details of activation processes.

Such experiments have been carried out for the LRP systems **b**, **d**, **e** and **f** in Scheme 2. The results confirm that each of these systems is driven virtually by a single mechanism, which is the thermal dissociation for the nitroxide system **b** (*61*), the degenerative transfer for both the iodide system **d** (*62*) and the dithiocompound system **e** (*63*), and the atom transfer mechanism for the bromide/CuBr system **f** (*64*).

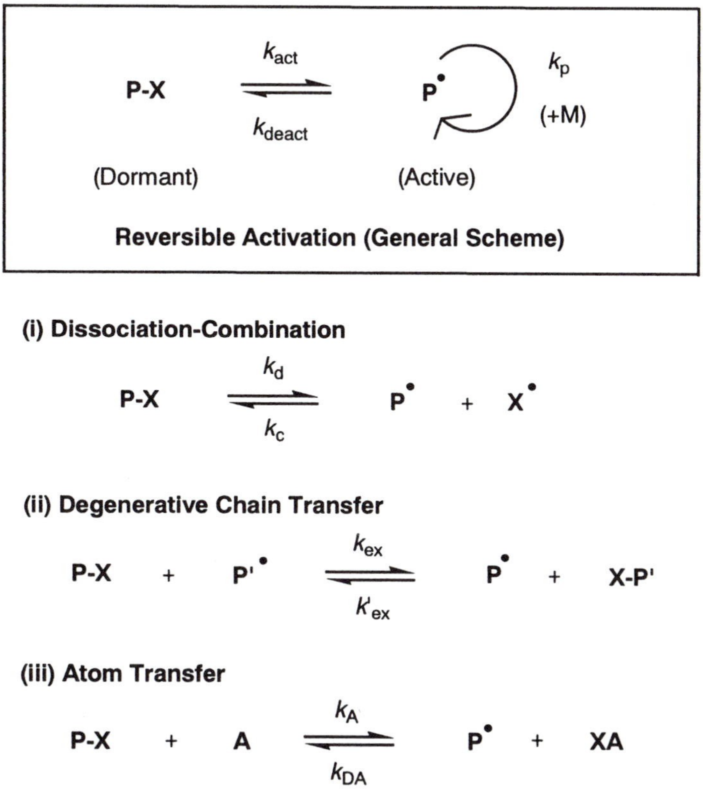

Scheme 1. A general scheme and the three main mechanisms (i, ii, and iii) of reversible activation.

Table 1. Activation Rate Constants

P-X	$10^3 \, k_{act} \, / \, s^{-1}$	$T \, / \, ^{\circ}C$	Ref.
1. PS-TEMPO	1.0	120	60
2. PS-DBN	42	120	65
3. PS-DEPN	11	120	65
4. PBA-DBN	1.0	120	66
5. PMA-Co/Porphyrin	$4.0 \pm 2.0^{b)}$	25	67
6. PS-I	$0.22^{a)}$	80	62
	$(C_{ex} = 3.6)$		
7. PS-SCSMe	$9.6^{a)}$	80	63
	$(C_{ex} = 160)$		
8. PS-SCSPh	$360 \pm 120^{a,b)}$	80	63
	$(C_{ex} = 6000 \pm 2000)$		
9. MMA-macromonomer	$0.013^{a)}$	80	68
	$(C_{ex} = 0.22)$		
10. PS-Br/CuBr	$23^{a)}$	110	64
	$(k_A = 0.45 \, M^{-1} \, s^{-1})$		

[a] Value approximately estimated for $R_p = 4.8 \times 10^{-4} \, M \, s^{-1}$. (Systems 1-5 are independent of R_p.)
[b] Preliminary results.

Table 1 lists the values of k_{act}, C_{ex}, and k_A for various systems (60,62-68). Since the k_{act}, for systems 6 through 10, is proportional to R_p, the listed values for those systems are referred to a standard value of $R_p = 4.8 \times 10^{-4} \, mol \, L^{-1} \, s^{-1}$, while the k_{act} for the other systems is independent of R_p. In comparison among the nitroxide systems 1 through 3, TEMPO, DBN and DEPN attached to the same polymer (PS) give large differences in k_{act}. The open-chain nitroxides DBN and DEPN give larger k_{act} than the ring-chain nitroxide TEMPO, in agreement with other indirect experimental (69) and theoretical (69,70) results. It is interesting to note that DBN gives a larger k_{act} than DEPN (at 120 °C). In comparison of the structures of the two nitroxides (Scheme 3), one may intuitively expect from an entropic point of view that DEPN would give a larger k_{act} than DBN, since the former has a much bulkier side group than the latter. However, Table 1 shows that it is not the case (at 120 °C). This suggests that the activation energy for PS-DEPN dissociation is larger than that for PS-DBN dissociation, even though the entropy of activation for the former reaction is larger than that for the latter. This in fact has been observed, as shown in Table 2 (65).

Table 1 also suggests that the DBN-mediated polymerization of *t*-butyl acrylate (BA) may be as well controlled as the TEMPO-mediated polymerization of styrene. Actually, however, the thermal degradation of the active chain-end of PBA-DBN

(a) "Iniferter" polymerization (Otsu et al., 1982[12])

PS-SCSNEt$_2$

(b) Nitroxide (Solomon et al., 1985[15]; Georges et al., 1993[16])

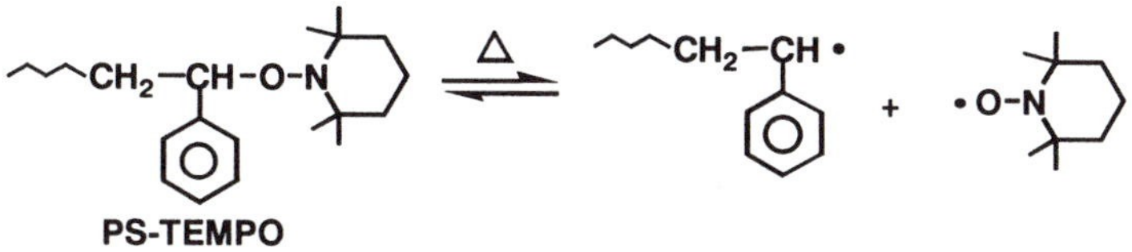

PS-TEMPO

(c) Cobalt/Porphyrin Complexes (Wayland et al., 1994[45])

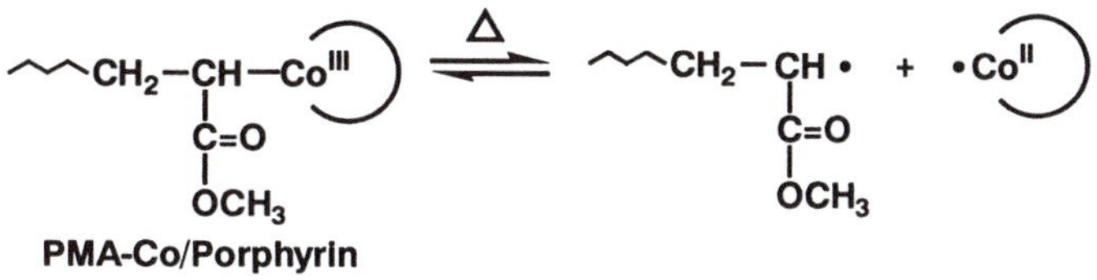

PMA-Co/Porphyrin

(d) "Iodine Transfer" (Tatemoto et al., 1991[34]; Sawamoto et al., 1994[35], Matyjaszewski et al., 1995[36])

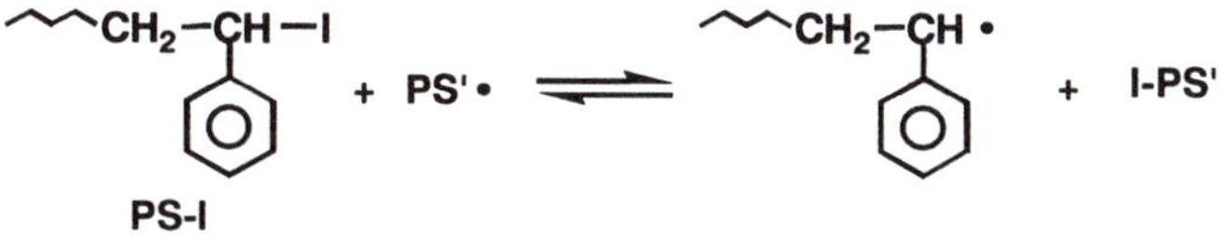

PS-I

(e) "RAFT" (Moad et al., 1995[49] & 1998[14])

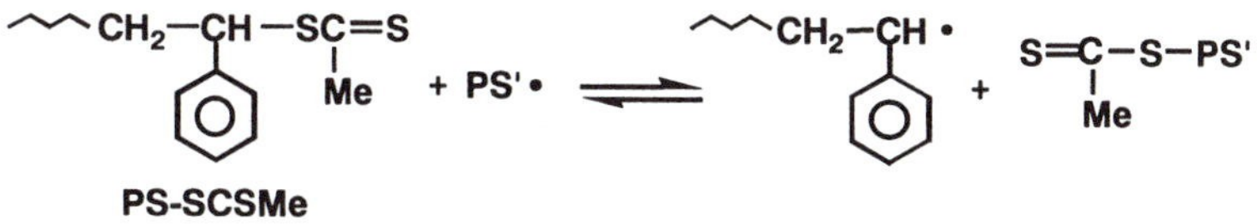

PS-SCSMe

(f) "ATRP" (Sawamoto et al., 1995[37]; Matyjaszewski et al., 1995[38])

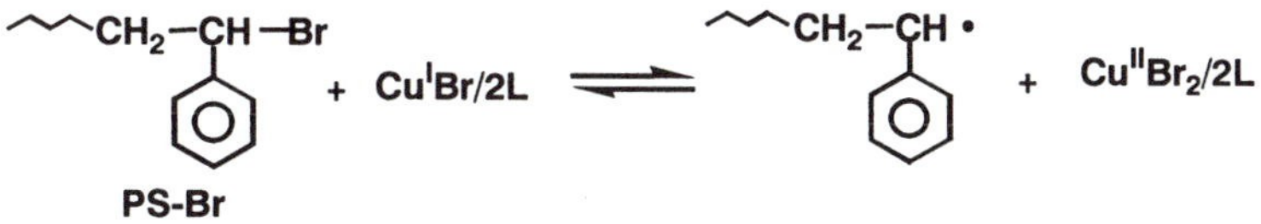

PS-Br

Scheme 2. Examples of LRP systems.

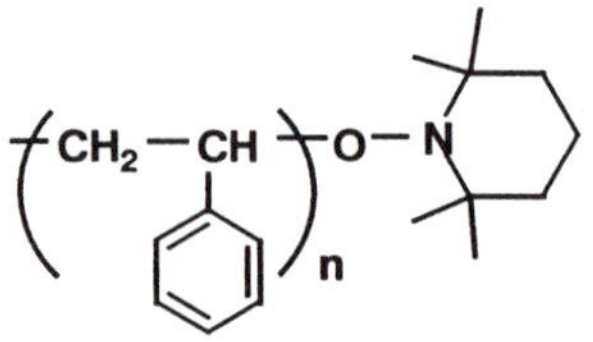

1: PS-TEMPO

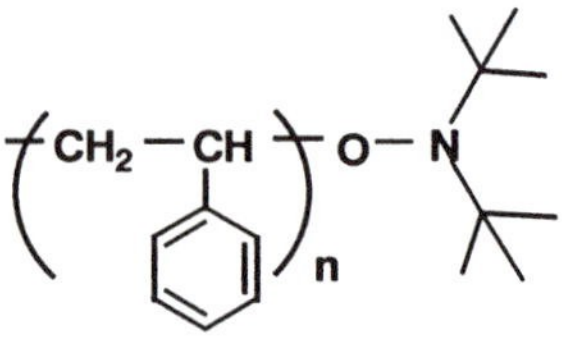

2: PS-DBN

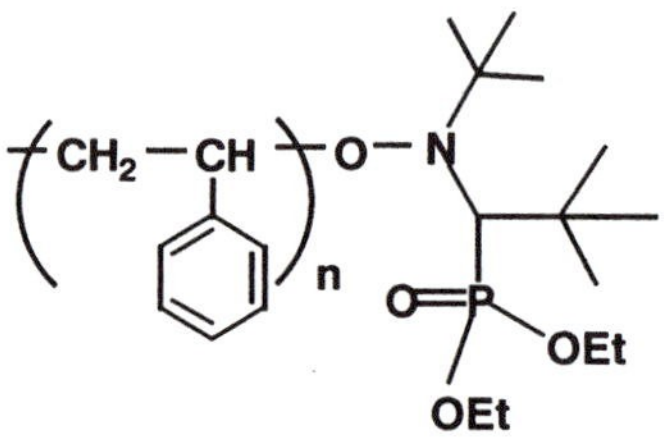

3: PS-DEPN

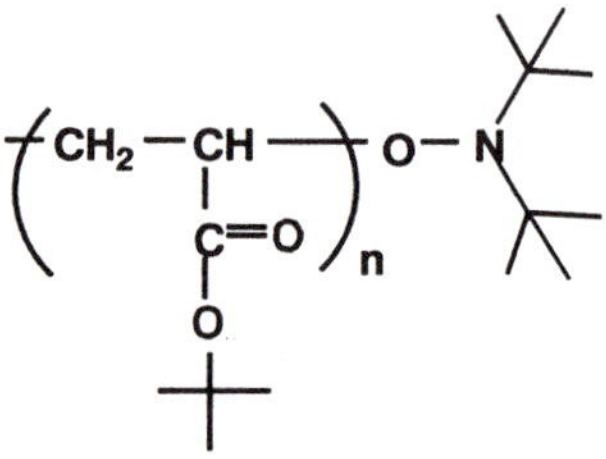

4: PBA-DBN

Scheme 3. Structures of polymer-nitroxide adducts.

occurs rather seriously at the high temperature, not allowing the polydispersity to be lowered as in the TEMPO/styrene system (66). Clearly, a large k_{act} is a necessary but not a sufficient condition for a high performance LRP. In this regard, the good achievements of the DEPN/BA and some other nitroxide/acrylate systems demonstrated by Benoit et al. (31,71) are noteworthy.

The k_{act} value for the RAFT (reversible addition-fragmentation chain transfer) system 8 with PS-SCSPh is surprisingly large, about 40 times as large as that for the system 7 with PS-SCSMe. The k_{act} value for the ATRP (atom transfer radical polymerization) system 10 is also large enough to explain the experimental observations that the system provides low-polydispersity polymers *from an early stage of polymerization*. (A large k_{act} gives a large y_n in spite of a small t, hence a low polydispersity: cf. equations 1-3.)

Table 2. Arrhenius parameters for the k_{act} of alkoxyamines

P-X	A_{act}[a] $/ s^{-1}$	E_{act}[b] $/ kJ \, mol^{-1}$
1. PS-TEMPO	3.0×10^{13}	124 ± 2
2. PS-DBN	3.8×10^{14}	120 ± 2
3. PS-DEPN	2.0×10^{15}	130 ± 2

[a] At 95 % confidence level, A_{act} = $(1.6\text{-}6.4) \times 10^{13}$, $(2.5\text{-}6.4) \times 10^{14}$, and $(1.1\text{-}3.7) \times 10^{15}$ for PS-TEMPO, PS-DBN, and PS-DEPN, respectively. [b] At 95 % confidence level, statistical error smaller than 2 kJ mol⁻¹.

Polymerization Rates

Here we first discuss the polymerization rates of the LRP systems of the dissociation-combination type (Scheme 1a), in which $X^•$ is a stable free radical. In the presence of both *conventional* initiation (rate = R_i) and termination (rate constant = k_t), the radical concentrations will follow the equations

$$d[P^•]/dt = k_d[P\text{-}X] - k_c[P^•][X^•] + R_i - k_t[P^•]^2 \tag{7}$$
$$d[X^•]/dt = k_d[P\text{-}X] - k_c[P^•][X^•] \tag{8}$$

Here a *conventional initiation* refers to the initiation induced thermally or by the decomposition of a conventional initiator like benzoyl peroxide, and should not be confused with the activation of an initiating dormant species $P_0\text{-}X$. We assume here that $P_0\text{-}X$ and P-X are kinetically identical. We are particularly interested in the case where dead chains are very small in number compared to "living" chains, i.e., the equality [P-X] = A_0 holds approximately, where A_0 is the concentration of the initiating dormant species ($A_0 = [P_0\text{-}X] = [P\text{-}X]_0$). Also it usually holds that $[X^•]$ » $[P^•]$ or $d[X^•]/dt$ » $d[P^•]/dt$. Under these conditions, equations 7 and 8 can be approximately solved for $[X^•]$ for the time range in which the quasi-equilibrium

$$[P^•][X^•] = K[P\text{-}X] \; (= KA_0) \qquad (K = k_d/k_c) \tag{9}$$

holds. (Fischer has shown this to be the case excepting the very initial stage of polymerization and noted the importance of the time range where the quasi-equilibrium holds (*73*).) The solution can be cast into the form

$$\ln\{[(1+x)/(1-x)]•[(1-x_0)/(1+x_0)]\} - 2(x-x_0) = at \tag{10}$$
$$x = (R_i/k_tK^2A_0^2)^{1/2}[X^•] \tag{11}$$
$$a = 2R_i^{3/2}/(k_tK^2A_0^2)^{1/2} \tag{12}$$

and x_0 is the value of x at $t = 0$ (R_i is assumed to be constant. For hints to the derivation of equations 10-12, see reference (*11*)). The $P^•$ concentration and hence R_p = $k_p[P^•][M]$ follow from equations 10-12 with $[P^•] = KA_0/[X^•]$ (equation 9).

Two special cases have been treated elsewhere (*72,73*). One is the case in which $R_i > 0$, $[X^•]_0 = 0$, *and t* is sufficiently large ($a•t$ » 1). In this limit, equation 10 simply reduces to $x = 1$ (the "stationary state") or equivalently

$$[X^•] = (KA_0)/(R_i/k_t)^{1/2} \qquad \text{(stationary state)} \tag{13}$$

or

$$[P^•] = (R_i/k_t)^{1/2} \qquad \text{(stationary state)} \tag{14}$$

Thus R_p is independent of the reversible activation reaction and identical with the stationary-state rate of polymerization of the nitroxide-free system. This has been experimentally observed (*72,74*).

This stationary-state kinetics is expected to hold also for degenerative transfer-type systems (Scheme 1b), since the radical concentration is basically unchanged by a transfer reaction unless it is a retarding or degradative one. Stationary-state kinetics was in fact observed for the styrene polymerizations with a PS-I (62) and a PS-SCSMe degenerative transfer agent (63).

The other case that has been discussed is the one with $R_i = 0$ ($a = 0$), where equation 10 simplifies to (11)

$$[X^\bullet]^3 - [X^\bullet]_0^3 = 3k_t(KA_0)^2 t \tag{15}$$

Equation 15 with $[X^\bullet]_0 = 0$ is the case discussed firstly by Fischer (73) and subsequently by us (75), which gives the characteristic power-law behavior of the conversion index $\ln([M]_0/[M])$:

$$\ln([M]_0/[M]) = (3/2)k_p(KA_0/3k_t)^{1/3}t^{2/3} \tag{16}$$

This behavior has been observed in part by experiments (75-77).

Equation 10 is applicable to more general cases in which, for example, R_i is nonzero but so small that the stationary state is reached only after a long time or never reached at all in the duration of a typical experiment. It also describes the case in which, for example, R_i is nonzero, *and* $[X^\bullet]_0$ is considerably larger than the stationary concentration given by equation 13. What happens in this case would be that the polymer radicals produced by the conventional initiation combine with the extra $X^\bullet$ radicals to produce extra adducts P-X until $[X^\bullet]$ decreases down to the stationary concentration (equation 13 with A_0 so modified as to include the extra adducts produced from the extra $X^\bullet$.) Correspondingly, the R_p in such a system would be small or virtually zero when t is small because of the extra $X^\bullet$, and gradually increase up to the stationary value given by equation 14. Such intermediate behaviors are clearly different from those for the two special (limiting) cases discussed above. The behavior that one would experimentally observe depends basically on the magnitudes of the two parameters x_0 and a, as equation 10 shows.

There are several causes that can introduce deviations from equation 10. One may be the inadequacy of the assumptions on which equation 10 is based, in particular, the approximation of $[P-X] = A_0$ (= constant). When $R_i = 0$, $[X^\bullet]$ increases and $[P-X]$ necessarily decreases with time because of termination (cf. equation 15). The magnitude of error introduced by this cause would be on the same order as that involved in the approximation $[P-X] = A_0$. For example, if $[P-X]$ is smaller than A_0 by 10 % at time t, the $[X^\bullet]$ (hence R_p) estimated from equation 15 would be in error by about 10 %. The second cause may be side reactions other than initiation and termination. For example, alkoxyamines are known to undergo thermal degradation (78,79). We also note that the conventional initiation can be accelerated by the presence of a nitroxide (80). The third possible cause of deviations may be the dependence of the kinetic parameters on chain length and/or polymer concentration. In the analysis of their ATRP experiments (81), Shipp and Matyjaszewski (82) have in fact noted the importance of taking the chain length dependence of k_t into account.

Remember that in a living polymerization, chain length and conversion or polymer concentration are directly related to each other. Also note that the above equations for nitroxide systems are basically applicable to ATRP systems by the reinterpretations of $X^{\bullet} = AX$, $k_d = k_A[A]$, and $k_c = k_{DA}$ ($K = k_A[A]/k_{DA}$), as suggested previously (*11,77*).

With these discussions in mind, the simple analytical equations given here will be hopefully useful in understanding the fundamental features, and enable consideration of the more sophisticated aspects, of LRP.

Abbreviations

a	reduced rate of initiation (equation 11)
A_0	initial concentration of dormant species ($A_0 = [P_0\text{-}X] = [P\text{-}X]_0$)
C	monomer conversion
C_{ex}	degenerative transfer constant (equation 5)
K	equilibrium constant (equation 9)
k_A	activation rate constant in ATRP (Scheme 1)
k_{act}	pseudo-first-order activation rate constant (Scheme 1)
k_c	combination rate constant (Scheme 1)
k_d	dissociation rate constant (Scheme 1)
k_{DA}	deactivation rate constant in ATRP (Scheme 1)
k_{deact}	deactivation rate constant (Scheme 1)
k_{ex}	degenerative transfer rate constant (Scheme 1)
k_p	propagation rate constant
k_t	termination rate constant
M_n	number-average molecular weight
M_w	weight-average molecular weight
$P^{\bullet}$	polymer radical
$P\text{-}X$	dormant species
$P_0\text{-}X$	initiating dormant species
R_i	(conventional) initiation rate
R_p	propagation rate
w_K	weight fraction of the subchain K (K= A or B) (equation 1)
x	reduced stable radical concentration (equation 10)
$X^{\bullet}$	stable free radical
Y	polydispersity factor ($Y = M_w/M_n - 1$)
y_n	average number of activation deactivation cycles that a chain experiences during polymerization time t ($y_n = k_{act}t$)
z_n	number-average degree of polymerization

Acknowledgments

We thank Professor K. Matyjaszewski, Dr. G. Moad, Dr. E. Rizzardo, Professor B. B. Wayland, and Professor H. Fischer for collaboration and/or valuable discussions.

References

1. Moad, G.; Rizzardo, E.; Solomon, D. H. In *Comprehensive Polymer Science*; Eastmond, G. C., Ledwith, A., Russo, S., Sigwalt, P., Eds.: Pergamon: London, 1989; Vol. 3, p 141.
2. Georges, M. K.; Veregin, R. P. N.; Kazmaier, P. M.; Hamer, G. K. *Trends Polym. Sci.* **1994**, *2*, 66.
3. Matyjaszewski, K.; Gaynor, S.; Greszta, D.; Mardare, D.; Shigemoto, T. *J. Phys. Org. Chem.* **1995**, *8*, 306.
4. Moad, G.; Solomon, D. H. *The Chemistry of Free Radical Polymerization*; Pergamon: Oxford, UK, 1995; p 335.
5. Davis, T. P.; Haddleton, D. M. In *New Methods of Polymer Synthesis*; Ebdon, J. R., Eastmond, G. C., Eds.; Blackie: Glasgow, UK, 1995; Vol. 2, p 1.
6. Hawker, C. J. *Trends Polym. Sci.* **1996**, *4*, 183.
7. Colombani, D. *Prog. Polym. Sci.* **1997**, *22*, 1649.
8. Matyjaszewski, K. In *Controlled Radical Polymerization*; Matyjaszewski, K., Ed.; ACS Symposium Series 685; American Chemical Society: Washington, DC, 1998; Chapter 1.
9. Sawamoto, M.; Kamigaito, M. In *Polymer Synthesis*; Materials Science and Technology Series; VCH-Wiley: Weinheim 1998, Chapter 1.
10. Otsu, T.; Matsumoto, A. *Adv. Polym. Sci.* **1998**, *136*, 75.
11. Fukuda, T.; Goto, A.; Ohno, K. *Macromol. Rapid Commun.* **2000**, *21*, 151.
12. Otsu, T.; Yoshida, M. *Makromol. Chem., Rapid Commun.* **1982**, *3*, 127.
13. Ajayaghosh, A.; Francis, R. *Macromolecules* **1998**, *31*, 1436.
14. Le, T. P. T.; Moad, G.; Rizzardo, E.; Thang, S. H. *International Pat. Appl. PCT/US97/12540, WO9801478* (Chem Abstr. **1998**, *128*, 115390).
15. Solomon, D. H.; Rizzardo, E.; Cacioli, P. *Eur. Pat. Appl. EP135280* (Chem. Abstr. **1985**, *102*, 221335q).
16. Georges, M. K.; Veregin, R. P. N.; Kazmaier, P. M.; Hamer, G. K. *Macromolecules* **1993**, *26*, 2987.
17. Greszta, D.; Matyjaszewski, K. *Macromolecules* **1994**, *27*, 638.
18. Hawker, C. J. *J. Am. Chem. Soc.* **1994**, *116*, 11185.
19. Li, I.; Howell, B. A.; Matyjaszewski, K.; Shigemoto, T.; Smith, P. B.; Priddy, D. B. *Macromolecules* **1995**, *28*, 6692.
20. Catala, J. -M.; Bubel, F.; Hammouch, S. O. *Macromolecules* **1995**, *28*, 8441.
21. Fukuda, T.; Terauchi, T.; Goto, A.; Tsujii, Y.; Miyamoto, T.; Shimizu, Y. *Macromolecules* **1996**, *29*, 3050.
22. Puts, R. D.; Sogah, D. Y. *Macromolecules* **1996**, *29*, 3323.
23. Baldoví, M. V.; Mohtat, N.; Scaiano, J. C. *Macromolecules* **1996**, *29*, 5497.
24. Yoshida, E.; Sugita, A. *Macromolecules* **1996**, *29*, 6422.
25. Steenbock, M.; Klapper, M.; Müllen, K.; Pinhal, N.; Hubrich, M. *Acta Polym.* **1996**, *47*, 276.
26. Schmidt-Naake, G.; Butz, S. *Macromol. Rapid Commun.* **1996**, *17*, 661.
27. Bon, S. A. F.; Bosveld, M.; Klumperman, B.; German, A. L. *Macromolecules* **1997**, *30*, 324.

28. Kothe, T.; Marque, S.; Martschke, R.; Popov, M.; Fischer, H. *J. Chem. Soc., Perkin Trans.* **1998**, *2*, 1553.
29. Bouix, M.; Gouzi, J.; Charleux, B.; Vairon, J. P.; Guinot, P. *Macromol. Rapid Commun.* **1998**, *19*, 209.
30. Grimaldi, S.; Lemoigne, F.; Finet, J. P.; Tordo, P.; Nicol, P.; Plechot, M. WO 96/24620.
31. Benoit, D.; Grimaldi, S.; Finet, J. P.; Tordo, P.; Fontanille, M.; Gnanou, Y. In *Controlled Radical Polymerization*; Matyjaszewski, K., Ed.; ACS Symposium Series 685; American Chemical Society: Washington, DC, 1998; Chapter 14.
32. Yamada, B.; Miura, Y.; Nobukane, Y.; Aota, M. In *Controlled Radical Polymerization*; Matyjaszewski, K., Ed.; ACS Symposium Series 685; American Chemical Society: Washington, DC, 1998; Chapter 12.
33. Fukuda, T.; Goto, A.; Ohno, K.; Tsujii, Y. In *Controlled Radical Polymerization*; Matyjaszewski, K., Ed.; ACS Symposium Series 685; American Chemical Society: Washington, DC, 1998; Chapter 11.
34. Yutani, Y.; Tatemoto, M. *Eur. Pat. Appl.* **1991**, *0489370A1*.
35. Kato, M.; Kamigaito, M.; Sawamoto, M.; Higashimura, T. *Polym. Prepr., Jpn.* **1994**, *43*, 225.
36. Matyjaszewski, K.; Gaynor, S.; Wang, J.-S. *Macromolecules* **1995**, *28*, 2093.
37. Kato, M.; Kamigaito, M.; Sawamoto, M.; Higashimura, T. *Macromolecules* **1995**, *28*, 1721.
38. Wang, J.-S.; Matyjaszewski, K. *J. Am. Chem. Soc.* **1995**, *117*, 5614.
39. Percec, V.; Barboiu, B. *Macromolecules* **1995**, *28*, 7970.
40. Granel, C.; Dubois, Ph.; Jérôme, R.; Teyssié, Ph. *Macromolecules* **1996**, *29*, 8576.
41. Haddleton, D. M.; Jasieczek, C. B.; Hannon, M. J.; Scooter, A. J. *Macromolecules* **1997**, *30*, 2190.
42. Leduc, M. R.; Hayes, W.; Fréchet, J. M. J. *J. Polym. Sci. Part A.: Polym. Chem.* **1998**, *36*, 1.
43. Jankova, K.; Chen, X.; Kops, J.; Batsberg, W. *Macromolecules* **1998**, *31*, 538.
44. Angot, S.; Murthy, K. S.; Taton, D.; Gnanou, Y. *Macromolecules* **1998**, *31*, 7218.
45. Patten, T. E.; Matyjaszewski, K. *Adv. Mat.* **1998**, *10*, 901.
46. Patten, T. E.; Matyjaszewski, K. *Acc. Chem. Res.* **1999**, *32*, 895.
47. Matyjaszewski, K. *Chem. Eur. J.* **1999**, *5*, 3095.
48. Sawamoto, M.; Kamigaito, M. *CHEMTECH* **1999**, *29*(6), 30.
49. Wayland, B. B.; Poszmik, G.; Mukerjee, S. L.; Fryd, M. J. *J. Am. Chem. Soc.* **1994**, *116*, 7943.
50. Mardare, D.; Matyjaszewski, K. *Macromolecules* **1994**, *27*, 645.
51. Arvanitopoulos, L. D.; Grenel, M. P.; Harwood, H. J. *Polym. Prepr. (Am. Chem. Soc., Div. Polym. Chem.)* **1994**, *35*(2), 549.
52. Druliner, J. D. *Macromolecules* **1991**, *24*, 6079.
53. Krstina, J.; Moad, G.; Rizzardo, E.; Winzor, C. L. *Macromolecules* **1995**, *28*, 5381.
54. Yang, W.; Ranby, B. *Macromolecules* **1996**, *29*, 3308.

38

55. Chung, T. C.; Janvikul, W.; Lu, H. L. *J. Am. Chem. Soc.* **1996**, *118*, 705.
56. Colombani, D.; Steenbock, M.; Klapper, M.; Müllen, K. *Macromol. Rapid Commun.* **1997**, *18*, 243.
57. Detrembleur, C.; Lecomte, Ph.; Caille, J.-R.; Creutz, S.; Dubois, Ph.; Teyssié, Ph.; Jérôme, R. *Macromolecules* **1998**, *31*, 7115.
58. Fukuda, T.; Goto, A. *Macromol. Rapid Commun.* **1997**, *18*, 682: the factor -2 appearing in eq. 4 in this reference is a misprint for $C-2$.
59. Müller, A. H. E.; Zhuang, R.; Yan, D.; Litvinenko, G. *Macromolecules* **1995**, *28*, 4326.
60. Goto, A.; Terauchi, T.; Fukuda, T.; Miyamoto, T. *Macromol. Rapid Commun.* **1997**, *18*, 673.
61. Goto, A.; Fukuda, T. *Macromolecules* **1997**, *30*, 5183.
62. Goto, A.; Ohno, K.; Fukuda, T. *Macromolecules* **1998**, *31*, 2809.
63. Goto, A.; Sato, K.; Fukuda, T.; Moad, G.; Rizzardo, E.; Thang, S. H. *Polym. Prepr. (Div. Polym. Chem., Am. Chem. Soc.)* **1999**, *40(2)*, 397.
64. Ohno, K.; Goto, A.; Fukuda, T.; Xia, J.; Matyjaszewski, K. *Macromolecules* **1998**, *31*, 2699.
65. Goto, A.; Fukuda, T. to be published.
66. Goto, A.; Fukuda, T. *Macromolecules* **1999**, *32*, 618.
67. Goto, A.; Fukuda, T.; Wayland, B. B. to be published.
68. Moad, C. L.; Moad, G.; Rizzardo, E.; Thang, S. H. *Macromolecules* **1996**, *29*, 7717.
69. Moad, G.; Rizzardo, E. *Macromolecules* **1995**, *28*, 8722.
70. Kazmaier, P. M.; Moffat, K. A.; Georges, M. K.; Veregin R. P. N.; Hamer, G. K. *Macromolecules* **1995**, *28*, 1841.
71. Benoit, D.; Chaplinski, V.; Braslau, R.; Hawker, C. J. *J. Am. Chem. Soc.* **1999**, *121*, 3904.
72. Fukuda, T.; Terauchi, T.; Goto, A.; Ohno, K.; Tsujii, Y.; Miyamoto, T.; Kobatake, S.; Yamada, B. *Macromolecules* **1996**, *29*, 6393.
73. Fischer, H. *Macromolecules* **1997**, *30*, 5666.
74. Catala, J. M.; Bubel, F.; Hammouch, S. O. *Macromolecules* **1995**, *28*, 8441.
75. Ohno, K.; Tsujii, Y.; Miyamoto, T.; Fukuda, T.; Goto, M.; Kobayashi, K.; Akaike, T. *Macromolecules* **1998**, *31*, 1064.
76. Kothe, T.; Marque, S.; Martschke, R.; Popov, M.; Fischer, H. *J. Chem. Soc., Perkin Trans.* **1998**, *2*, 1553.
77. Fischer, H. *J. Polym. Sci. Part A.: Polym. Chem.* **1999**, *37*, 1885.
78. Li, I.; Howell, B. A.; Matyjaszewski, K.; Shigemoto, T.; Smith, P. B.; Priddy, D. B. *Macromolecules* **1995**, *28*, 6692.
79. Ohno, K.; Tsujii, Y.; Fukuda, T. *Macromolecules* **1997**, *30*, 2503.
80. Moad, G.; Rizzardo, E.; Solomon, D. *Polym. Bull.* **1982**, *6*, 589.
81. Matyjaszewski, K.; Patten, T. E.; Xia, J. *J. Am. Chem. Soc.* **1997**, *119*, 674.
82. Shipp, D. A.; Matyjaszewski, K. *Macromolecules* **1999**, *32*, 2948.

Chapter 3

Initiation and Termination Rates Associated with Free-Radical Polymerization in Extended Ranges of Temperature and Pressure

Michael Buback

**Institut für Physikalische Chemie der Georg-August-Universität,
Tammannstrasse 6, D–37077 Göttingen, Germany**

An almost constant value of chain-length averaged termination rate coefficient, k_t, is observed during the initial period of free-radical homo- and copolymerizations in bulk and in solution with the extension of this plateau region depending on monomer type and on the particular polymerization conditions. The plateau region of k_t is assigned to termination being controlled by the diffusivity of a small number of segments at the free-radical site. This size of k_t is determined by steric shielding and by intra-coil viscosity due to the monomer and, if present, the solvent. Also a chain-length dependence of k_t is seen in the initial polymerization range. The second part of the paper deals with the spectroscopic measurement of the decomposition of organic peroxides, with particular emphasis on the influence of substituents on kinetics and selectivity. For alkyl peroxyesters, the impact of substituents on the decomposition mechanism, via single-bond or via concerted two-bond scission, and thus on the formation of either oxygen-centered or carbon-centered radicals is illustrated.

The advent of pulsed-laser-assisted techniques has enormously improved the quality of rate coefficient measurements in free-radical polymerization. As has been detailed in a contribution to the ACS volume developed from the 1997 San Francisco symposium on "Controlled Radical Polymerization" (*1*), two such methods are of particular importance: The PLP-SEC method (*2*) uses pulsed-laser polymerization (PLP) in conjunction with size-exclusion chromatographic (SEC) analysis of the polymeric material. The pulsed laser light almost instantaneously creates an intense burst of free radicals. Varying the laser pulse repetition rate allows for producing packages of primary free radicals at pre-selected time intervals. The pulsing translates into a characteristic pattern of the molecular weight distribution (MWD). Analysis of the MWD enables the unambiguous determination of the propagation rate coefficient, k_p, as far as certain consistency criteria are met. Within the second method, SP-PLP, polymerization induced by a single laser pulse (SP) is measured via infrared or near-infrared spectroscopy with a time resolution of microseconds (3). From the measured trace of monomer conversion vs. time (after applying the laser pulse), the ratio of termination to propagation rate coefficients, k_t/k_p, and thus (with k_p being available from PLP-SEC) also individual k_t are obtained. Following the IUPAC recommendations, k_t values presented in the subsequent text refer to the overall termination rate law: $d\,c_R/d\,t = -\,2\,k_t \cdot c_R{}^2$, where c_R is free-radical concentration.

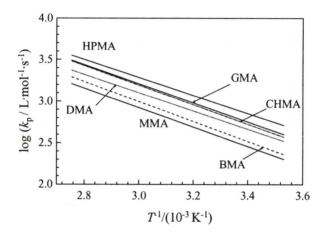

Figure 1: Temperature dependence of the propagation rate coefficient k_p at ambient pressure for methyl methacrylate (MMA), butyl methacrylate (BMA), dodecyl methacrylate (DMA), cyclohexyl methacrylate (CHMA), glycidyl methacrylate (GMA), and hydroxypropyl methacrylate (HPMA). The data is from References (5,6,7,8); see text.

The PLP-SEC strategy is recommended by the IUPAC Working Party "Modeling of Polymerisation Kinetics and Processes" as the method of choice for reliable k_p determination. The compilation of benchmark k_p data by this Working Party has started (4,5). Figure 1 shows an Arrhenius plot of ambient pressure k_p data from PLP-SEC investigations into several methacrylates: methyl methacrylate (MMA), butyl methacrylate (BMA), dodecyl methacrylate (DMA), cyclohexyl methacrylate (CHMA), glycidyl methacrylate (GMA), and hydroxypropyl methacrylate (HPMA). The fitted lines for MMA, BMA, and DMA have already been presented in (6). The k_p data for CHMA, for GMA, and for HPMA are from (7) and (8). A family-type behavior of the methacrylates is clearly seen in that the lines are almost parallel which indicates nearly identical activation energy. IUPAC-recommended benchmark k_p values of MMA have already been published (5). The compilation of the corresponding Arrhenius fits for BMA, DMA, CHMA, GMA, and HPMA by the IUPAC Working Party is currently underway.

With the PLP-SEC method being available, a very satisfactory situation with respect to accurately measuring k_p of free-radical homo- and copolymerizations has emerged. Even high-pressure k_p data from PLP-SEC have been reported for several monomers (9). As k_p refers to a chemically controlled (radical–molecule) process, the coefficients from PLP-SEC experiments in the initial polymerization stage are valid up to fairly high conversion (10), mostly throughout the entire conversion range of technical polymerizations. To model polymerization kinetics and polymer properties, additional rate coefficients need to be accurately known. Among them, the termination rate coefficient, k_t, and the rate coefficient of initiator decomposition, k_d, deserve special attention. It is the purpose of this contribution to briefly outline recent developments in studying k_t and k_d. The essential problem of investigations into k_t is associated with this (radical–radical) process being diffusion controlled. Because of cage effects, also the overall initiator decomposition rate coefficient (k_d) may depend on diffusive processes. Thus both k_t and k_d need to be correlated to transport properties of the reaction medium. k_t, in addition, may vary with the chain length of the particular set of terminating radicals. Section I of this paper deals with the measurement and interpretation of k_t and Section II refers to k_d studies.

Termination Rate Coefficients from SP-PLP Experiments

Free-radical bulk polymerizations to high degrees of monomer conversion exhibit quite different types of k_t vs. conversion behavior: In MMA and also in styrene homo-polymerization, k_t decreases by several orders of magnitude, whereas the variation with conversion of k_t is less pronounced, e.g., for butyl acrylate (1). As has been described in Ref.(1), the latter type of behavior is typical for so-called type A monomers which are relatively high in k_p and yield polymer of significant segmental mobility. Type B monomers, such as MMA and styrene, are associated with low values of k_p and with a reduced segmental mobility of the associated polymer. Irrespective of the enormous differences in k_t vs. conversion behavior between type A and B monomers that are seen upon inspection of the entire conversion range, both types of monomers experience a plateau region of approximately constant k_t in the initial polymerization period. The subsequent discussion will be restricted to this initial plateau region of k_t which may be rather extended as is demonstrated in Fig. 2. It should be noted that in the first part of this section k_t refers to a chain-length averaged value of termination rate coefficient. The determination of microscopic chain-length dependent termination coefficients will be addressed further below.

Shown in Figure 2 are k_t values as a function of monomer conversion during bulk homo-polymerizations of MMA, BMA, and DMA at 40 °C and 1000 bar (11). Experimental data points are only given for DMA, whereas the measured k_t data for MMA and BMA are represented by dashed lines fitted to measured data in the initial polymerization period. For MMA, the conversion range of (almost) constant k_t extends only up to about 15 to 20 % with this limiting conversion being strongly dependent on the polymerization conditions of a particular experiment. With the reaction conditions being chosen such that very high molecular weight material is produced, the initial plateau range is smaller. Above 20 % and up to 50 % monomer conversion, k_t of MMA rapidly decreases upon further conversion, by approximately one order of magnitude per each 10 % of (further) conversion (1,10). The extent by which k_t decreases during this gel effect region depends on the history of the preceding polymerization, in particular on the type of polymeric material that has been produced during the plateau period and during the early gel effect region. As is shown in Figure 2, the k_t plateau value of BMA is significantly smaller than the corresponding MMA value, but the region of constant k_t (indicated by the length of the dashed line) is far more extended than in MMA. These trends appear to be associated with the size of the alkyl group of the ester moiety. In DMA bulk polymerization, k_t is further decreased and the plateau region of constant k_t extends up to about 50 %.

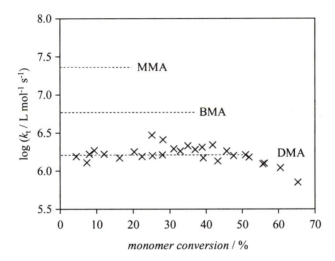

Figure 2: Termination rate coefficients of methyl methacrylate (MMA), butyl methacrylate (BMA), and dodecyl methacrylate (DMA) homo-polymerizations as a function of monomer conversion at 40 °C and 1000 bar (11).

Two questions result from the data in Figure 2: (1) Why is the low and moderate conversion k_t insensitive toward conversion and thus toward bulk viscosity which changes roughly by one order of magnitude with each 10 % conversion? (2) Why are the MMA and DMA k_t's different by a factor of 16, whereas the associated difference in the monomer viscosities is only by about a factor of 6?

Within the acrylate family, the differences in plateau k_t values between the methyl and dodecyl members, methyl acrylate (MA) and dodecyl acrylate (DA), are even larger than between MMA and DMA. This is demonstrated in Figure 3 where, for the bulk homo-polymerizations of MMA, BMA, DMA, MA, BA, and DA, k_p is plotted vs. the associated plateau k_t value with all numbers referring to 40 °C and 1000 bar. A family-type behavior of k_p occurs for both acrylates and methacrylates, whereas no indication of any family-type behavior of k_t can be seen, e.g., k_t of DA is remarkably close to the DMA value, but significantly different from k_t of MA.

To better understand the origin of the significant differences in plateau k_t values, rate coefficients of copolymerization, $k_{t,copo}$, have been measured for several binary (meth)acrylate mixtures. The entire body of results has been detailed in the *Ph.D. Thesis* of Kowollik (*13*) with part of the data being given in Ref. (*12*) and in Feldermann´s *Diploma Thesis* (*14*). Before presenting the essential evidence from these studies, one experimental aspect of the SP-PLP studies used for the investigations into $k_{t,copo}$ will be briefly mentioned.

To allow for a consistent k_t analysis, it is highly desirable to use the same technique for measuring k_t of the copolymerizations and the (limiting) homo-polymerization systems. The SP-PLP technique yields conversion vs. time traces of excellent signal-to-noise quality for monomers which are low in k_t and high in k_p. Thus, among the monomers summarized in Figure 3, DA is an optimum candidate for k_t analysis via SP-PLP.

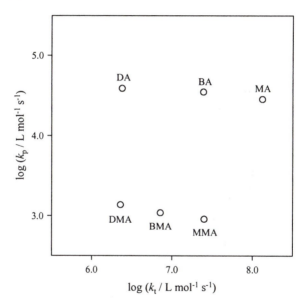

Figure 3: Propagation rate coefficients, k_p, and termination rate coefficients, k_t, of methyl methacrylate (MMA), butyl methacrylate (BMA), dodecyl methacrylate (DMA), methyl acrylate (MA), butyl acrylate (BA), and dodecyl acrylate (DA) homo-polymerizations at 40 °C and 1000 bar (The sources for the numbers in Figure 3 are given in Ref. 12).

As is shown in the upper part of Figure 4, a single laser pulse (of about 15 ns width), within one second after firing the laser pulse, induces a monomer conversion that exceeds 2 %. Among the monomers depicted in Figure 3 an unfavorable SP-PLP situation must be expected for MMA which is high in k_t and is lowest in k_p. By co-adding around 100 "true" SP signals, however, even for MMA a conversion vs. time trace can be obtained (see lower part of Figure 4) that is sufficient for k_t

determination. Literature k_t data for high pressure are not available. As is however shown in Ref. (*15*), extrapolation of the MMA k_t value from SP−PLP (Figure 5) toward ambient pressure yields satisfactory agreement with published data. The experimental procedure of time-resolved measurement of such extremely small pulsed-laser-induced conversions as with MMA is outlined in References (*13*) and (*15*).

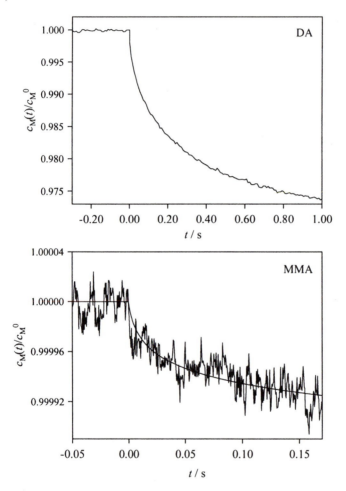

Figure 4: Time-resolved measurement of monomer concentration vs. time (after applying the laser pulse) for DA (upper part) and MMA (lower part) homo-polymerizations. Both SP-PLP experiments have been carried out in the initial period of reactions at 40 °C and 1000 bar. Plotted is the relative concentration, $c_M(t)/c_M^O$ with c_M^O referring to the monomer concentration prior to applying the laser pulse; for more details see text or reference (15).

Shown in Figure 5 are copolymerization k_t data for the systems MA–DA and DA–DMA (*12*). The particular interest in the MA–DA system is due to the fact that the homo-polymerization k_p's are almost the same whereas the k_t's largely differ, by a factor of 55. Within the DA–DMA system, on the other hand, the homo-k_t's are virtually identical, but the k_p's differ by a factor of 29. (The numbers refer to the 40 °C and 1000 bar conditions under which the majority of homo- and copolymerization experiments has been carried out so far; see also Figure 3). Particularly remarkable from Figure 5 is the linear log $k_{t,copo}$ vs. monomer mole fraction, f, correlation by which the associated homo-termination rate coefficients are linked together. This result is not overly surprising for DA–DMA as the homo-termination k_t's are more or less identical, but is remarkable for MA–DA where the homo-polymerization k_t's are clearly different.

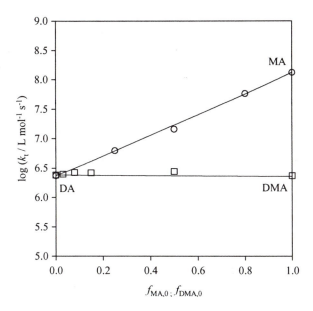

Figure 5: Termination rate coefficients in the plateau region of DA–MA and DA–DMA copolymerizations at 40 °C and 1000 bar plotted vs. f_{MA} or f_{DMA}, the mole fraction of MA or DMA in the monomer mixture with DA. The data is taken from Reference (12).

As has been outlined in more detail in (*12*), among the existing models for copolymerization k_t, summarized in Fukuda et al.'s article (*16*), the treatment which considers penultimate units on the free-radical terminus and uses the geometric mean approximation to estimate (penultimate) cross-termination rate coefficients, provides the best representation of the log $k_{t,copo}$ vs. f data in Figure 5. The suitability of this penultimate model to fit $k_{t,copo}$ strongly suggests that steric effects are controlling termination behavior in the initial plateau region of (meth)acrylates. Consideration of the terminal and the penultimate segments on the macroradical chain appears to sufficiently account for the shielding action at the free-radical site.

Assuming a rate control by steric effects may also explain the similarity in the homo-k_t's of DA and DMA: As the shielding effect of the dodecyl group will largely exceed the corresponding action of a methyl group, the addition of a CH_3 group to DA, to yield DMA, will not extensively vary the shielding situation. Thus k_t of DA should be close to the DMA value, which is indeed what has been observed (Figure 3). Shielding considerations also explain why the k_t's of the methyl members of the two families, MA and MMA, in which steric effects on k_t should be less pronounced, are significantly larger than dodecyl (meth)acrylate k_t. Moreover, addition of a methyl group to MA, to yield MMA, should have a stronger effect on k_t than the corresponding change from DA to DMA. This expectation is also backed by the experimental data in Figure 3.

It should further be noted that plotting the Ito and O'Driscoll ambient pressure $k_{t,copo}$ data for the MMA–BMA and MMA–DMA systems (17) on a log $k_{t,copo}$ vs. f scale also yields a linear correlation (12). Such very simple dependences of k_t on monomer mixture composition are not generally valid, but will presumably be restricted to pairs of monomers belonging to the same family, e.g., alkyl acrylate–alkyl acrylate or alkyl methacrylate–alkyl methacrylate mixtures, or to mixtures of monomers which are identical in their k_t value and underly the same mechanism of termination rate control. In DA–MMA and in MA–DMA mixtures, on the other hand, where both the homo-k_t's and the homo-k_p's are significantly different, non-linear correlations of log $k_{t,copo}$ vs. f have been obtained (13,14). Copolymerization systems without a linear log $k_{t,copo}$ vs. f behavior are also reported in Fukuda et al.'s review (16). The systems in (16) are still adequately represented by the penultimate unit $k_{t,copo}$ model applied in conjunction with the geometric mean approximation to estimate cross-termination (penultimate) rate coefficients. Thus steric shielding is expected to play a major role in controlling termination also in these systems.

Under conditions where k_t is significantly affected by such steric effects, the termination reaction is not diffusion-controlled in the sense that a close proximity of the two free-radical sites necessarily results in reaction. In case of strong shielding, even after formation of a short-lived contact or an entangled free-radical pair the two species may diffuse apart before termination has taken place. The significant lowering (and control) of k_t by steric hindrance due to the alkyl substituents on the (meth)acrylic acid alkyl esters is considered to be responsible for the low sensitivity of k_t toward center-of-mass diffusivity of the macroradicals and thus toward bulk viscosity in the initial polymerization period. k_t should, however, depend on intra-coil viscosity. Lowering this viscosity should enhance the mobility of chain segments including the end-segments which carry the free-radical site, thereby increasing encounter frequencies and thus also k_t. That such an effect indeed plays a major role, is strongly indicated by the observed large difference of k_t values for polymerization in bulk and in low-viscosity supercritical CO_2 solution.

Figure 6 shows such data for DA at 40 °C and 1000 bar (18,19). In the solution experiments with supercritical CO_2, k_t is found to be approximately one order of magnitude above the corresponding bulk k_t value. Figure 6 clearly displays two remarkable observations: (i) Toward increasing monomer conversion and thus toward enhanced bulk viscosity (with this viscosity increase being particularly pronounced in bulk polymerization) k_t is not significantly changed whereas: (*ii*) A variation in the monomer-solvent viscosity appreciably affects k_t.

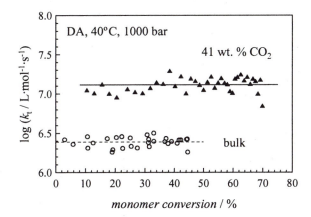

Figure 6: Homo-polymerization k_t as a function of monomer conversion of dodecyl acrylate at 40 °C and 1000 bar in bulk and in solution containing 41 wt.% CO_2.

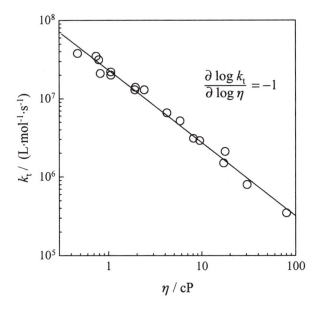

Figure 7: Dependence of MMA homo-polymerization k_t on the viscosity of the monomer-solvent mixture (measured at zero conversion prior to polymerization). The experiments have been carried out at 0 °C and ambient pressure. The data points are taken from Ref. 20.

A strong influence on solvent viscosity of low conversion k_t has already been reported by Fischer et al. (20). Their data for k_t of MMA homo-polymerizations carried out in a variety of solvent surroundings in which the solvent viscosity (measured as the viscosity of the monomer-solvent mixture prior to polymerization) has been varied by a factor of 170 are given in Figure 7. k_t nicely scales with the inverse of viscosity of the monomer-solvent mixture. It goes without saying that, according to this argument, part of the differences seen in homo-polymerization k_t values of the (meth)acrylates (Figure 3) will be due to differences in viscosity.

As has been found by Beuermann et al. (18,19), the difference in plateau k_t values for bulk and solution (in CO_2) polymerization depends on the type of monomer. E. g., no significant difference between bulk and solution k_t is seen for MA. This observation may be explained by the poor shielding with MA which results in termination being close to diffusion controlled. Enhancement of micro-viscosity by adding CO_2 thus should not be associated with a much higher termination rate as the fraction of unsuccessful free-radical collisions is small in MA bulk polymerization.

Finally, the chain-length dependence (CLD) of the initial plateau values of (meth)acrylate k_t will be briefly addressed. An overview of the extended literature about the CLD of k_t has very recently been provided by de Kock (21). The following expression is frequently used to represent the dependence on chain length i of the rate coefficient for termination of two free radicals of identical size:

$$k_t = 2\, k_t^\circ(1,1)\cdot i^{-\alpha}$$

where $k_t^\circ(1,1)$ refers to the termination rate coefficient of very small free radicals. SP-PLP experiments are particularly well suited for studies into the CLD of k_t according to this equation. The chain length of radicals is proportional to the time after applying the laser pulse (unless chain-transfer processes interfere at larger chain lengths). The SP-PLP trace thus provides access to the termination behavior of free radicals which are of approximately the same size with the chain length of the two radicals being proportional to the time after applying the laser pulse.

The exponent α has been estimated by Mahabadi (22) to be 0.15 for MMA and 0.24 for styrene. Olaj and Vana (23) report α of both monomers to be around 0.17. In a very recent paper by Olaj et al. (24) the exponent α is deduced via a novel procedure to be close to 0.2. A study into styrene k_t at high pressure (25) also considered a weak CLD with the measured rate data being consistent with a value of α around 0.2.

Attempts to deduce α from SP-PLP experiments for acrylates were not immediately successful despite of the high signal-to-noise quality of conversion vs. time traces in these polymerizations. As is shown in more detail in Ref. 13, the analysis is rather difficult because the photo-initiator 2,2-dimethoxy-2-phenylacetophenone (DMPA) used in these experiments decomposes into a propagating benzoyl radical and into a non-propagating α,α-dimethoxybenzyl radical. The latter primarily participates in radical–radical reactions (26). The simultaneous formation of a propagating and of an inhibiting radical from DMPA results in a rather peculiar type of behavior seen with the SP-PLP data. An example is given in Figure 8.

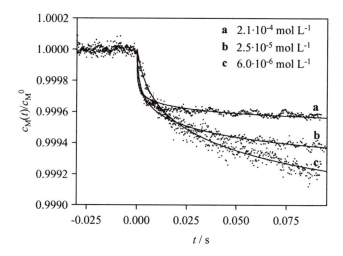

Figure 8: Relative monomer concentration vs. time traces measured for methyl acrylate homo-polymerizations. The SP-PLP studies were carried out in the initial period of polymerizations at 40 °C and 1000 bar, but at significantly different DMPA photo-initiator concentrations (given in the figure as a, b, and c).

The SP-PLP data in Figure 8 are from MA polymerizations at 40 °C and 1000 bar with all experimental parameters being kept constant with the exception of the initial DMPA concentration which was varied considerably, between $2.1 \cdot 10^{-4}$ and $6.0 \cdot 10^{-6}$ mol·L^{-1}. A very remarkable result is seen in that the resulting conversion vs. time traces intersect. Taking the photo-dissociation dynamics of DMPA into account (26), the detailed kinetic analysis shows that the observed crossings only occur if k_t decreases with chain length. The quantitative study of this effect, which has been carried out via the simulation program PREDICI®, yields $\alpha = 0.15$ for MMA and $\alpha = 0.16$ for styrene (13,27) with both these numbers being very close to the above-mentioned literature values.

A reasonable explanation for an exponent of this size in reactions between polymeric species may be provided by excluded volume arguments, as have been put forward by Khokhlov (28). An overview on effects of excluded volume is given in (21). Applying the scaling theory of polymer solutions (28) yields an exponent of $\alpha = 0.16$ in cases where the reactive sites are at the chain ends of large polymeric species whereas higher values of α occur when reaction occurs between the central part of one species and the chain end of a second polymeric species ($\alpha = 0.28$). An even higher value of α ($\alpha = 0.43$) is predicted for reaction between two central positions of the long-chain species (28). It should be mentioned that higher exponents ($\alpha = 0.32$) are derived from SP-PLP traces measured for methyl and dodecyl acrylate (13). De Kock very recently also determined a number of about this size, $\alpha = 0.36 \pm 0.1$, for several acrylates at chain lengths exceeding $i = 50$ (21). Further work is certainly required in order to be able to draw firm conclusions about the origin of these higher α values.

In summary: SP-PLP experiments on (meth)acrylate homo- and co-polymerizations and on styrene homo-polymerization demonstrate the occurrence of plateau k_t values in the initial polymerization stage. The insensitivity of k_t toward bulk viscosity, which for DA and DMA holds to rather high conversion, is assigned to kinetic control by steric shielding effects. Even under these conditions, the termination rate coefficient is, however, not independent of viscosity. Comparison of bulk and solution (in $scCO_2$) data suggests that the viscosity of the monomer (or of the monomer-CO_2 mixture in the solution polymerizations) strongly affects k_t. The termination mechanism in the plateau region is controlled by the segmental diffusivity of a small number of segments at the free-radical site. This segmental motion occurs under friction conditions provided by the monomer (mixture) inside the macroradical coils. The extent of steric shielding is determined by the substituents on the chain. The more pronounced the shielding contributions are, the larger is the plateau region of almost constant k_t. A specific plateau k_t value, e.g. in homo-polymerizations, thus refers to the mobility of a few segments in the more or less unchanged surrounding that is made up by the monomer (plus the solvent in solution polymerization experiments). According to this argument, the temperature and pressure dependence of plateau k_t is expected to be close to the temperature and pressure dependence of the inverse of monomer viscosity. This expectation indeed holds for styrene (25). The plateau k_t values exhibit some weak chain-length dependence which most likely is due to excluded volume effects of the reacting polymeric chains.

It should be noted that the preceding arguments do not rule out that very small free radicals, the action of which is not easily seen in the SP-PLP experiments, may show a different type of termination behavior. In particular, for such radicals shielding contributions to k_t may be different.

The preceding discussion has been restricted to the initial plateau region of k_t. A pronounced "non-plateau" k_t behavior occurs at higher conversions where other mechanisms such as translational diffusion control or reaction diffusion control apply. Under gel effect conditions where translation diffusion is dominant, strong dependences of k_t on chain length may be found. Moreover, the concentration and microstructure of the polymer (produced during the preceding reaction) may play an important role. The termination kinetics thus may depend on the history of the particular polymerization, whereas no such influence, at least no major one, is seen in the plateau region in which segmental motion under intra-coil viscosity conditions appears to be rate controlling. Improving the knowledge of the chain-length dependence of k_t at high conversion is of particular importance for detailed studies into the kinetics of controlled radical polymerization.

Decomposition Rates of Peroxide Initiators

Organic peroxides are extensively used as initiators in free-radical polymerizations at pressures and temperatures up to 300 °C and 3000 bar. In addition to the technically oriented interest in these reactions which is motivated by the requirement of knowing decomposition rate coefficients, k_d, for modeling purposes, a significant academic interest focuses on the detailed understanding of kinetics and selectivity of these unimolecular processes. The reaction mechanism needs to be known in order to characterize the type of free radicals produced during the primary dissociation event. Whether, e.g. in peroxyester decomposition, one-bond scission or concerted two-bond scission takes place, has immediate consequences for initiator efficiency, as will be outlined below for alkyl peroxyesters of the type $R(CO)OOC(CH_3)_2R^*$.

$$R\!-\!\overset{O}{\underset{O-O}{C}}\!\!\!\diagup\!\!\!\!\overset{CH_3}{\underset{CH_3}{C}}\!\!-\!R^*$$

If one-bond homolysis (of the O–O bond) occurs, an acyloxy and an alkoxy free radical are formed, whereas concerted two-bond scission, associated with the formation of CO_2, produces an alkyl and an alkoxy radical. Recombination after single-bond scission thus may restore the peroxide whereas cage recombination after concerted two-bond scission leads to the formation of an ether which is not capable of yielding free radicals within a further dissociation step. The efficiency will thus depend on the dissociation mechanism. Moreover, the type of primary free radicals, oxygen-centered or carbon-centered, is different for the two mechanisms. This may have consequences for the chain-transfer activity of these species. The type of primary radicals is also of interest with respect to applications in controlled radical polymerization.

The selectivity considerations associated with the immediate formation of CO_2 are related to the particular type of alkyl group R on the "acid side" of the alkyl peroxyester. It should be noted that, in addition to concerted two-bond scission, also rapid β-scission of the acyloxy radical gives CO_2 (29). Variation of the R* group on the "alcohol side" of the peroxyester may also influence the type of primary free-radical species as, depending on R*, a β-scission reaction of the C–R* bond subsequent to the primary dissociation event may convert the oxygen-centered alkoxy radical into a carbon-centered radical (plus acetone).

Peroxide decomposition rates were measured in solution of *n*-heptane via on-line FT-IR spectroscopy at temperatures and pressures up to 200 °C and 2500 bar, respectively. Three types of experimental procedures have been applied, two of them discontinuous and one continuous (30,31). The continuous procedure is used at higher temperatures where the reaction is fast. A tubular reactor which is a high-pressure capillary tube, typically of 10 m length and 0.5 mm internal diameter, allows to choose residence times as low as a few seconds. The IR analysis is carried out in a high-pressure optical cell which is positioned directly behind the tubular reactor, but outside the heat bath into which the reactor is embedded. To determine k_d at each set of p and T reaction conditions, between five and ten solution flow rates, corresponding to different residence times, were adjusted.

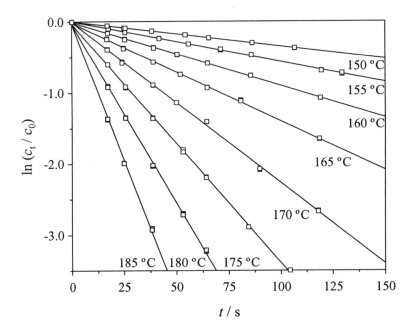

Figure 9: Dependence of relative peroxide concentration on residence time in tert-butyl peroxypropionate decomposition reactions at 500 bar and several temperatures (32). The reactions have been carried out in a tubular reactor. c_o is the initial concentration of the peroxide, typically around 0.01 mol·L⁻¹.

Shown in Figure 9 are plots of the logarithm of relative peroxide concentration, ln (c_t/c_o), vs. residence time t for *tert*-butyl peroxypropionate decompositions at 500 bar and several temperatures. c_o refers to the initial peroxide concentration. Data at a constant temperature up to several half lives closely fit to a linear ln(c_t/c_o) vs. t correlation indicating first-order decomposition kinetics. The slope to each of these lines yields the associated overall first-order rate coefficient k_d. Linear fits as in Figure 9 have been obtained for a wide variety of *tert*-butyl peroxyesters with the C-atom in R linked to the (CO)OO moiety being a primary, secondary, or tertiary one (31,32).

The substitution at this particular C-atom is of key importance for decomposition kinetics and selectivity. The reactions of *tert*-butyl peroxyacetate (TBPA) and of *tert*-butyl peroxypivalate (TBPP) which have been studied in quite some detail (31), may be considered as limiting situations of primary and tertiary carbon centers in R, respectively. Shown in Figure 10 is the pressure dependence of TBPA and of TBPP decomposition rate coefficients at 120 °C. At identical p and T, TBPP decomposes much faster than TBPA. The pressure dependence is also remarkably different: A weak decrease of k_d with pressure, corresponding to an activation volume of about 3 cm³·mol⁻¹, is seen for TBPP whereas a significantly larger pressure dependence occurs with TBPA and the activation volume itself depends on pressure. This pressure dependence of activation volume in TBPA decomposition is not easily seen on the extended log k_d scale of Figure 10, but clearly emerges if a more suitable scale is chosen (*31*). As is demonstrated in more detail in (*31*), the experimental material for TBPA and TBPP clearly shows that these two peroxides may be considered as the archetypes of decomposition behavior via single-bond and via concerted two-bond scission, respectively.

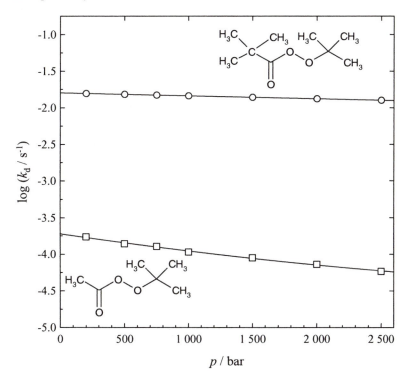

Figure 10: Pressure dependence of the overall decomposition rate coefficient, k_d, of tert-butyl peroxypivalate and of tert-butyl peroxyacetate at 120 °C; solvent: n-heptane.

The systematic variation of the decomposition rate coefficient of *tert*-butyl peroxyesters with the type of the particular C-atom in R, whether it is primary, secondary or tertiary, is illustrated by Arrhenius plots of k_d at 2000 bar for TBPA, *tert*-butyl peroxy-*iso*-butyrate (TBP*i*B), and TBPP in Figure 11. In passing from the C-atom in R being primary (TBPA), to secondary (TBP*i*B), and to tertiary (TBPP), the decomposition rate coefficient increases and, as can be seen from the slope to the straight lines in Figure 11, the activation energy slightly decreases. The activation energy associated with k_d linearly scales with the dissociation energy of the R–H bond in cases where the C-atom in the R substituent is either secondary or tertiary (*31,32*). This finding together with the observed pressure dependence of k_d strongly indicates that concerted two-bond scission accompanied by an immediate CO_2 release occurs with *tert*-butyl peroxyesters where the particular C-atom in R is secondary or tertiary.

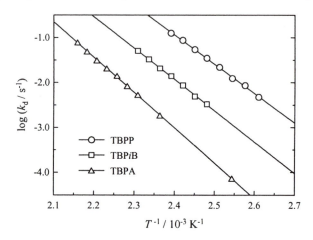

Figure 11: Arrhenius plots of the decomposition rate coefficient of TBPP, TBPiB, and TBPA dissolved in n-heptane. The data are for 2000 bar (31,32).

For each of the three peroxyesters (TBPA, TBPiB, and TBPP) an Arrhenius-type equation for $k_d(p,T)$ has been derived from an extended set of data measured in solution of *n*-heptane at initial peroxide concentrations of 0.01 M. The resulting three equations read:

TBPP $(110 \leq T / °C \leq 145; 100 \leq p/ \text{bar} \leq 2500)$

$$k_d\ (p,\ T)/\text{s}^{-1} = 6.10\ \cdot\ 10^{14}\ \cdot\ \exp\text{-}\left(\frac{15011 + 0.0367\ \cdot\ p\ /\ \text{bar}}{T\ /\ \text{K}}\right)$$

TBPiB $(130 \leq T / °C \leq 160, 200 \leq p\ /\ \text{bar} \leq 2500)$

$$k_d\ (p,T)/\text{s}^{-1} = 4.03 \cdot 10^{14} \cdot \exp\text{-}\left(\frac{15809 + 0.0267 \cdot p\ /\ \text{bar}}{T\ /\ \text{K}}\right)$$

TBPA $(120 \leq T / °C \leq 190; 100 \leq p/ \text{bar} \leq 2500)$

$$k_d\ (p,\ T)/\text{s}^{-1} = 6.78 \cdot\ 10^{15}\ \cdot\ \exp\text{-}\left(\frac{17714\ +\ (0.2471\ \text{-}3.336\ \cdot\ 10^{-5}\ \cdot\ p\ /\ \text{bar}) \cdot p\ /\ \text{bar}}{T\ /\ \text{K}}\right)$$

Also indicated are the temperature and pressure ranges of the underlying experiments *(31,32)*.

It is beyond the scope of this article to detail the consequences on kinetics and selectivity of varying the R* substituent on the "alcohol side" of the peroxyester, e.g., in going from *tert*-butyl, to *tert*-amyl, to 1,1,3,3-tetramethylbutyl, and to tetramethylpropyl peroxyesters. The R* substituent has only a minor effect on k_d, but strongly influences the distribution of products originating from the intermediate OR* radical (*33*). Thus the ratio of alcohol to ketone, produced from the intermediate OR* radical in *n*-heptane solution, strongly varies with substituent R*. The stability of the free radical formed upon β-scission of the OR* species plays a dominant role in controlling the relative amounts of oxygen-centered and carbon-centered radicals that are obtained from OR* in *n*-heptane.

Insight into the dynamics of photo-induced decomposition of organic peroxyesters, in particular into the unambiguous assignment of one-bond vs. two-bond-scission processes, is available from highly time-resolved studies into the evolution of CO_2 after applying ultra-short laser pulses. Picosecond UV-pump/IR-probe experiments along these lines have already been carried out on specially prepared peroxides (*34*). With respect to technical applications, the measurement of initiation rate and of initiation efficiency during the course of high-temperature high-pressure polymerizations is of particular importance. Such investigations are underway.

Acknowledgments

The author is grateful to the *Deutsche Forschungsgemeinschaft* for supporting the work underlying this survey within the Graduiertenkolleg "Kinetik und Selektivität chemischer Prozesse in verdichteter fluider Phase" and within the SFB 357 ("Molekulare Mechanismen unimolekularer Reaktionen"). The cooperation with AKZO NOBEL (in particular with Dr. B. Fischer, Dr. J. Meijer, Dr. A. van Swieten, and R. Gerritsen) in the field of peroxide decomposition is gratefully acknowledged, as is the provision of the peroxide samples. The author is grateful to Dr. Sabine Beuermann, Christopher Kowollik, and Johannes Sandmann for their help in finalizing this paper.

References

1. Beuermann, S.; Buback, M. *Controlled Radical Polymerization;* K. Matyjaszewski, Ed.; ACS Symposium Series 685: Washington, DC, 1998, p 84.
2. Olaj, O.F.; Bitai, I.; Hinkelmann, F. *Makromol. Chem.* **1987**, *188*, 1689; Olaj, O.F.; Schnöll-Bitai, I. *Eur. Polym. J.* **1989**, *25*, 635.
3. Buback, M.; Hippler, H.; Schweer, J.; Vögele, H.-P. *Makromol. Chem., Rapid Commun.* **1986**, *7*, 261.
4. Buback, M.; Gilbert, R.G.; Hutchinson, R.A.; Klumperman, B.; Kuchta, F.-D.; Manders, B.G.; O'Driscoll, K. F.; Russell, G. T.; Schweer, J. *Macromol. Chem. Phys.* **1995**, *196*, 3267.
5. Beuermann, S.; Buback, M.; Davis, T.P.; Gilbert, R.G.; Hutchinson, R.A.; Olaj, O.F.; Russell, G.T.; Schweer, J.; van Herk, A.M. *Macromol. Chem. Phys.* **1997**, *198*, 1545.

56

6. Beuermann, S.; Buback, M. *Pure & Appl. Chem.*, **1998**, *70*, 1415.
7. Buback, M.; Kurz, C. H. *Macromol. Chem. Phys.* **1998**, *199*, 2301.
8. Hutchinson, R. A.; Beuermann, S.; Paquet, D. A., Jr.; McMinn, J. H.; Jackson, C. *Macromolecules* **1998**, *31*, 1542.
9. Buback, M.; Geers, U.; Kurz, C. H. *Macromol. Chem. Phys.* **1997**, *198*, 3451; Buback, M.; Kuchta, F.-D. *Macromol. Chem. Phys.* **1995**, *196*, 1887; Beuermann, S.; Buback, M.; Russell, G. T. *Macromol. Rapid Commun.* **1994**, *15*, 351; Buback, M.; Kurz, C. H.; Schmaltz, C. *Macromol. Chem. Phys.* **1998**, *199*, 1721.
10. Buback, M. *Makromol. Chem.* **1990**, *191*, 1575.
11. Buback, M.; Kowollik, C. *Macromol. Chem. Phys.* **1999**, *200*, 1764.
12. Buback, M.; Kowollik, C. *Macromolecules* **1999**, *32*, 1445.
13. Kowollik, C. *Ph.D. Thesis*, Göttingen 1999.
14. Feldermann, A. *Diploma Thesis*, Göttingen 1999.
15. Buback, M.; Kowollik, C. *Macromolecules* **1998**, *31*, 3221.
16. Fukuda, T.; Ide, N. ; Ma, Y.-D. *Macromol. Symp.* **1996**, *111*, 305.
17. O'Driscoll, K. F.; Ito, K. *J. Polym. Sci.* **1979**, *17*, 3913.
18. Beuermann, S.; Buback, M.; Nelke, D. in preparation for publication.
19. Nelke, D. *Ph.D Thesis*, Göttingen 2001
20. Fischer, J. P.; Mücke, G.; Schulz, G. V. *Ber. Bunsenges. Phys. Chem.* **1970**, *73*, 1077.
21. de Kock, J. B. L. *Chain-Length Dependent Bimolecular Termination in Free-Radical Polymerization*, Universiteitsdrukkerij TUE, Eindhoven 1999.
22. Mahabadi, H. K. *Macromolecules* **1985**, *18*, 1319.
23. Olaj, O. F.; Vana, P. *Macromol. Rapid Commun.* **1998**, *19*, 433; Olaj, O. F.; Vana, P. *Macromol. Rapid Commun.* **1998**, *19*, 533.
24. Olaj, O. F.; Vana, P.; Kornherr, A.; Zifferer, G. *Macromol. Chem. Phys.* **1999**, *200*, 2031.
25. Buback, M.; Kuchta, F.-D. *Macromol. Chem. Phys.* **1997**, *198*, 1455.
26. Fischer, H.; Baer, R.; Hany, R.; Verhoolen, I.; Walbiner, M. *J. Chem. Soc. Perkin Trans.* **1990**, *2*, 787.
27. Buback, M.; Busch, M.; Kowollik, C. submitted to *Macromol. Theory Simul.*
28. Khokhlov, A. R. *Macromol. Chem., Rapid Commun.* **1981**, *2*, 633.
29. Moad, G.; Solomon, D. H. *Azo and Peroxy Initiators*; Eastmond, G. C.; Ledwith, A.; Russo, S.; Sigwalt, P., Eds.; Pergamon: London; Vol.3, p 97.
30. Buback, M.; Hinton, C. in: *High-pressure techniques in chemistry and physics: a practical approach*, Isaacs, N.; Holzapfel W. B., Eds.; Oxford Univ. Press, 1997.
31. Buback, M.; Klingbeil, S.; Sandmann, J.; Sderra, M.-B.; Vögele, H.-P., Wackerbarth, H.; Wittkowski, L. *Z. Phys. Chem. (Munich)* **1999**, *210*, 199.
32. Sandmann, J. *Ph.D. Thesis*, Göttingen 2000; Buback, M.; Sandmann, J., submitted to *Z. Phys. Chem. (Munich)*.
33. Nelke, D. *Diploma Thesis*, Göttingen 1998.
34. Aschenbrücker, J.; Buback, M.; Ernsting, N. P.; Schroeder, J.; Steegmüller, U. *Ber. Bunsenges. Phys. Chem.* **1998**, *102*, 965.

Chapter 4

Stereochemistry in Radical Polymerization of Vinyl Esters

Yoshio Okamoto[1], Kazunobu Yamada[2, 4], and Tamaki Nakano[3]

[1]Department of Applied Chemistry, Graduate School of Engineering,
Nagoya University, Furo-cho, Chikusa-ku, Nagoya 464–8603, Japan
[2]Joint Research Center for Precision Polymerization (JRCPP)-
Japan Chemical Innovation Institute (JCII), Nagoya University,
Furo-cho, Chikusa-ku, Nagoya 464–8603, Japan
[3]Graduate School of Materials Science, Nara Institute of Science and Technology
(NAIST), Takayama-cho 8916-5, Ikoma, Nara 630–0101, Japan

This chapter describes a new stereoregulation method for free-radical polymerization of vinyl esters using fluoroalcohols as solvents. The stereochemistry of the vinyl ester polymerization was remarkably affected by the fluoroalcohols such as $(CF_3)_3COH$ and $(CF_3)_2CHOH$. The polymerization of vinyl acetate (VAc) gave a polymer rich in syndiotacticity (up to $rr = 50\%$, $r = 72\%$), while the polymerization of vinyl propionate (VPr), vinyl isobutyrate (ViBu), vinyl pivalate (VPi), vinyl 2,2-dimethylbutylate (VDMB), and vinyl 2,2-dimethylvalerate (VDMV) gave polymers rich in heterotacticity (up to $mr = 61\%$). A polymer rich in isotacticity (up to $mm = 33\%$) was obtained by the polymerization of vinyl benzoate (VBz). This stereochemistry represents the highest level of content for the radical polymerization of vinyl esters reported so far. The stereochemical effects observed in this study are due to a hydrogen-bonding interaction between the fluoroalcohol molecules and the ester groups of the vinyl ester monomer and the growing polymer. Conversely, the stereochemistry of the polymerization of bulky 2,2-bis(trifluoromethyl)propionate (VF6Pi) in fluoroalcohols was similar to that of the bulk polymerization presumably due to the weak hydrogen-bonding interaction between the monomer and the fluoroalcohols.

Control of free-radical polymerization is drawing much current attention and several living polymerization methods have been developed. Because polymer properties are significantly influenced not only by the molecular weight but also by the main-chain tacticity, stereocontrol of the polymerization is also an important topic in macromolecular chemistry. However, stereoregulation by free-radical polymerization has been achieved only in limited cases so far (*1*).

[4]Present address: R & D Center, Unitika Ltd., 23 Kozakura, Uji, Kyoto 611–0021, Japan.

Vinyl esters afford polymers only by radical polymerization and solvolysis of the obtained polymers leads to poly(vinyl alcohol) (PVA). The thermal and mechanical properties of PVA are significantly influenced by a slight change in the main-chain tacticity, and increasing syndiotacticity of PVA improves the properties (2). However, commercially available PVA, which is produced through the polymerization of vinyl acetate (VAc), has an atactic structure.   In order to increase the syndiotactic specificity of the vinyl ester polymerization, various bulky monomers including vinyl trifluoroacetate (3,4), vinyl pivalate (VPi) (4,5), vinyl diphenylacetate (6), and vinyl 2,2-bis(trifluoromethyl)propionate (VF6Pi) (7) have been designed.   Although the monomer-design method is effective, it lacks versatility and often requires expensive monomers.   In the search for a versatile method which can be applied to conventional monomers, we investigated solvent effects on the stereochemistry of the vinyl ester polymerization.

The solvent effect on the stereochemistry of VAc polymerization is generally very small. However, the polymerization in phenol has been reported exceptionally to afford a polymer rich in syndiotacticity (rr = 32–33%) (8).   This effect has been ascribed to the increased bulkiness of the monomer's side group due to hydrogen-bond formation between the solvent and the acetyl group. This work stimulated us to investigate the radical polymerization of vinyl esters in a wide range of protic solvents that would efficiently interact with the ester groups of the monomers and the growing species   through   the   hydrogen-bonding.   We   were   especially   interested   in fluoroalcohols as solvents because they are more acidic than the corresponding parent alcohols   and   the   monomers   containing   fluoroalkyl   groups   are   known   to   give syndiotactic-rich polymers (3,7).   The solvents used for the polymerization in this study   include CF_3CH_2OH, $(CF_3)_2CHOH$,   and $(CF_3)_3COH$.   The structures of the monomers employed here are shown below.   Unambiguous solvent effects on the polymerization stereochemistry were confirmed as discussed hereafter. However, the mechanism of the effects is still not completely clear, and the discussion on this aspect contains speculations because the reaction systems are quite complex.

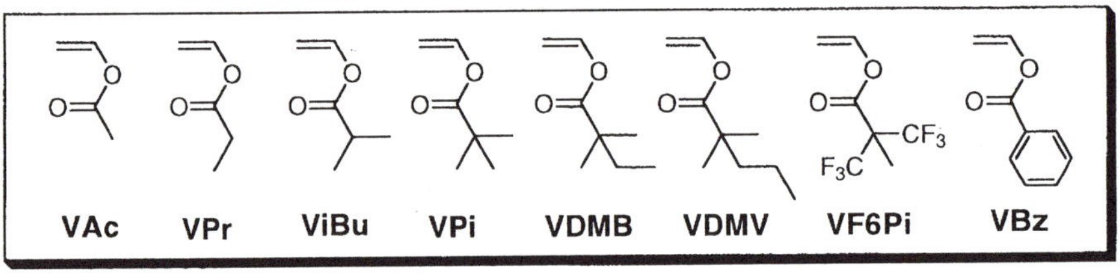

VAc   VPr   ViBu   VPi   VDMB   VDMV   VF6Pi   VBz

Polymerization of Vinyl Esters Using Fluoroalcohols As Solvents

Polymerization of VAc

Table I shows the results of the polymerization of VAc in various solvents.   The VAc polymerization in fluoroalcohols gave polymers with higher syndiotacticity (rr) than that obtained in bulk polymerization or in methanol. The stereochemical effect was larger with a solvent having a smaller pK_a and higher bulkiness and at lower temperature. Syndiotactic specificity of the VAc polymerization in $(CF_3)_3COH$ at -78 °C reached an rr of 50% ($r = 72\%$) (run 8), which is the highest for the radical polymerization of vinyl esters (9a). This effect may be due to the steric or electrostatic repulsion between the apparently bulky side groups of the monomer and the growing chain-end hydrogen-bound by the fluoroalcohol as will be discussed later.

Table I. Radical Polymerization of VAc in Alcoholic Solvents[a]

Run	Solvent	pK_a of Solvent	Temp. °C	Time h	Yield[b] %	$\bar{M}_n \times 10^{-4}$ [c]	$\bar{M}_w/\bar{M}_n$ [c]	Tacticity[d] /%			
								mm	mr	rr	r [e]
1	None	—	20	1	71	5.5	1.9	22.6	48.9	28.5	52.9
2	$(CH_3)_3COH$	19	20	24	73	6.4	1.6	20.9	49.5	29.5	54.3
3	CH_3OH	16	20	24	58	1.1	1.9	22.2	49.2	28.6	53.2
4	CF_3CH_2OH	12.4	20	24	62	1.3	1.9	19.8	49.9	30.3	55.2
5	$(CF_3)_2C(CH_3)OH$	9.6	20	24	64	1.9	2.7	17.9	50.0	32.1	57.1
6	$(CF_3)_2CHOH$	9.3	20	24	81	1.7	2.0	18.6	49.1	32.3	57.7
7	$(CF_3)_3COH$	5.2	20	24	94	6.2	1.8	13.0	49.4	37.6	62.3
8	$(CF_3)_3COH$	5.2	-78	168	50	8.3	1.5	5.4	44.9	49.8	72.2

[a] $[VAc]_0 = 2.2$ M (20 vol%). Polymerizations at 20°C and –78°C were initiatied with AIBN (0.15 M) under UV light irradiation and with $(nBu)_3B$ (0.20 M) in the presence of a small amount of air, respectively.
[b] Et_2O-insoluble part.
[c] Determined by GPC of original polymers using standard polystyrenes in THF.
[d] Determined by ^1H or ^{13}C NMR of PVA in DMSO-d_6.
[e] Calculated on the basis of triad tacticity ($r = rr + mr/2$).
SOURCE: Adapted from reference 9b.

The differences in activation enthalpy ($\Delta H^{\ddagger}$) and activation entropy ($\Delta S^{\ddagger}$) between isotactic and syndiotactic-specific propagation for the VAc polymerization are summarized in **Table II** along with the data for the bulk polymerization of VF6Pi. The data were obtained from the results of polymerization at different temperatures. The large positive values of $\Delta H_i^{\ddagger} - \Delta H_s^{\ddagger}$ and the small positive values of $\Delta S_i^{\ddagger} - \Delta S_s^{\ddagger}$ in the VAc polymerization in the fluoroalcohols indicate that the syndiotactic-specific propagation was favored by enthalpy and disfavored by entropy. These results are similar to those for the VF6Pi polymerization. This suggests that the stereochemical mechanism of the VAc polymerization in the bulky fluoroalcohols may be similar to that for the VF6Pi polymerization. Bulkiness of VAc and the growing chain hydrogen-bound by the bulky solvents may have similar steric effects to that of the bulky VF6Pi monomer and its growing radical.

Polymerization of Monomers Bulkier than VAc

Polymerizations of the monomers bulkier than VAc including VPr, ViBu, VPi, VDMB, and VDMV were also examined using the fluoroalcohols as solvents. In contrast to the syndiotacticity-enhancing effect of the fluoroalcohols in the VAc polymerization, the solvents appeared to be effective for the polymerization of these monomers in increasing heterotactic triad content (mr) and at the same time meso diad content (*m*) of the polymers (**Table III**). The polymerization of VPi in $(CF_3)_3COH$ at –40 °C gave an *mr* of 61% (run 11), which is the highest reported for the radical homopolymerization of vinyl monomers (*9b*). The mechanism of the heterotactic-specific polymerization will be discussed later.

Table II. Activation Parameters for the Polymerization of VAc and VF6Pi

Monomer	Solvent	[Solvent]$_0$ vol%	$\Delta H_i^{\ddagger} - \Delta H_s^{\ddagger}$ cal/mol	$\Delta S_i^{\ddagger} - \Delta S_s^{\ddagger}$ cal/deg•mol	Reference
VAc	None	0	-10 ± 40	-0.3 ± 0.2	9b
	PhOH	10[a]	28	-0.35	8
	CH_3OH	80	80 ± 70	0.0 ± 0.3	9b
	CF_3CH_2OH	80	200 ± 70	0.3 ± 0.3	9b
	$(CF_3)_2CHOH$	80	520 ± 70	1.1 ± 0.3	9b
	$(CF_3)_3COH$	80	550 ± 50	0.9 ± 0.2	9b
VF6Pi	None	0	460	0.2	7

[a] wt %.

Table III. Radical Polymerization of Bulkier Monomers than VAc[a]

Run	Monomer	Solvent	Temp. °C	Time h	Conv.[b] %	$\bar{M}_n \times 10^{-4}$ [c]	$\bar{M}_w/\bar{M}_n$ [c]	Tacticity[d] mm	mr	rr
1	**VPr**	None	20	2	31	4.3	1.9	19.5	49.9	30.5
2	**ViBu**	None	20	4	58	1.7	2.1	16.7	49.8	33.6
3	**VPi**	None	20	1	65	7.1	1.6	13.7	49.2	37.1
4	**VPi**	None	-40	24	52	15.3	2.1	11.1	47.9	41.0
5	**VDMB**	None	20	4	76[e]	8.3	1.6	14.5	48.7	36.8
6	**VDMV**	None	20	4	45[e]	21.7	3.1	12.9	48.4	38.8
7	**VF6Pi**	None	20	9	72	4.0	1.4	11.2	44.5	44.4
8	**VPr**	$(CF_3)_3COH$	20	24	83	4.7	1.8	14.6	53.4	32.1
9	**ViBu**	$(CF_3)_3COH$	20	24	70	2.9	1.7	16.9	57.0	26.1
10	**VPi**	$(CF_3)_3COH$	20	24	73	2.4	2.0	20.7	57.5	21.8
11	**VPi**	$(CF_3)_3COH$	-40	24	84	4.2	1.8	21.3	61.0	17.7
12	**VDMB**	$(CF_3)_3COH$	20	24	63[e]	1.9	3.0	21.0	57.7	21.4
13	**VDMV**	$(CF_3)_3COH$	20	24	53[e]	1.9	1.6	20.3	56.0	23.7
14	**VF6Pi**	$(CF_3)_3COH$	20	24	25	1.2	1.9	12.9	45.7	41.4

[a] Polymerizations at 20 °C and -40 °C were initiated with AIBN (0.15 M) under UV light irradiation and with $(nBu)_3B$ (0.20 M) in the presence of a small amount of air, respectively.
[b] Determined by ^1H NMR of reaction mixture in acetone-d_6.
[c] Determined by GPC of original polymers using standard polystyrenes in THF.
[d] Determined by ^1H or ^{13}C NMR of PVA in DMSO-d_6. [e] MeOH/H$_2$O (=1/1)-insoluble part.
SOURCE: Adapted from references 9b and 10.

Polymerization of VF6Pi

The bulk polymerization of VF6Pi is known to be syndiotactic-specific due to steric repulsion between the bulky side groups (7). Contrary to the aforementioned polymerization of vinyl esters in fluoroalcohols, the stereochemistry of the VF6Pi polymerization was little affected by the fluoroalcohol solvents (runs 7 and 14 in Table III). In order to understand the effect of the bulkiness of the monomers on the polymerization stereochemistry, the triad tacticities of the polymer are plotted versus syndiotactic specificity (r) in the bulk polymerization at 20 °C ($r_{20^\circ C}$) (**Figure 1**). The diad r in the bulk polymerization can be taken as an index of the bulkiness of the monomers. With increasing bulkiness of the monomer ($r_{20^\circ C}$ value), rr increased and mm decreased in the bulk polymerization (Figure 1A), while the stereochemistry of the polymerization in (CF$_3$)$_3$COH was more complex; mm and mr reached maximum values and rr reached a minimum value when $r_{20^\circ C}$ comes close to those of VPi, VDMB, and VDMV (Figure 1B). These results suggest that the hydrogen-bonding interaction between VF6Pi, which is much bulkier than VPi, and the fluoroalcohol is much weaker than that between other monomers and the solvent due to steric reasons. The bulkiness of VPi, VDMB, and VDMV may be a critical one beyond which the hydrogen bonding with the fluoroalcohol becomes relatively weak.

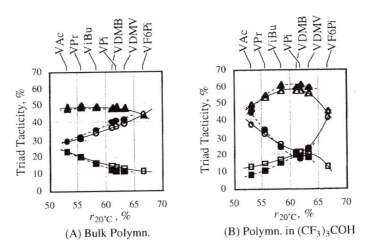

Figure 1. Relationship between $r_{20^\circ C}$ and the triad tacticity of the polymers obtained by the bulk polymerization (A) and polymerization in (CF$_3$)$_3$COH (B). Triad tacticity: mm (□, ■), mr (△, ▲), rr (○, ●). Polymerization temperature: 20 °C (open symbols), –40 °C (filled symbols). (Reproduced from reference 10 under copyright permission from 1999 John Wiley & Sons, Inc.).

Polymerization of VBz

It has been reported that the polymerization of VBz derivatives affords slightly isotactic-rich polymers (up to mm = 30% for VBz (*11*), and mm = 32% for vinyl *m*-bromobenzoate (*12*)), and this was attributed to the stacking effect of the monomer's aromatic ring (*13*). In the present study, we found that the polymerization of VBz in CF_3CH_2OH and $(CF_3)_2CHOH$ gave polymers with higher isotacticity (mm) than the bulk polymerization (**Table IV**). It is notable that the polymerization of VBz in $(CF_3)_2CHOH$ at –20 °C led to an mm of 33% (run 6), which is the highest for the polymerization of vinyl esters, while $(CF_3)_3COH$ caused only a slight effect. In the polymerization of VBz, the π-stacking effect, which may be enhanced using polar alcoholic solvents, and the hydrogen-bonding effect making the monomer apparently bulkier can coexist. In the VBz polymerization in CF_3CH_2OH and $(CF_3)_2CHOH$, the π-stacking effect may be dominant over the hydrogen-bonding effect. On the other hand, in the polymerization in $(CF_3)_3COH$, the syndiotacticity-enhancing effect based on the hydrogen-bonding interaction is considered to cancel out the isotacticity-enhancing effect based on the π-stacking, resulting in the slight change in the polymerization stereochemistry.

Table IV. Radical Polymerization of VBz in Fluoroalcohols[a]

Run	Solvent	$[M]_0$ mol/L	Temp. °C	Time h	Yield[b] %	$\bar{M}_n \times 10^{-4}$ [c]	$\bar{M}_w/\bar{M}_n$ [c]	Tacticity[d] mm mr rr
1	None	7.2	20	48	32	1.8	1.9	26.7 50.8 22.5
2	CF_3CH_2OH	1.4	20	48	43	1.2	1.8	29.0 50.7 20.3
3	$(CF_3)_2CHOH$	1.4	20	48	71	1.1	1.8	30.3 51.0 18.7
4	$(CF_3)_3COH$	1.4	20	48	79	1.2	1.8	27.0 50.4 22.6
5	$(CF_3)_2CHOH$	1.4	-20	48	51	1.4	1.8	31.8 51.6 16.5
6	$(CF_3)_2CHOH$	0.7	-20	48	36	1.2	1.8	32.7 50.5 16.8

[a] Polymerizations at 20°C and –20°C were initiated with AIBN (0.15M) under UV light irradiation and with ($nBu)_3B$ (0.10 M) in the presence of a small amount of air, respectively.
[b] MeOH-insoluble part.
[c] Determined by GPC of original polymers using standard polystyrenes in THF.
[d] Determined by ^{1}H NMR of PVA in DMSO-d_6.
SOURCE: Adapted from reference 10.

Hydrogen-Bonding between Fluoroalcohols and Vinyl Esters

Interaction between VAc and Alcohols

As already described, the interaction between the vinyl esters and the fluoroalcohols can affect the stereochemistry of the polymerization. In order to learn about the nature and the function of the interaction, solution systems of VAc containing CH_3OH, CF_3CH_2OH, $(CF_3)_2CHOH$, and $(CF_3)_3COH$ were studied in $CDCl_3$ at 20 °C by means of ^1H and ^{13}C NMR spectroscopy. In the presence of an alcohol, the signal of the carbonyl carbon of VAc in the ^{13}C NMR shifted downfield, and the OH signal of an alcohol in the ^1H NMR also showed a downfield shift in the presence of VAc. The extent ($\Delta\delta$) of the downfield shift was higher when an alcohol with a smaller pK_a was used for the experiment (**Figure 2A**). These results indicate that VAc and an alcohol interact through hydrogen-bonding between the carbonyl oxygen of VAc and the hydroxyl group of an alcohol, and the interaction becomes stronger when the pK_a of the alcohol is smaller.

Interaction between Vinyl Esters and $(CF_3)_3COH$

Influence of the ester side groups of the monomers on the hydrogen-bonding formation was examined with $(CF_3)_3COH$. **Figure 2B** shows the changes in the ^1H NMR chemical shift of the hydroxyl group of $(CF_3)_3COH$ in the presence of the vinyl esters (*9b*, *10*). The downfield shift of the OH signal due to hydrogen bonding was found in the presence of the vinyl esters. The magnitude of the shift was similar when VBz, VAc, and VPi were used, in spite of their different bulkiness, while it was much smaller for VF6Pi. This accounts for the small stereochemical effect of $(CF_3)_3COH$ on the VF6Pi polymerization.

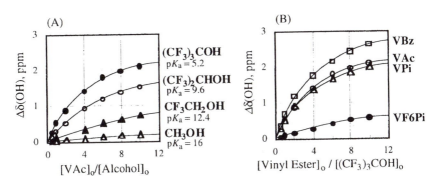

Figure 2. *Changes in the chemical shift of the OH proton of the alcohols in the presence of vinyl esters. [Alcohol]$_o$ = 50 mM, 400 MHz, CDCl$_3$, 20 ± 1 °C. ((A): Reproduced from reference 9b. Copyright 1998 American Chemical Society. (B): Reproduced from reference 10 under copyright permission from 1999 John Wiley & Sons, Inc.).*

Stereochemical Statistics and Stereoregulation Mechanism

Stereochemical Statistics

With the triad, tetrad, and the pentad stereochemical information obtained from the ^{13}C NMR spectra of PVA, stereochemical statistics including a Bernoulli model and a first-order Markov model were tested (*9b*). The tacticity fraction of the syndiotactic PVA derived from the VAc polymerization in (CF$_3$)$_3$COH (run 8 in Table I, r = 72%) was well described by a first-order Markov model with $P(m/r)$ = 0.81 and $P(r/m)$ = 0.31, where $P(m/r)$ means the probability that a monomer adds in an r-fashion to ~~~mM• species. The stereochemistry of the bulk VPi polymerization giving a syndiotactic polymer (run 4 in Table III, rr = 41%, r = 65%) could be described by a Bernoulli model with $P(m)$ = 0.35 or a first-order Markov model with $P(m/r)$ = 0.68 and $P(r/m)$ = 0.37. This suggests that the steric repulsion between the entering VAc and the chain-end (penultimate and/or penpenultimate) monomeric units hydrogen-bound by the solvent influences the stereochemistry of the syndiotactic-specific polymerization. However, the stereochemistry of the polymerization in (CF$_3$)$_3$COH giving a heterotactic polymer (run 11 in Table III, mr = 61%) could not be described by either a Bernoulli or a first-order Markov model or even by a second-order Markov model, indicating that the heterotactic-specific propagation is based on a more complicated mechanism than that for the syndiotactic-specific propagation.

Mechanism of Heterotactic-Specific Propagation

In order to explain the stereochemistry, we proposed the models of the growing species shown in **Figure 3**; i.e., the two kinds of growing radicals, ~~~mM• and ~~~rM•, having m and r chain-end configurations, respectively. The former adds VPi in an r-fashion and the latter in an m-fashion to achieve heterotactic-specific polymerization. These different stereoselectivities may be related to the different strength of the interaction between the monomeric units and the alcohol molecules; ~~~rM• species may have a stronger interaction than ~~~mM• species due to steric reasons. Accordingly, the ~~~rM• species may have alcohol molecules attached to the chain-terminal and the penultimate unit (Figure 3A), whereas the ~~~mM• species may not have an alcohol at the penultimate unit (Figure 3B). Steric repulsion between the entering monomer bound by an alcohol and the alcohol-bound chain-terminal unit of ~~~mM• may explain the monomer addition in an r-fashion. The monomer addition of ~~~rM• in an m-fashion may be attributable to be the higher crowdedness in the vicinity of the chain-and of ~~~rM• than that of ~~~mM•. The growing species may assume a temporary conformation (partial helix) leading to m-addition induced by the steric repulsion. The helical conformation is known for a highly crowded growing chain in the bulky triarylmethyl methacrylates polymerization giving isotactic polymers (*14*).

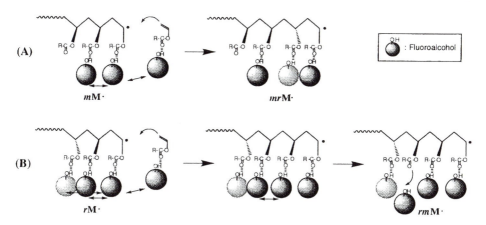

Figure 3. Stereoregulation mechanism in the heterotactic-specific free-radical polymerization. (Reproduced from reference 9b. Copyright 1998 American Chemical Society).

Melting Point of PVA

Stereoregular PVA is expected to have superior thermal and mechanical properties, and the melting point (T_m) can be a criterion for the heat-resistance property of PVA. Recently, it was reported that the polymerization of VPi in *n*-hexane and saponification of the obtained polymer afforded a PVA with $r = 69\%$ and T_m of 258°C (*5b*). **Figure 4** shows the relationship between syndiotacticity and T_m of the PVA's obtained through the polymerization of VAc and VPi in $(CF_3)_3COH$ in the present study along with those obtained by the VPi polymerization in *n*-hexane (*5b,15,16*). With an increase in syndiotacticity, T_m increased and the PVA with $r = 72\%$ showed a T_m of 269°C. This value is higher than that of the PVA with $r = 69\%$ obtained through the polymerization of VPi. On the other hand, T_m of the heterotactic PVA obtained through the VPi polymerization in $(CF_3)_3COH$ is lower than that of an atactic PVA. This may be attributed to the decrease in syndiotactic sequences which may contribute to enhancement of T_m through the intermolecular hydrogen-bonding of PVA.

The stereochemical effect of the fluoroalcohols on the vinyl ester polymerization and the origin of the effects can be summarized as shown in **Table V**. Based on the hydrogen-bonding to alcohols or π-stacking of aromatic groups of the monomer, the stereochemistry of the vinyl ester polymerizations can thus be controlled.

We recently found that the solvent effects of fluoroalcohols on the stereochemistry of the vinyl ester polymerizations is applicable to the radical polymerization of different classes of monomers including methacrylates (*17*).

66

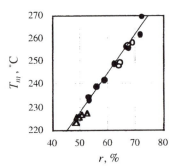

Figure 4. *Relationship between diad syndiotacticity (r) and melting point (Tm) of PVA obtained through the polymerization of VAc in (CF₃)₃COH (●), VPi in (CF₃)₃COH (△), and VPi in n-hexane (○). (Reproduced from reference 9b. Copyright 1998 American Chemical Society).*

Table V.   Summary of the Stereochemical Effect of Fluoroalcohols on the Vinyl Ester Polymerization

Monomer	Fluoroalcohol	H-Bonding	π-Stacking	Stereoeffect
VAc	$(CF_3)_3COH$	+	−	rr ↑
VPi	$(CF_3)_3COH$	+	−	mr ↑
VF6Pi	$(CF_3)_3COH$	±	−	−
VBz	$(CF_3)_2CHOH$	±	+	mm ↑
VBz	$(CF_3)_3COH$	+	+	−

Acknowledgements

This work was supported by the New Energy and Industrial Technology Development Organization (NEDO) under the Ministry of International Trade and Industry (MITI), Japan, through the grant for "Precision Catalytic Polymerization" in the Project "Technology for Novel High-Functional Materials" (1996–2000).

Literature Cited

1. Nakano, T; Okamoto, Y. In *ACS Symp. Ser. 685*, Matyjaszewski, K. Ed.; ACS: Washington D.C., 1998, p. 451–462.
2. For reviews: (a) Finch, C. A. Ed. *Polyvinyl Alcohol -Developments*; Wiley: Chichester, 1992. (b) Fujii, K. *J. Polym. Sci., Macromol. Rev.* **1971**, *5*, 431. (c) *PVA no Sekai (The World of PVA) (in Japanese)*; Kobunshi Kankokai: Kyoto, 1992.

3. Matsuzawa, S.; Yamaura, K.; Noguchi, H.; Hayashi, H. *Makromol. Chem.* **1973**, *165*, 217.
4. Nozakura, S.; Sumi, M.; Uoi, M.; Okamoto, T.; Murahashi, S. *J. Polym. Sci., Polym. Chem.* **1973**, *11*, 279.
5. (a) Fukae, R.; Kawakami, K.; Yamamoto, T.; Sangen, O.; Kato, T.; Kamachi, M. *Polym. J.* **1995**, *27*, 1257. (b) Fukae, R.; Yamamoto, T.; Fujita, Y.; Kawatsuki, N.; Sangen, O.; Kamachi, M. *ibid.* **1997**, *29*, 293.
6. Nakano, T.; Makita, K.; Okamoto, Y. *Polym. J.* **1998**, *30*, 681.
7. Yamada, K.; Nakano, T.; Okamoto, Y. *Polym. J.* **1998**, *30*, 641.
8. Imai, K.; Shiomi, T.; Oda, N.; Otsuka, H. *J. Polym. Sci., Part A: Polym. Chem.* **1986**, *24*, 3225.
9. (a) Yamada, K.; Nakano, T.; Okamoto, Y. *Proc. Jpn. Acad.* **1998**, *74(B)*, 46. (b) Yamada, K.; Nakano, T.; Okamoto, Y. *Macromolecules* **1998**, *31*, 7598.
10. Yamada, K.; Nakano, T.; Okamoto, Y. *J. Polym. Sci., Part A: Polym. Chem.* **1999**, *37*, 2677.
11. Imai, K.; Shiomi, T.; Tezuka, Y.; Kawanishi, T.; Jin, T. *J. Polym. Sci., Part A: Polym. Chem.*, **1988**, *26*, 1961.
12. Imai, K.; Shiomi, T.; Tezuka, Y.; Fujioka, N.; Hosokawa, T.; Ueda, N.; Fujita, K. *Kobunshi Ronbunshu*, **1989**, *46*, 261; *Chem. Abstr.*, **111**, 98048 (1989).
13. Yamamoto, T.; Matsumoto, T.; Sugimoto, M.; Sangen, O.; Isono, T.; Kamachi, M. *Prepr. 34th IUPAC Int. Symp. MACRO AKRON*, 1994, p. 94 .
14. Kamachi, M. In *Catalysis in Precision Polymerization*; Kobayashi, S. Ed.; Wiley: Chichester, 1997, p. 239–240.
15. Fukae, R.; Yamamoto, T.; Sangen, O.; Saso, T.; Kato, T.; Kamachi, M. *Polym. J.* **1990**, *22*, 636.
16. Lyoo, W. S.; Ha, W. S. *J. Polym. Sci., Part A: Polym. Chem.*, **1997**, *35*, 55.
17. Isobe, Y; Yamada, K.; Nakano, T.; Okamoto, Y. *Macromolecules* **1999**, *32*, 5979.

Chapter 5

Electron Paramagnetic Resonance Study of Conventional and Controlled Radical Polymerizations

Atsushi Kajiwara[1], Krzysztof Matyjaszewski[2], and Mikiharu Kamachi[3]

[1]Department of Materials Science, Nara University of Education,
Takabatake-cho, Nara 630–8528, Japan
[2]Department of Chemistry, Carnegie Mellon University, 4400 Fifth Avenue,
Pittsburgh, PA 15213
[3]Department of Applied Chemistry, Fukui University of Technology,
3–6–1 Gakuen, Fukui 910–0028, Japan

Electron paramagnetic resonance (EPR) spectroscopy has been applied to both conventional and controlled radical polymerization systems. In the study of conventional radical polymerization, well-resolved signals of radical species in the polymerization systems were clearly observed by careful optimization of experimental conditions. Precise information on the structures of the primary propagating radicals could be obtained. EPR spectroscopy and kinetic analysis were applied to controlled radical polymerization (atom transfer radical polymerization, ATRP) systems of styrene and (meth)acrylates. Although only a copper (I) species was added to the system initially, EPR signals of copper (II) species were clearly observed in the polymerization mixtures. As the polymerization proceeded, the concentration of copper (II) increased gradually until a nearly steady state was reached. The correlation between time dependence of concentration of copper (II) species and kinetics of polymerization is discussed for various ATRP systems.

Electron paramagnetic resonance (EPR) spectroscopy is a very useful tool to investigate paramagnetic species.[1] Structures, concentrations, and dynamics of paramagnetic compounds can be obtained from EPR measurements in the study of radical polymerizations.[2,3]

EPR has been applied to the investigation of elementary processes of radical polymerization. Initiation, propagation, and termination rate constants have been estimated by EPR methods. If the steady state of the radical concentration can be measured by EPR spectroscopy, the propagation rate constant (k_p) can be determined directly from

$$R_p \ (= -d[M]/dt) = k_p[P_n\bullet][M] \qquad (1)$$

or its integrated form

$$\ln([M]_1/[M]_2) = k_p[P_n\bullet](t_2\text{-}t_1) \qquad (2)$$

where R_p and $[P_n\bullet]$ are the rate of polymerization and the concentration of the propagating radical, respectively, and $[M]_1$ and $[M]_2$ are the monomer concentrations at time t_1 and t_2, respectively. The estimated rate constants are very important for understanding elementary processes of radical polymerizations.

EPR has also been applied to mechanistic studies of atom transfer radical polymerization systems. ATRP has proved to be a method for obtaining polymers with low polydispersities and predictable molecular weights.[4] EPR spectroscopy can be used to investigate the chemistry of paramagnetic metal complexes of ATRP systems. EPR can potentially yield information on the local structure, coordination structure, aggregated structure, symmetry, and concentrations of paramagnetic copper (II) species.[5]

Correlation of the concentration of copper (II) species with the kinetics of polymerizations in ATRP systems was also investigated.

Conventional Radical Polymerization

Direct Detection of Propagating Radicals in Polymerization Systems

Direct detection of propagating radicals of radical polymerization by EPR spectroscopy has been very difficult, mainly due to both the labile nature and the extremely low concentration of the propagating radicals. About 20 years ago, Kamachi *et al.* observed EPR spectra of propagating radicals of methacrylates under similar conditions to conventional polymerization by means of a specially designed flat cell and cavity.[1] Recently, better resolved EPR spectra of propagating radicals have been observed using a commercially available cavity and normal sample cell, due to both improvement in EPR spectrometers and careful optimization of preparation of the sample.[6] Well-resolved spectra provide detailed information about the structures and chemical properties of the radicals.

EPR spectra of the propagating radical of styrene were clearly observed in a bulk polymerization at various temperatures under irradiation. A typical example of the EPR signal of the propagating radical of styrene in a conventional radical polymerization is shown in Figure 1a.   A well-resolved signal with good S/N ratio was obtained by careful choice of experimental conditions.   Simulation of the spectrum fit with the observed spectrum of the propagating radical of styrene. GPC measurements showed that polymers of molecular weight above 10000 were formed in the polymerization systems, so the signal was considered to be due to a polymeric radical.

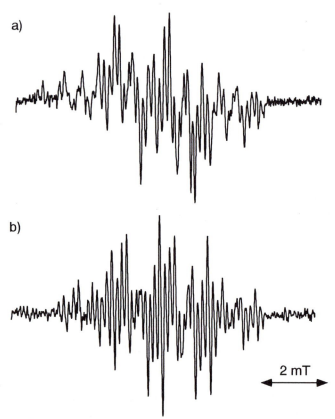

Figure 1   a) EPR spectrum of the normal propagating radical of styrene initiated by tBPO under irradiation at 0 °C.
b) EPR spectrum of the primary propagating radical of styrene initiated by MAIB under irradiation at 0 °C.

When azo initiator (dimethyl 2,2'-azobis(isobutyrate), MAIB) was used in photo-initiated polymerization of styrene, an EPR signal was also observed (Fig. 1b).   The spectrum was very similar to that of the normal propagating radicals, but the precise values of the hyperfine splitting constants (hfc's) were different from those of the propagating radical. Results from various experiments suggested that the signal is due to a primary propagating radical, which is formed by an addition of a fragment of the initiator to one monomer (Fig. 2). In that case, the electron density of the protons of the styrene unit is changed.

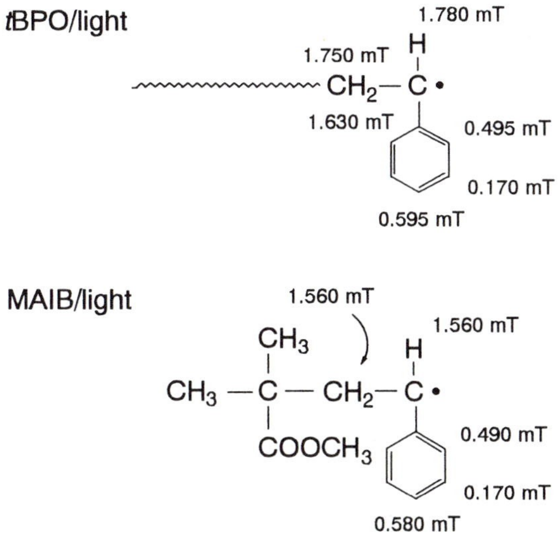

Figure 2   Structures and hyperfine splitting constants of normal and primary propagating radicals of styrene.

Two kinds of information could be obtained from the results of the EPR measurements of the azo-initiated polymerization. One is concerned with the chain length of the radicals. The result showed that spectra of very short radicals (primary propagating radical) can be distinguished from those of longer propagating radicals by the precise EPR measurements. That finding will be helpful in choosing the polymerization conditions for estimation of rate constants of propagating radicals. Another finding is important information which is concerned with a penultimate effect. In the normal propagating radical of styrene, the penultimate unit is a styrene unit. However, in the case of a primary propagating radical, the penultimate unit has an MMA structure. A difference in hfc's between normal and primary propagating radicals is considered to be caused by the different electronic interaction of a penultimate MMA unit compared to a penultimate styrene unit. The postulated penultimate effect for copolymerization systems has been investigated. The difference in hfc's between the well-resolved EPR spectra of the normal and primary propagating radicals indicates that the qualitative difference of the electronic state of the propagating radicals was clearly observed, which is probably the origin of the penultimate effect on the copolymerization.[7] Detection of the well-resolved spectra of propagating radicals in EPR spectroscopy provides further information on the polymerizations.

72

Estimation of Propagation Rate Constants

The steady state radical concentration can be calculated by double integration of the EPR spectrum. Values of k_p were calculated from eq. 1 or 2. Previously, k_p values have been estimated for styrene, methacrylates, and dienes by the EPR methods.[2,3,8-10] For example, the k_p's were ca. 400 (at 70 °C)[8], ca. 150 (at 5 °C)[2], and ca. 400 (at 20 °C)[10] for styrene, butadiene, and dodecyl methacrylate (DMA), respectively. In some cases, (e.g. DMA and styrene) the values showed good agreement with those determined by the Pulsed Laser Polymerization (PLP) method.[10], some (e.g. butadiene) did not. The disagreement might be caused by a difference in both the chain length of propagating radicals and the extent of oligomer formation. When considerable amounts of soluble oligomers wre formed in a polymerization system, the amount of consumption of monomers will be over estimated to give larger k_p value.

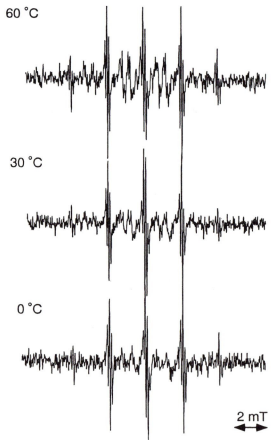

Figure 3   EPR spectra of propagating radicals of nBMA at various temperatures.

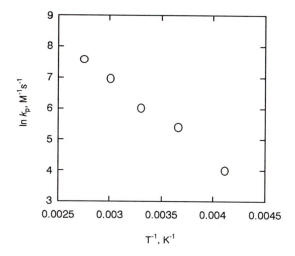

Figure 4   Arrhenius plot for k_p values for nBMA.

Table 1   Estimated k_p values for Various Monomers

monomer	temperature, °C	k_p	ref.
n-butyl methacrylate	90	2000 ± 200	11
	60	1070 ± 50	11
	30	410 ± 20	11
	0	220 ± 10	11
	-30	55 ± 7	11
dodecyl methacyrlate	40	714	10
	10	321	10
benzyl methacrylate	40.1	899 ± 51	9
	10	480 ± 87	9
styrene	70	450 ± 70	8
butadiene	5	150 ± 40	2
isoprene	5	125 ± 30	2
2-methyl-1,3-pentadiene	5	35 ± 10	2
1,3-hexadiene	5	20 ± 10	2
2,4-hexadiene	5	16 ± 12	2

Recently, k_p values for n-butyl methacrylate (nBMA) were estimated by the EPR method. In order to calculate the steady state radical concentration, EPR spectra of propagating radicals of nBMA were observed at various temperatures (Figure 3). As shown in the figure, the features of the spectra changed gradually

74

with temperature. We interpret this change in terms of the dynamics of the propagating chain end affecting the spectra. This is a characteristic feature of the EPR spectra of methacrylates. Steady state radical concentrations were calculated by double integration of the spectra and k_p values were calculated by eq. 1 or 2. An Arrhenius plot of the results is shown in Figure 4. The activation energy (E_a) and preexponential factor (A) for the photo-initiated radical polymerization of nBMA were estimated from the slope and the intercept of these plots to be (22.0 ± 1.0) kJ/mol and (3.40 ± 0.4) x 10^6 $M^{-1}s^{-1}$, respectively. The results was compared with those from PLP method to investigate a precision and an accuracy of the obtained values, which have been independently reported to be 23.3 kJ/mol and 3.44 x 10^6 $M^{-1}s^{-1}$,[12] (20.6 ± 0.3) kJ/mol and (1.81 ± 0.3) x 10^6 M^{-1} s^{-1},[13] 21.9 kJ/mol and 2.65 x 10^6 $M^{-1}s^{-1}$,[24] and 23.6 kJ/mol and 4.78 x 10^6 $M^{-1}s^{-1}$,[15] respectively. The E_a showed fairly good agreement among EPR and PLP method. The A is also in fair agreement with an average value of these exponential factors obtained by PLP method. Detailed analysis for both analyses should be considered for more accurate determination in the future. Values of k_p for benzyl methacrylate were also estimated by the EPR method and already reported.[9]

Controlled Radical Polymerization

EPR spectroscopy has been applied to investigations of ATRP systems.

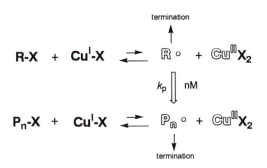

X = halogen atom
M = styrene, MMA, acrylonitrile, and MA

Scheme 1 shows a typical process of ATRP. In the scheme, the initiating radical, propagating radical, and copper (II) species are paramagnetic and EPR observable. In principle, all of the paramagnetic species could be observed by EPR spectroscopy; however, unfortunately (or fortunately) only the paramagnetic copper (II) species can be observed due to its high concentration relative to the initiating and propagating radicals. The concentration of the organic radicals in these systems is usually in the range of $10^{-8} - 10^{-7}$ mol/L. The concentrations of copper (II) species in this system are above 10^{-3} mol/L which is $10^4 - 10^5$ times

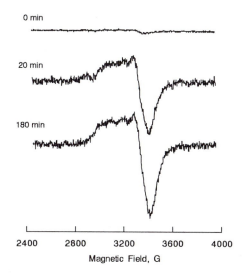

Figure 5   EPR spectra of the polymerization mixture measured at 25 °C after 0, 20, and 180 min at 110 °C, for a system of (styrene/1-phenylethyl bromide/CuBr/dNbipy = 100/1/1/2 in toluene (50 %)).

higher than the initiating and propagating radicals, according to a persistent radical effect resulting from irreversible radical termination.   Thus, the copper (II) species is the predominant species observed by EPR in the ATRP system.

ATRP systems of styrene and (meth)acrylates were investigated by EPR spectroscopy.   Polymerizations were carried out in an EPR tube at 110 °C (styrene) or 90 °C ((meth)acrylate) and spectra were recorded at room temperature (styrene) or 20 K ((meth)acrylates).   For example, the result of the time dependence of the EPR signal of copper (II) species in the ATRP of styrene is shown in Figure 5.   The spectrum which was observed after 20 min of heating is considered to be a typical axially symmetric copper (II) signal and can correspond to either trigonal bipyramidal or square pyramidal structures.   Frozen state signals of copper (II) species might provide more detailed information about the structure of the copper complexes (vide infra).

The concentration of copper (II) species was estimated by double integration of the spectra.   The time dependence of the concentrations of these species for various polymerization conditions of styrene and (meth)acrylates is shown in Figure 6.   Concentrations of copper (II) species in ATRP systems of styrene, MMA, and methyl acrylate (MA) were ca. 2.5 − 3 mM, ca. 0.7 − 0.9 mM, and ca. 0.6 mM, respectively.   The percentage of copper (II) formed from copper (I) was calculated to be ca. 5 − 6 %, ca. 5 − 6 %, and ca. 3 % for styrene, MMA, and MA, respectively.   In the case of a typical styrene ATRP system (styrene/1-phenylethyl bromide/CuBr/dNbipy = 100/1/1/2 in toluene (50 %)), approximately 5-6 % of the copper (I) was converted to copper (II) species during the polymerization, leaving 94 – 95 % of the initial amount of copper (I) still in the monovalent state.[16] These values obviously depend on the initial concentrations of the initiator and copper (I) species and should correspond to the particular concentrations used for experiments presented in the Figures.

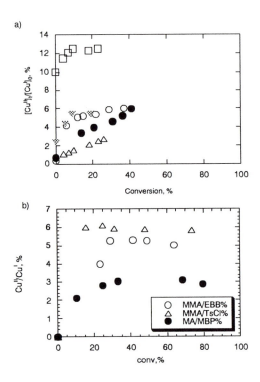

Figure 6   a) Plots of time dependence of proportion of copper (II) species formed from copper (I) species by ATRP initiated by 1-phenylethyl bromide (open circle) , 1-phenylethyl chloride (open triangle), and benzyl bromide (filled circle) in toluene solution and bulk polymerization initiated by 1-phenylethyl bromide (filled triangle) as well as in the presence of externally added CuBr$_2$ (open square).
b)   Plots of time dependence of proportion of copper (II) species formed from copper (I) species by ATRP initiated by ethyl 2-bromoisobutyrate (open circle) and p-toluenesulfonyl chloride (open triangle) in diphenyl ether solution and MA ATRP initiated by methyl 2-bromopropionate (filled circle).

EPR signals at a polymerization temperature of 110 °C have also been measured and investigated.[17]   The spectra are very similar to those measured at room temperature (25 °C).[16] Estimated concentrations of copper (II) species are almost the same as those at room temperature. These results show that the EPR data at room temperature can be used for analysis of the mechanism of ATRP.

Correlation of Copper (II) Concentrations and Kinetics of ATRP.

A correlation of the copper (II) concentration and kinetics of the same polymerization system was examined (Figures 7 and 8). This correlation is useful for a discussion of the mechanism of ATRP.[17,18]

In the case of a 1-phenylethyl bromide initiated and CuBr catalyzed system (rapid initiation system), plots of copper (II) concentration, M_n, and M_w/M_n as a function of monomer conversion are shown in Figure 7. The copper (II) concentration reached a pseudo-steady state within 30 min. M_n increased linearly with conversion and showed good agreement with the theoretical prediction (solid line). Polydispersities decreased with conversion and remained low, $M_w/M_n <$ 1.15, during the polymerization.

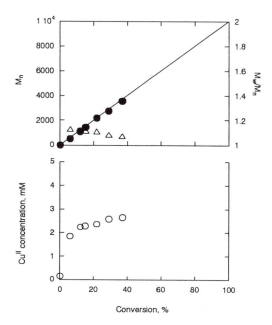

Figure 7   Molecular weight, M_n, and molecular weight distribution, M_w/M_n, dependence on monomer conversion for a polymerization system of (styrene/1-phenylethyl bromide/CuBr/dNbipy = 100/1/1/2 in toluene (50 %)) as well as copper(II) concentration dependence on conversion.

On the other hand, in a 1-phenylethyl chloride initiated and CuCl catalyzed styrene ATRP system, the copper (II) concentration increased slowly and did not reach a steady state (Figure 8). Copper (II) concentration was more than 1 mM above 20 % conversion. At less than 10 % conversion, M_n was larger than the theoretical prediction, perhaps due to an insufficient rate of deactivation. Above 20 % conversion, M_n showed a linear relationship with conversion and agreed with the theoretical prediction. The higher than predicted M_n in the initial stage of polymerization indicates incomplete initiation and slow deactivation. The copper (II) concentrations are relatively low in this polymerization system compared to the polymerization process which showed a steady M_n. This may mean that the concentration of copper (II) species is too low to control the radical polymerization. After a considerable amount of copper (II) species is formed, the system starts to become self-controlled. This is indicated by the observation that with the continuous increase of the copper (II) concentration, polydispersities decreased dramatically (relative to other cases) from M_w/M_n = 1.7-1.2.

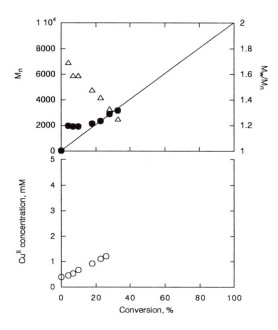

Figure 8   Molecular weight, M_n, and molecular weight distribution, M_w/M_n, dependence on monomer conversion for a polymerization system of (styrene/1-phenylethyl chloride/CuCl/dNbipy = 100/1/1/2 in toluene (50 %)) as well as copper(II) concentration dependence on conversion.

Insight into the Structure of the Copper (II) Complex

Frozen state signals of copper (II) species in the ATRP systems might provide more detailed information about the structure of the copper complexes.

Frozen state EPR spectra of ATRP systems of styrene, MMA, MA, and acrylonitrile (in toluene solution) at 20 K showed almost identical spectra. The spectrum in the case of a MA-ATRP system is shown in Figure 9. This finding indicates that the structure of the copper (II) complex in these systems should be almost the same.

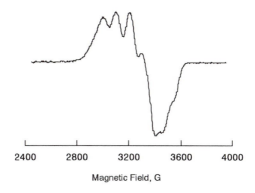

Magnetic Field, G

Figure 9 Frozen state EPR spectra of MA ATRP systems (monomer/1-phenylethyl bromide/CuBr/dNbipy = 100/1/1/2 in toluene (50 %, v/v)) at 20 K.

Spectroscopic simulation of the frozen state signal may provide structural information for the copper (II) complex. The signal was reasonably simulated to be a highly anisotropic pattern which has three kinds of coupling constants (190.4 G, 203.5 G, and 421.3 G for A_x, A_y, and A_z, respectively) and three kinds of g values (2.0079, 2.1564, and 2.1809 for g_x, g_y, g_z, respectively). EPR spectra of various kinds of copper (II) complexes have been investigated in the past.[5] Similar EPR signals have already been reported for some kinds of copper (II) bis(bipyridine) complexes.[5, 19] Furthermore, crystal structures of these complexes ([Cu^{II}(bipy)$_2$X]$^+$X$^-$, (bipy = 2,2'-bipyridine, X = Cl, Br, and PF$_6$)) have determined that they have a trigonal bipyramidal structure.[18] Therefore, the complex in the ATRP systems is considered to have a trigonal bipyramidal structure consistent with their characteristic three different A values and g values. This speculation is supported by EXAFS and X-ray crystallographic analysis of the complexes in ATRP systems.[20]

Experimental

EPR measurements

EPR spectra of propagating radicals in conventional radical polymerizations were recorded on a JEOL JES RE-2X spectrometer operating in the X-band, utilizing a 100 kHz field modulation, and a microwave power of 0.2 mW. A TE_{011} mode cavity was used. Temperature was controlled by a JEOL DVT2 variable-temperature accessory.

EPR spectra in ATRP systems were recorded on a Bruker ESP-300 X-band EPR spectrometer. A 0.2 mL sample was taken from the polymerization systems and put into an EPR tube (o.d. 4 mm) under argon. The sample was degassed 3 times by freeze-pump-thaw cycles and sealed under vacuum. Spectra were recorded at room temperature after polymerization at controlled temperature for a given time. It is recognized that the concentration of growing radicals is much higher at polymerization temperature ($[P\bullet] \sim 10^{-7}$ M at 110 °C) than at room temperature ($[P\bullet] \ll 10^{-8}$ M). However, this does not affect the concentration of copper (II) species, which can change by less than 0.01 %, since [copper (II)] $>10^{-3}$ M.

Concentrations of copper (II) species were estimated by double integration of spectra. Spectra of Cu^{II}(trifluoroacetylacetonate) in the same media under the same conditions were used as standards.

Materials: Monomers, initiators, copper salts, and ligand were purified in the usual manner.

Polymerization: The general procedures for the polymerization reactions can be obtained from ATRP literature.[21]

Characterization: Monomer conversion was determined from the concentration of residual monomer, with THF as internal standard, using a Shimadzu GC-14 gas chromatograph equipped with a J&W Scientific 30 m DB-WAX column with a Shimadzu CR501 Chromatopac. Molecular weights and molecular weight distributions were measured using a Waters 712 WISP autosampler and the following Phenogel GPC columns: guard, linear, 1000 Å and 100Å. Polystyrene standards were used to calibrate the columns.

References

1) Kamachi, M. *Adv. Polym. Sci.*, **1981**, *38*, 56.
2) Kamachi, M.; Kajiwara, A. *Macromolecules*, **1996**, *29*, 2378
3) Kamachi, M.; Kajiwara, A.; Saegusa, K.; Morishima, Y. *Macromolecules*, **1993**, *26*, 7369.

4)   Wang, J. S.; Matyjaszewski, K.   *J. Am. Chem. Soc.*, **1995**, *117*, 5614.

5) a) Hathaway, B.; Billing, D. E.   *Coordination Chemistry Reviews*, **1970**, *5*, 143.   b) Hathaway, B.; Duggan, M.; Murph, A.; Mullane, J.; Power, C.; Walsh, A.; Walsh, B.   *Coordination Chemistry Reviews*, **1981**, *36*, 267.

6) Kamachi, M.   *ACS Symposium Series 685*, American Chemical Society, San Francisco, **1997**, 145 and references cited there.

7) Fukuda, T.; Ma, Y. –D. ; Inagaki, H.   *Makromol. Chem., Suppl.*, **1985**, *12*, 125.

8) a) Kamachi, M.; Kajiwara, A.   *Macromol. Chem. Phys.*, **1997**, *198*, 787.   b) Tonge, M. P.; Kajiwara, A.; Kamachi, M; Gilbert, R. G.   *Polymer*, **1998**, *39*, 2305.

9) Noda, T; Morishima, Y.; Kamachi, M; Kajiwara, A.   *Macromolecules*, **1998**, *31*, 9078.

10) Buback, M.; Kowollik, C.; Kamachi, M.; Kajiwara, A.; *Macromolecules*, **1998**, *31*, 7208.

11) Kajiwara, A.; Kamachi, M.   *Macromol. Chem. Phys.*, submitted.

12) Davis, T.P., O'Driscoll, K.F., Piton, M.C., Winnik, M.A., *Macromolecules*, **1990**, *23*,2113.

13) Hatchinson, R.A., Paquet,Jr., D.A., McMinn, J.H., Fuller, R.E., *Macromolecules*, **1995**, *28*,4023.

14) Buback, M.; Geers, U.; Hurz, C. H.   *Macromol. Chem. Phys.* **1997**, *198*, 3451.

15) Zammit, M. D.; Coote, M. L.; Davis, T. P.; Willett, G. D.   *Macromolecules*, **1998**, *31*, 955.

16) Matyjaszewski, K.; Kajiwara, A. *Macromolecules*, **1998**, *31*, 548.

17) Kajiwara, A; Matyjaszewski, K.; Kamachi, M.   *Macromolecules*   **1998**, *31*, 5695.

18) Kajiwara, A.; Matyjaszewski, K.   *Polym. J.*   **1999**, *31*, 70.

19) Barclay, G. A.; Hoskins, B. F.; Kennard, C. H. L.   *J. Chem. Soc.*, **1963**, 5691.

20) Kickelbick, G; Rinoehl, U; Ertel, T. S.; Bertagnolli, H.; Matyjaszewski, K.   *Polym. Prep.*, **1999**, *40*, 334.

21) Matyjaszewski, K.; Patten, T. E.; Xia, J. *J. Am. Chem. Soc.*, **1997**, *119*, 674.

Chapter 6

Models for Free-Radical Copolymerization Propagation Kinetics

Michelle L. Coote[1,3], Thomas P. Davis[1,4], and Leo Radom[2]

[1]Centre for Advanced Macromolecular Design, School of Chemical Engineering
and Industrial Chemistry, University of New South Wales,
Sydney, New South Wales 2052, Australia
[2]Research School of Chemistry, Australian National University,
Canberra, ACT 0200, Australia

The alternative models for copolymerization kinetics are critically examined using recent experimental and theoretical results —both from our own work, and from other recent studies— with the aim of identifying a suitable replacement for the terminal model as the basis of copolymerization kinetics. Based on the combined results of these studies, it is concluded that the explicit penultimate model should be adopted as the basis of free-radical copolymerization kinetics, and that the origin of this penultimate unit effect is likely to be the result of a number of factors, including polar interactions, radical stabilization effects, entropic effects and direct interactions. The implications of these results for copolymerization kinetics are discussed.

Copolymerization models are used to predict the overall propagation rate of a copolymerization, and the composition and sequence distribution of the resulting copolymer, as a function of the feed ratio of the comonomers and a small set of characteristic constants. In order to derive these models it is necessary to make simplifying assumptions as to the factors influencing the rate of the propagation step, and the types of side-reactions that may occur.

One of the simplest models is the terminal model (1-3), in which it is assumed that side-reactions are not significant, and that the reactivity of the propagation reaction is governed only by the nature of the monomer and the terminal unit of the polymer radical. For many years it was thought that this model could describe the

[3]Current address: IRC in Polymer Science, University of Durham, Durham DH1 3LE,
United Kingdom.
[4]Corresponding author.

82

majority of copolymerization systems and it was thus the basis of copolymerization kinetics. Those systems that did not obey the terminal model were regarded as exceptions, with the failure of the terminal model being attributed to the particular chemical properties of these monomers, which rendered them susceptible to additional system-specific influences. For these 'exceptional' systems, alternative models were proposed. These took into account the influence of additional units of the polymer radical and/or the interference of side-reactions (such as complex formation, monomer partitioning and depropagation).

Although the terminal model was widely accepted as the basis of copolymerization kinetics, critical testing of this model was, until relatively recently, rarely undertaken. Instead the model was merely fitted to the available copolymer composition data by selecting appropriate values for its characteristic parameters (known as reactivity ratios). When, in 1985, Fukuda et al. (4) used these characteristic parameters to predict the terminal model propagation rate for the system styrene with methyl methacrylate, they found that the predicted values and their measured values were completely different. Subsequent studies have since demonstrated the almost general failure of the terminal model to describe simultaneously the composition and propagation rate in free-radical copolymerization. (For a review of this work, see for example Fukuda et al. (5) and references cited therein.) To replace the terminal model as the basis of copolymerization kinetics, workers have adopted some of the alternative models which had been originally reserved for exceptional systems —suggesting that these models may have more general significance than previously assumed. The most popular of these models is the implicit penultimate model, in which it is assumed that both the terminal and penultimate units of the polymer radical affect its reactivity, but only the terminal unit affects its selectivity. While this model can successfully describe the available experimental data, other models can also be fitted to the same data (6,7). Some of these alternative models include the explicit penultimate model, in which it is assumed that both the terminal and penultimate units of the radical may affect both the reactivity and selectivity of the radical, and the bootstrap model, in which it assumed that some form of monomer partitioning occurs. Discrimination between these and other alternative models has not yet been possible and thus a new basis model for copolymerization kinetics remains to be established.

In this article, the alternative models for copolymerization kinetics are critically examined using recent experimental and theoretical results —both from our own work, and from other recent studies— with the aim of identifying a suitable replacement for the terminal model as the basis of copolymerization kinetics.

Possible Models

In the terminal model it is assumed that the only factors affecting the rate of the propagation reaction are the reactivity of the monomer and the terminal unit of the radical. The failure of the terminal model indicates that some additional factor(s) must be important, which could include additional reactions (such as depropagation, monomer partitioning, or complex formation) which may compete with or interfere with the propagation reaction, and/or additional units of the polymer radical (notably the penultimate unit) which may also affect radical reactivity.

In the present work, we discount additional reactions as being important in a *basis model* for copolymerization kinetics (though obviously these factors are important in certain exceptional systems) because experimental evidence suggests that these factors are not important (at normal operating conditions) in the 'fruit-fly' system, the bulk copolymerization of styrene with methyl methacrylate. For this system, previous studies indicate that: (a) (to our knowledge) there is no spectroscopic evidence for complex formation; (b) toluene (a solvent with a similar dielectric constant to the comonomers) does not alter the copolymerization propagation kinetics (*8*); (c) depropagation is not significant at ordinary reaction temperatures, styrene having a ceiling temperature of 310°C and methyl methacrylate a ceiling temperature of 220°C (*9*); (d) thermodynamic data indicate that these comonomers are ideally mixed (*10*) and that bulk preferential solvation of the polymers by one of the monomers is unlikely (*7,11*); (e) studies (see for example Semchikov (*12*) and references cited therein) have found that the composition of styrene with methyl methacrylate copolymers is chain-length independent; and (f) although the bootstrap model can be simultaneously fitted to the composition and propagation rate coefficients, the parameters obtained are physically unrealistic (*13,14*). Thus, for this particular system, models based on complex formation can be ruled out by (a)–(b), models based on depropagation can be ruled out by (c), and the bootstrap model can be ruled out by (d)–(f). Hence, these models appear to be incapable of accounting for the failure of the terminal model in this common copolymerization system, and thus do not appear to be suitable as a basis model for copolymerization kinetics.

Instead we consider variants of the penultimate model. Two different versions of the penultimate model have been proposed: the *explicit* (*15*) and the *implicit* (*4*). In the explicit penultimate model, it is assumed that only the terminal and penultimate units of a polymer radical affect its reactivity. In the copolymerization model derived from this assumption, the composition, sequence distribution and propagation rate equations all deviate from the terminal model. In the implicit penultimate model it is again assumed that both the terminal and penultimate units of the polymer radical affect its reactivity, but that the magnitude of this penultimate unit effect is independent of the coreactant. In the copolymerization model derived from this assumption, the composition and sequence distribution collapse to their terminal model forms, and only the propagation rate equation deviates from the terminal model. It is thus important to establish whether or not penultimate unit effects are implicit or explicit, as this has important implications for the large numbers of terminal model monomer reactivity ratios that have been and continue to be published, and the terminal-model based empirical schemes for predicting them.

Model Discrimination Approach

An obvious way of discriminating between the alternative models is to examine their ability to describe experimental data. However, such studies have been largely inconclusive as the alternative models contain various monomer and radical reactivity ratios that are not independently measured but are instead treated as adjustable parameters as part of the model-fitting procedure. By selecting appropriate values of these adjustable parameters, any number of different models (regardless of their physical validity) can be made to fit the same set of data (*7*). For some systems, it has been possible to discriminate between the implicit and explicit models by examining the ability of the terminal model to describe simultaneously the composition and sequence distribution. However, for other systems (such as the copolymerization of styrene with methyl methacrylate), the triad/pentad fraction data that is used in the model testing cannot be extracted from the measured peak fraction data without first assuming a model and 'measuring' the coisotacticity parameter as an adjustable parameter.

Given these problems, we have attempted to discriminate between the alternative models by examining the physical validity of their assumptions. In both models it is assumed that the penultimate unit affects the rate of the propagation step, while in the implicit model it is additionally assumed that the magnitude of this effect is independent of the coreactant. Whether or not this second assumption is likely to be valid, depends on which factors cause penultimate unit effects. A number of causes have been proposed, which include the following:

- *Polar Model* (*16,17*). The penultimate unit affects the stability of the charge-transfer configurations of the transition structure, and this causes an explicit effect in the barrier.
- *Radical Stabilization Model* (*18*). The penultimate unit affects the stability of the propagating radical and, provided the Evans-Polanyi rule (*19,20*) holds, this results in an implicit effect in the barrier.
- *Entropic Model* (*21*). The penultimate unit affects the frequency factor of the propagation reaction and, depending on the relative steric differences between the coreactants, this results in an implicit or explicit effect in the frequency factor.

In this article, we thus assess the alternative models by looking for direct evidence for penultimate unit effects, and by studying their cause (from which we can conclude whether they will be implicit or explicit).

Experimental Results

Temperature Effects in STY-MMA Copolymerization

In a pulsed laser polymerization study of styrene with methyl methacrylate at several temperatures, it was found that the penultimate unit effect in this system was temperature dependent, though it was impossible to establish with any certainty the extent of this temperature dependence (*13,14*). Thus, it appeared that there was a penultimate unit effect in the barrier, though it was impossible to establish its cause. However, other experimental studies (*22-24*) have noted that the monomer reactivity ratios of this system are correlated with the dielectric constant of the reaction medium, and this suggests that polar interactions are important in the transition structure of the propagation reactions. Since the existence of significant polar interactions is known (*25*) to undermine the Evans-Polanyi rule (*19,20*) —an important assumption of the radical stabilization model— and since polar interactions are themselves associated with explicit penultimate unit effects in the reaction barrier (see below), this would suggest that the enthalpic penultimate unit effects observed in this system are explicit rather than implicit.

Validity of $r_1r_2 = s_1s_2$

A prediction of the radical stabilization model is that, provided there are no penultimate unit effects in the frequency factors, the product of the monomer and radical reactivity ratios should be equal (i.e. $r_1r_2 = s_1s_2$). When there is no penultimate unit effect (i.e. $s_1 = s_2 = 1$), the radical reactivity ratio product equals unity ($s_1s_2 = 1$), and thus the monomer reactivity ratio product should also equal unity (i.e. $r_1r_2 = s_1s_2 = 1$). Thus, the radical stabilization model can be assessed by testing the terminal model, in systems for which $r_1r_2 \neq 1$. Early studies found that those systems obeying the terminal model, such as the copolymerization of p-methoxystyrene with styrene (26), did indeed have values of $r_1r_2 = 1$, while systems in which penultimate unit effects were observed had values of $r_1r_2 \neq 1$. (Fukuda et al. (27) provide a listing of some of these systems.) However, in a recent pulsed laser polymerization study of two other para-substituted styrene copolymerization systems, p-chlorostyrene with styrene and with p-methoxystyrene, it was found that (even when uncertainties in the parameter estimates were taken into account) $r_1r_2 \neq s_1s_2$. Thus the radical stabilization model was shown to be invalid in these systems, and it was argued that this failure was probably caused by polar interactions, which rendered invalid the Evans-Polanyi rule (19,20), one of the model's principal assumptions.

Penultimate Unit Effects in Small Radicals

The penultimate unit effect can be studied directly by examining the effect of the γ substituent on the properties of small radicals. A number of studies (16,28-30) have provided direct evidence for γ-substituent effects on the selectivity of small radicals toward various alkenes, and the trends in these results indicate a polar origin for these effects. However, this direct evidence for explicit penultimate unit effects is limited to systems involving polar substituents such as nitrile groups, and no studies of a small-radical model of the styrene with methyl methacrylate system have as yet been performed.

In addition, ESR studies (31,32) indicate that the penultimate unit can affect the conformation and stability of small radicals. Although this supports one of the primary assumptions of the radical stabilization model (namely, that there is a penultimate unit effect on radical stability), this evidence is again limited to systems involving polar substituents such as nitrile groups —systems for which, as noted above, there is direct evidence for explicit polar penultimate unit effects.

Terpolymerization Studies

Pulsed laser polymerization studies of terpolymerization systems may provide a more sensitive means of discriminating between the implicit and explicit penultimate models as, under the implicit (but not the explicit) model, the terpolymerization behavior should be entirely predictable on the basis of parameters estimated from the constituent copolymerization systems (33,34). In a recent study of the terpolymerization of styrene with methyl methacrylate and para-methoxystyrene (*35*), it was found that the propagation rate coefficients could indeed be predicted by the implicit penultimate model, using parameters estimated from the constituent copolymerization systems. However, while this would appear to provide strong evidence for the implicit penultimate model in both the terpolymerization and constituent copolymerization systems, it is possible that the implicit penultimate model was not critically tested in this system as one of the constituent copolymerizations (styrene with p-methoxystyrene) obeyed the terminal model (*26*). Further terpolymerization studies, involving monomers for which penultimate unit effects are present in all constituent copolymerizations, are thus required in order to confirm this result.

Theoretical Results

Ab initio molecular orbital calculations have also been used to study the penultimate unit effect. They enable penultimate unit effects in the frequency factor and barrier, and other related quantities (such as radical stabilization energies), to be studied individually. However, in order to treat the reactions at an adequate level of theory, the size of the systems (and thus the radical and monomer substituents) that can be studied is somewhat limited. Nevertheless, in the systems that can be studied, such calculations provide a very powerful tool for probing the origin of the penultimate unit effect.

Penultimate Unit Effects in the Barrier

In an early study by Heuts et al. (*36,37*) it was found that the reaction barrier was relatively insensitive to the nature of the γ substituent (X) in the reactions of 3X-propyl radicals (X = H, F, NH_2, CN) with ethylene. However, in recent studies we have extended these calculations to include a wider range of radicals and monomers and found direct evidence for explicit penultimate unit effects in the reaction barrier. The results of these calculations may be summarised as follows.

Polar Penultimate Unit Effects
In a study (*17*) of the addition of 3X-propyl radicals (X = H, F, NH_2, CN) to various alkenes (CH$_2$=CHY; Y = F, NH_2, CN, CHO), it was found that significant penultimate unit effects in the barrier were possible for reactions with some of the alkenes, the magnitude and direction of the penultimate unit effects on the reaction barrier were strongly dependent on the nature of the alkene (i.e. they were strongly explicit), and the origin of these effects was likely to be polar.

Radical Stabilization Effects
Since penultimate unit effects on radical stability were not significant for the 3X-propyl radicals considered above, the addition reactions of 1F,3X-propyl and 1CN,3X-propyl radicals (X = H, F, CN) with ethylene were next studied (*38*). For these radicals, large penultimate unit effects on radical stability were observed. However,

these effects did not carry over to the barrier to any great extent and, where penultimate unit effects in the barrier were observed, it was impossible to rule out polar interactions (and thus explicit effects). Furthermore, it was argued on the basis of these results that *in general* radical stabilization effects were unlikely to result in significant penultimate unit effects in the barrier without the simultaneous occurrence of explicit polar effects. The reasons for this are as follows. Under the radical stabilization model, the penultimate unit effect in the barrier is predicted to be proportional to the penultimate unit effect on radical stability. The size of the proportionality constant is governed by how early or late the transition structure is. The small proportionality constants observed in the model propagation reactions of this study indicate an early transition structure —a result which is likely to be general, given the highly exothermic nature of the propagation step in common copolymerization systems. As result, it would be predicted that, in general, for such effects to lead to significant penultimate unit effects in the barrier, radical stabilization effects would need to be very large and hence would generally involve radical substituents that are strong electron donors or acceptors. However, such substituents are simultaneously associated with strong polar interactions —especially when they appear on the monomer, as they would in the cross-propagations of a free-radical copolymerization.

Conformational Effects

A representative sample of the above reactions was then examined for conformational effects (*39*). It was found that the penultimate unit effect in the barrier was strongly dependent on the conformation of the reacting radical and transition structure and thus the penultimate unit effect in the overall reaction was strongly dependent on the relative contributions of these different pathways. These results indicated that: (a) in certain, more crowded conformations, direct interactions involving the radical and monomer substituents were likely to contribute to the penultimate unit effect; (b) entropic factors would probably play a role in the penultimate unit effect in the overall reaction barrier (and vice versa); (c) the penultimate unit effects on the tacticity of the copolymer were likely to be important, a result that is supported by NMR studies (*40,41*). Finally, it was argued that, given (a) and (b), penultimate unit effects in the overall barrier of a free-radical propagation step were unlikely to be independent of the coreactant and hence were likely to be explicit rather than implicit. For, not only is it highly unlikely that direct interactions involving the coreactant would be independent of the nature of the coreactant, but it is also unlikely that the relative contributions of the different pathways to the overall reaction would be independent of the coreactant —since this would require that both the relative barriers *and* relative frequency factors of the individual pathways are independent of the monomer.

Penultimate Unit Effects in the Frequency Factor

Heuts et al. (*21*) used ab initio molecular orbital calculations to study the frequency factors in free-radical propagation reactions. Based on a comparison of the frequency factors for ethyl and propyl addition to ethylene, they argued that the penultimate unit would be able to hinder certain internal motions in the transition structure, which were important in determining the frequency factor. Since penultimate unit effects on these various motions would be multiplicative, they argued that entropic penultimate unit effects could affect the rate of propagation by a factor (range) of 0.1–10. However, it should be noted that this prediction is based on a comparison of ethyl and propyl propagation (i.e. a β-substituent effect), and has yet to be confirmed in studies of penultimate unit effects (i.e. γ-substituent effects).

Assessment: What is the Best Model for Copolymerization Propagation Kinetics?

Based on the above studies, it may be concluded that there is direct experimental and theoretical evidence for explicit penultimate unit effects in the reaction barrier, and these penultimate unit effects are likely to be polar in origin. In addition, the theoretical results also indicate that direct interactions —which would be predicted to result in explicit rather than implicit penultimate unit effects— are also likely to play a role. Furthermore, although both the experimental and theoretical studies provide direct evidence for penultimate unit effects on radical stability, these effects appear unlikely to result in significant penultimate unit effects in the barrier without the simultaneous occurrence of explicit polar effects. In other words, it would appear that, where penultimate unit effects in the barrier are significant, they are likely to be explicit rather than implicit. While the possibility of implicit penultimate unit effects in the frequency factor cannot be ruled out, if there is an explicit penultimate unit effect in the barrier, it follows that there is an explicit penultimate unit effect in the overall propagation rate.

Although the prediction that penultimate unit effects in the barrier are likely to be explicit rather than implicit is based on the results for a small number of systems (involving highly polar substituents such as nitrile groups), there appears to be reasonably strong theoretical grounds for predicting them to be general. Furthermore, although there is no direct evidence for this result for common copolymerization systems such as the bulk copolymerization of styrene with methyl methacrylate, there is indirect evidence. For, the study of temperature effects in this system indicates that there is a penultimate unit effect in the reaction barrier, while the correlations observed between its monomer reactivity ratios and the solvent's dielectric constant indicate that polar interactions are likely to be important in this system. There is also direct evidence against the radical stabilization model in two related copolymerization systems —the copolymerization of p-chlorostyrene with styrene and with p-methoxystyrene. Finally, while there is only circumstantial evidence for the explicit penultimate model in systems such as the bulk copolymerization of styrene with methyl methacrylate, there appears to be no evidence for the implicit penultimate model (and in particular its assumption that the magnitude of the penultimate unit effect is independent of the coreactant) in such systems. Hence, based on existing evidence, it appears that the explicit penultimate model should be favored over the implicit penultimate model as a more physically realistic description of copolymerization kinetics in systems for which the terminal model has been shown to fail.

Implications of these Results for Copolymerization Kinetics

The conclusion that the explicit (rather than implicit) penultimate model should be adopted as the basis of free-radical copolymerization kinetics has important implications. In particular it implies that the large numbers of terminal model monomer reactivity ratios that have been and continue to be published have limited physical meaning. Although these parameters are roughly correlated with average radical reactivity, they do not reflect their proposed physical meaning and can thus lead to false predictions of the other copolymerization properties which depend upon them —as indeed they are already known to do, in the case of propagation rate coefficients. Not only does this imply that such parameters should not be used in quantitative studies of radical reactivity, but it also suggests that the 'measured' sequence distribution data for systems such as styrene with methyl methacrylate which rely on fitted coisotacticity factors may be incorrect, since these have been fitted to the (incorrect) terminal model composition equation.

However, the adoption of the more physically realistic explicit penultimate model may not offer a substantive solution to the above problem. For, even in two-parameter fits of the terminal model to composition data, or of the implicit penultimate model (with fixed monomer reactivity ratios) to propagation rate coefficients, the uncertainties in the fitted model parameters are large. In the case of the radical reactivity ratios it has been shown that, even compiling the most extensive data-set to date, the uncertainties in the estimated radical reactivity ratios for the styrene with methyl methacrylate system were so large as to preclude the attachment of any physical meaning to their point estimates (*13,14*). Given this, it is clear that the *six* reactivity ratios estimated when fitting the explicit penultimate model to the data are likely to be indeterminate. Indeed, it has recently been shown that, under the explicit penultimate model, multiple sets of monomer reactivity ratios can describe with reasonable accuracy the composition data from the copolymerization of styrene with methyl methacrylate (*42*). It is important to note that this does not mean that these parameters are superfluous —merely that they should not be estimated by fitting models to data. Instead, these parameters should be measured directly via, for example, experimental or theoretical studies of small-radical models of the various types of the propagation reactions.

In conclusion, the present work indicates that the various reactivity ratios that have been estimated from copolymerization data using the terminal or implicit penultimate models are likely to convey only limited physical meaning. Provided that these parameters are only used to reproduce the data to which they have been fitted (and of course that the model is able to provide a reasonable fit to data), this is not a problem —although this job could be performed equally well using any third-order polynomial with "A", "B", "C" and "D" as constants. However, such parameters are not suitable for making independent predictions, or for quantitative studies of radical reactivity. Unfortunately, the parameters estimated by fitting the explicit penultimate model to the data are also likely to convey only limited physical meaning —owing to the enormous uncertainty in their point estimates— and thus the further development of direct means of measuring these parameters is important.

Acknowledgments We gratefully acknowledge a generous allocation of time on the Fujitsu VPP300 and SGI Power Challenge computers of the Australian National University Supercomputer Facility, support from the Australian Research Council, and the award (to MLC) of an Australian Postgraduate Award.

Literature Cited

1. Mayo, F. R.; Lewis, F. M. *J. Am. Chem. Soc.* **1944**, *66*, 1594.
2. Alfrey, T.; Goldfinger, G. *J. Chem. Phys.* **1944**, *12*, 205.
3. Jenkel, E. *Z. Phys. Chem. Abt. A* **1942**, *190*, 24.
4. Fukuda, T.; Ma, Y.; Inagaki, H. *Macromolecules* **1985**, *18*, 17.
5. Fukuda, T.; Kubo, K.; Ma, Y. *Prog. Polym. Sci.* **1992**, *17*, 875.
6. Schweer, J. *Makromol. Chem., Theory Simul.* **1993**, *2*, 485.
7. Maxwell, I. A.; Aerdts, A. M.; German, A. L. *Macromolecules* **1993**, *26*, 1956.
8. Fukuda, T.; Kubo, K.; Ma, Y.; Inagaki, H. *Polym. J. (Tokyo)* **1987**, *19*, 523.
9. Odian, G. *Principles of Polymerization*; John Wiley & Sons, Inc.: New York, 1991.
10. Egorochkin, G. A.; Semchikov, Y. D.; Smirnova, L. A.; Karyakin, N. V.; Kut'in, A. M. *Eur. Polym. J.* **1992**, *28*, 681.
11. Kratochvil, P.; Strakova, D.; Stejskal, J.; Tuzar, Z. *Macromolecules* **1983**, *16*, 1136.
12. Semchikov, Y. D. *Macromol. Symp.* **1996**, *111*, 317.
13. Coote, M. L.; Johnston, L. P. M.; Davis, T. P. *Macromolecules* **1997**, *30*, 8191.
14. Coote, M. L.; Zammit, M. D.; Davis, T. P.; Willett, G. D. *Macromolecules* **1997**, *30*, 8182.
15. Merz, E.; Alfrey, T., Jr.; Goldfinger, G. *J. Polym. Sci.* **1946**, *1*, 75.
16. Giese, B.; Engelbrecht, R. *Polymer Bulletin* **1984**, *12*, 55.
17. Coote, M. L.; Davis, T. P.; Radom, L. *J. Mol. Struct. (THEOCHEM)* **1999**, *461-462*, 91-96.
18. Fukuda, T.; Ma, Y.; Inagaki, H. *Makromol. Chem., Rapid Commun.* **1987**, *8*, 495.
19. Evans, M. G. *Disc. Faraday Soc.* **1947**, *2*, 271.
20. Evans, M. G.; Gergely, J.; Seaman, E. C. *J. Polym. Sci.* **1948**, *3*, 866.
21. Heuts, J. P. A.; Gilbert, R. G.; Maxwell, I. A. *Macromolecules* **1997**, *30*, 726.
22. Bonta, G.; Gallo, B.; Russo, S. *Polymer* **1975**, *16*, 429.
23. Ito, T.; Otsu, T. *J. Macromol. Sci. –Chem.* **1969**, *A3*, 197.
24. Fujihara, H.; Yamazaki, K.; Matsubara, Y.; Yoshihara, M.; Maeshima, T. *J. Macromol. Sci. —Chem.* **1979**, *A13*, 1081.
25. Wong, M. W.; Pross, A.; Radom, L. *J. Am. Chem. Soc.* **1994**, *116*, 6284.
26. Piton, M. C.; Winnik, M. A.; Davis, T. P.; O'Driscoll, K. F. *J. Polym. Sci: Part A: Polym. Chem.* **1990**, *28*, 2097.
27. Fukuda, T.; Ide, N.; Ma, Y.-D. *Macromol. Symp.* **1996**, *111*, 305.
28. Jones, S. A.; Prementine, G. S.; Tirrell, D. A. *J. Am. Chem. Soc.* **1985**, *107*, 5275.
29. Cywar, D. A.; Tirrell, D. A. *J. Am. Chem. Soc.* **1989**, *111*, 7544.

30. Busfield, W. K.; Jenkins, I. D.; van Le, P. *J. Polym. Sci. A. Polym. Chem.* **1998**, *A36*, 2169.
31. Tanaka, H.; Sasai, K.; Sato, T.; Ota, T. *Macromolecules* **1988**, *21*, 3534.
32. Sato, T.; Inui, S.; Tanaka, H.; Ota, T.; Kamachi, M.; Tanaka, K. *J. Polym. Sci. A Polym. Chem.* **1987**, *A25*, 637.
33. Coote, M. L.; Davis, T. P. *Polym. React. Eng.* **1999**, *7*, 347.
34. Olaj, O. F.; Schnöll-Bitai, I. *Macromol. Theory Simul.* **1995**, *4*, 577.
35. Coote, M. L.; Davis, T. P. *Polym. React. Eng.* **1999**,*7*, 363.
36. Heuts, J. P. A. Ph.D. thesis, University of Sydney, Sydney, Australia, 1996.
37. Heuts, J. P. A.; Gilbert, R. G.; German, A. L.; Radom, L. *unpublished.*
38. Coote, M. L.; Davis, T. P.; Radom, L. *Macromolecules* **1999**, *32*, 2935.
39. Coote, M. L.; Davis, T. P.; Radom, L. *Macromolecules* **1999**, *32*, 5270.
40. Moad, G.; Solomon, D. H.; Spurling, T. H.; Johns, S. R.; Willing, R. I. *Aust. J. Chem.* **1986**, *39*, 43.
41. Yamamoto, T.; Hanatani, M.; Matsumoto, T.; Sangen, O.; Isono, T.; Fukae, R.; Kamachi, M. *Polym. J.* **1994**, *26*, 417.
42. Kaim, A.; Oracz, P. *Macromol. Theory Simul.* **1997**, *6*, 565.1. Mayo, F. R.; Lewis, F. M. *J. Am. Chem. Soc.* **1944**, *66*, 1594.

Chapter 7

Topochemical Polymerization of Diene Monomers in the Crystalline State to Control the Stereochemistry of the Polymers

Akikazu Matsumoto and Toru Odani

Department of Applied Chemistry, Faculty of Engineering,
Osaka City University, Sugimoto, Sumiyoshi-ku, Osaka 558–8585, Japan

The topochemical polymerization of several 1,3-diene monomers in the crystalline state as a new method of polymer structure control via a radical polymerization mechanism is described in this chapter. The concept of topochemical polymerization, the features and mechanism of the polymerization of diethyl (Z,Z)-muconate (EMU), and the characterization of poly(EMU) obtained as polymer crystals are presented. We also mention the polymerization of the ammonium salts of 1,3-diene mono- and dicarboxylic acids including muconic and sorbic acid derivatives bearing different configurations. It has been demonstrated that the introduction of a naphthylmethylammonium moiety efficiently induces the topochemical polymerization of not only (Z,Z)-diene but also (E,E)-diene derivatives. The stereochemical structure of the polymers is discussed in relation to the configuration of the monomers and molecular packing in the crystals.

In recent years, many reports have been published on polymer structure control using radical polymerization, which is the most convenient and important process for the production of vinyl polymers because of the advantageous features of this polymerization (*1-4*). For example, the molecular weight and molecular weight distribution as well as both chain end structures are satisfactorily controlled by living radical polymerization. A well-defined branching structure of polymers such as graft, comb-like, star-shape, and hyperbranched polymers is also obtained by a macromonomer technique and a living radical polymerization process. In contrast, the control of tacticity is very difficult by free radical polymerization. As exceptional events, specially designed monomers are known to provide uniquely a stereoregular polymer through radical polymerization. In the radical polymerization of triphenylmethyl methacrylate and its related monomers, highly isotactic polymers with a helical chain structure are produced under substrate control even under free radical polymerization conditions (5,6). Chiral auxiliary control was demonstrated to be effective for the synthesis of an isotactic polyacrylamide (7), similarly to the stereochemical control in the radical reactions of small molecules (8). Several inclusion compounds also provide an optically active polymer of diene monomers (9). We have recently found the topochemical polymerization of several 1,3-diene monomers in the crystalline state as a new method of polymer structure control via a radical polymerization mechanism. The concept of topochemical polymerization, the features and mechanism of the polymerization, and the design and control of the polymerization, as well as the characterization of the polymers obtained as polymer crystals are described in this chapter.

Features of Polymerization in the Solid State

A large number of chemists are convinced that organic reactions are commonly carried out in a gaseous or liquid phase, but actually there are many examples of the reaction proceeding in the solid state. Recently, organic synthesis performed in the solid state is one of the most intriguing fields of chemistry because of the high selectivity of the reaction, the specific morphology of the products, and environmental aspects of the solvent-free process (10-12). When the reactants are crystalline by themselves or are included in host crystals, the rate and selectivity of the reaction are different from those observed in an isotropic reaction medium. The polymerization reactivity and the structure of the resulting polymers have been discussed for several types of polymerizations in organized media such as liquid crystals, interfaces, micelles, mono- and multi-layers, vesicles, inclusion compounds, and templates (13,14). The formation of polymers with highly controlled chain structures would be expected during the polymerization of molecular crystals consisting of the most organized structure (Figure 1).

Among solid-state polymerizations, the polymerization that provides a polymer with a specific crystal structure formed under the control of the crystal lattice of the monomer is distinguished as topotactic polymerization. Ring-opening polymerization of trioxane in the solid state is a typical example of topotactic polymerization, yielding a highly crystalline polymer. In contrast to this, topochemical polymerization is described as the reaction in which the crystallographic position and symmetry of the monomer as the reactant are retained in the resulting polymer as the product. During a topochemical polymerization process, the primary structure of polymer chains including regioselectivity and tacticity and higher order structures of the chains such as ordering and crystallinity are controlled, resulting in facile formation of polymer crystals as polymerized (15,16). [2+2] Photopolymerization of 2,5-distyrylpyradine derivatives (17) and thermal or radiation polymerization of diacetylenic compounds (18) are typical examples of topochemical polymerization. Although the discovery of these two polymerizations dates back to the late 1960s, few monomers had been found to undergo topochemical polymerization.

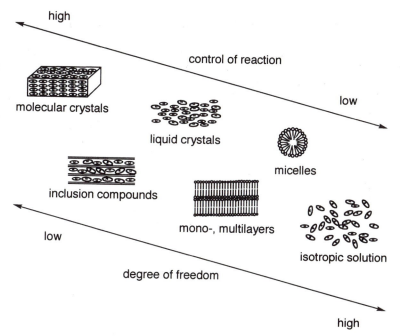

Figure 1. Various types of the organized structures of monomers for controlled polymerization.

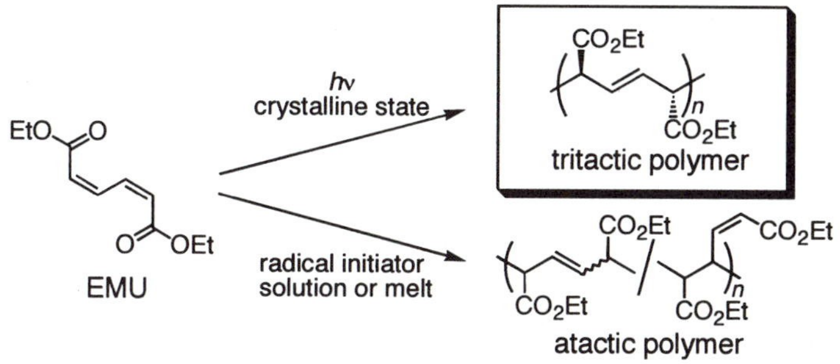

Scheme 1.

In 1994, we discovered that diethyl (Z,Z)-muconate (EMU, diethyl (Z,Z)-2,4-hexadienedioate) yields a tritactic polymer, i.e., a *meso*-diisotactic-*trans*-2,5-polymer, by photoirradiation of the monomer crystals, as shown in Scheme 1, in contrast to the formation of an atactic polymer by conventional radical polymerization in an isotropic state *(19)*. X-ray diffraction as well as NMR, IR, and ESR spectroscopic studies revealed that this is the first evident example of topochemical polymerization of 1,3-diene derivatives via a radical chain mechanism, regarded as the reaction-locus controlled polymerization *(20-22)*.

Topochemical Polymerization of EMU in the Crystalline State

The crystals of EMU provide the polymer on exposure to sunlight or UV light, irrespective of the circumstance of the crystals, i.e., in vacuo, in air, and even dispersed in water. In contrast to the polymerization of EMU, no polymerization occurred with the other alkyl esters of *(Z, Z)*-muconic acid and the ethyl esters of the *(E, Z)*- and *(E, E)*-isomers. The polymer obtained by the topochemical polymerization of EMU has been revealed to be of excellent stereoregularity, i.e., a *meso*-diisotactic-*trans*-2,5-structure, as shown in the NMR spectrum in Figure 2 (*20*). We can obtain poly(EMU) as needle-form crystals as polymerized under UV irradiation. It was revealed by X-ray diffraction using an imaging plate that the polymer chains were aligned along a specific axis of the crystals with completely extended conformation (Figure 3(a)) (*21*). The powder X-ray diffraction profiles were also examined to reveal the difference in the crystal lattice of the polymer from that of the monomer. It is clear that poly(EMU) has a crystalline structure resembling that of the monomer and that the polymerization reaction evidently proceeds via a crystal-to-crystal process as shown in Figure 3(b) (*21*). When melted, the polymer readily crystallized during the cooling process and spherulites were formed, indicating that the stereoregular poly(EMU) has high crystallinity.

Poly(EMU) prepared in the crystalline-state polymerization has a limited solubility because of its high stereoregularity; it is soluble in trifluoroacetic acid and 1,1,1,3,3,3-hexafluoro-2-propanol but insoluble in any other solvent. Therefore, gel permeation chromatography using hexafluoro-2-propanol as an eluent was employed to determine the molecular weight and molecular weight distribution (*23,24*). It has been clarified that the polymer is of ultrahigh molecular weight, at least more than 10^6 when the EMU monomer crystals prepared by conventional recrystallization are used, and that the molecular weight of the polymer keeps an approximately constant value irrespective of the polymer yield (*21*). It has been demonstrated that the size of the EMU crystals depends on the method used for crystal preparation, i.e., recrystallization, milling, freeze drying, and precipitation. Furthermore, the molecular weight of the resulting polymer decreases as the crystal size becomes small. Precipitation was the best method for the fabrication of microcrystals, the polymerization of which yields poly-(EMU) with well-controlled molecular weight and molecular weight distribution (*24,25*).

In the initial step of the polymerization, a radical species is produced from EMU excited by photoirradiation with the UV light of approximately 300 nm or less, which is consistent with the absorption band of EMU (λ_{max} = 259 nm, ε = 21,300 in cyclohexane). In the crystals, the diradical produced by the irradiation immediately reacts with neighboring monomer molecules, resulting in the formation of long-lived polymer radicals, which are easily detected by ESR spectroscopy (Figure 4). The decay of the radicals after interruption of the irradiation was extremely slow, indicating that the termination process of the propagating radicals occurs less frequently because the polymer chain produced cannot diffuse in the crystal.

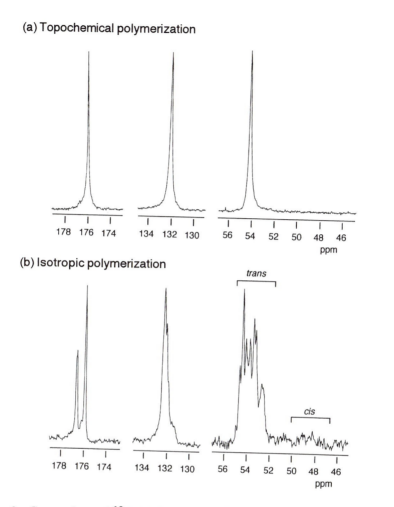

Figure 2. Comparison of ^{13}C NMR spectra of (a) tritactic poly(EMU) prepared by topochemical polymerization and (b) atactic poly(EMU) prepared by isotropic polymerization in the melt. Measurement solvent: trifluoroacetic acid-d (ref. 20).

When the polymerization process was monitored by IR spectroscopy during the polymerization under UV irradiation, the characteristic bands due to the monomer decreased and the bands of the polymer increased with some isosbestic points. From the kinetic analysis of the spectral change, it was revealed that the polymerization process apparently obeyed the first-order reaction with respect to the monomer concentration (21,22). The continuous X-ray irradiation also induced the polymerization of the EMU crystals. Therefore, we can follow the polymerization process during the X-ray diffraction measurement. The reflection due to the monomer crystals shifted and approached those of the polymer crystals; that is, the crystal lattice of the EMU monomer changed continuously to

(a)

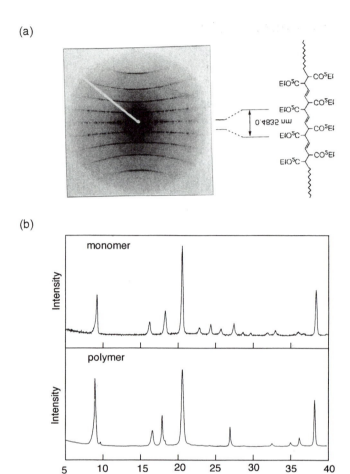

(b)

Figure 3. (a) Full-rotation X-ray photograph of the poly(EMU) crystal with an imaging plate detector. (b) Powder X-ray diffraction profiles of the EMU monomer and poly(EMU) crystals (ref. 21).

the lattice of the polymer. The large and continuous shift of the reflection indicates the generation of mechanical strain in the monomer crystal lattice followed by transformation to the lattice of the polymer crystal. Because the single crystal structure of the EMU monomer is very hard to determine by a conventional four-circle crystal structure analysis system or even by an imaging plate system, the monomer crystal structure was successfully analyzed using only a CCD camera system as a two-dimensional detector for rapid analysis within several tens of minutes or less. It has been clarified that the EMU monomer molecules form a columnar structure in the crystals and that the space group is

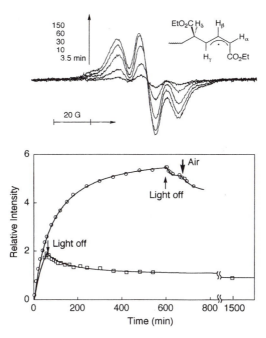

Figure 4.   Change in the concentration of propagating radicals during the crystalline-state polymerization of EMU under continuous UV irradiation and after interruption of the irradiation at 0 °C (ref. 21).

the same, $P2_1/c$, for the crystals of both monomer and polymer (26). From the comparison of the crystal structure of the monomer with that of the polymer, it was proved that the polymerization proceeded with minimum movement of the center of the molecular mass accompanying the change in chemical bonding from the 1,3-diene structure of the monomer to the successive *trans*-2,5-structure of the polymer.   These results indicate that the polymerization of EMU in the crystalline state is a typical topochemical reaction.

Topochemical Polymerization of Ammonium Muconates

As a result of the survey of the photoreaction behavior of some muconic acid derivatives including the esters, the amides, and the ammonium salts bearing various alkyl substituents, we have found that the topochemical polymerization is possible not only for EMU but also for several alkylammonium derivatives such as the benzyl- and *n*-alkylammonium salts of the acid, although most compounds are isomerized to the corresponding *(E,E)*-isomers or have no reactions under similar photoirradiation conditions, as summarized in Table I (27). The topochemical polymerization reactivity is very sensitive to the molecular structure because of the drastic changes in packing and orientation of the monomer

Table I. Photoreaction Behavior of Muconic Acid Derivatives Bearing Various Alkyl Substituents under UV Irradiation in the Crystalline State

R	RO_2C⌒⌒CO_2R	$RHNOC$⌒⌒$CONHR$	RH_3NO_2C⌒⌒CO_2NH_3R
Methyl	Isomerization	-	No reaction
Ethyl	Polymerization	No reaction	Isomerization
n-Propyl	Isomerization	-	Isomerization
n-Butyl	No reaction	Isomerization	Isomerization
Octyl	Isomerization	-	Isomerization
Decyl	Isomerization	Isomerization	Polymerization
Dodecyl	Isomerization	No reaction	Polymerization
Isopropyl	No reaction	Isomerization	Isomerization
Cyclohexyl	Isomerization	No reaction	Isomerization
Benzyl	Isomerization	No reaction	Polymerization
α-Methylbenzyl	-	No reaction	No reaction

molecules in the crystals. The difficulty of the crystal structure design prevented us from constructing a new topochemical polymerization system. To solve this problem, we chose the hydrogen bonding of primary ammonium carboxylates as a supramolecular synthon of the solid state (28,29) and started the search for topochemical polymerizable monomers on the basis of the recent rational design of the packing and orientation of organic molecules in the crystalline state, which has attracted much attention as crystal engineering. There are several merits for the ammonium salts of the muconic acid; facile monomer preparation, the high melting point of the crystals, the variety of the crystal structures depending on the kind of amine used, and the possibility of transformation to other derivatives

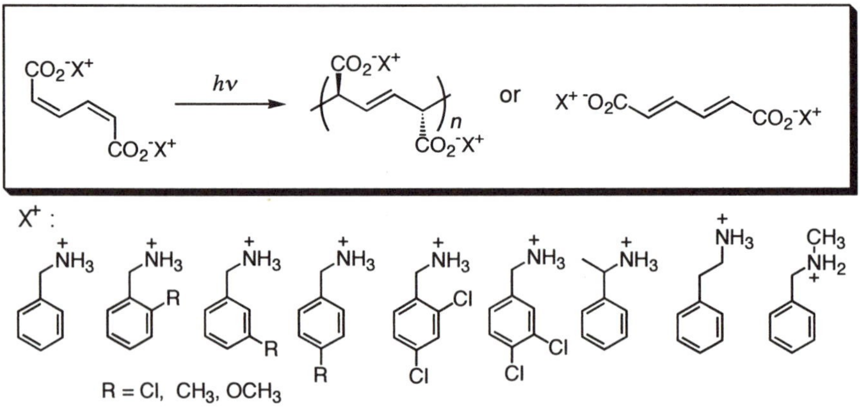

Scheme 2.

after polymerization (30). We have demonstrated the relationship between the photoreactivities and crystal structures of a series of benzylammonium muconates shown in Scheme 2 (31). X-ray structure analysis has revealed that molecular packing in the crystalline state controls the pathway of the photoreaction, the mode of the reaction and the stereochemistry of the photoproducts. The columnar-type structure formed by the appropriate stacking of the muconate anions is indispensable for the process of the topochemical polymerization. Namely, the photopolymerizable columnar structure implies the face-to-face arrangement of the muconates within a suitable distance of the double bonds. The stereoregularity of the resulting polymers is in good agreement with that expected from the crystal structure. The photopolymerization of the benzyl-ammonium muconates also belongs to a crystal lattice-controlled reaction, i.e., topochemical polymerization.

Design and Control of the Crystalline-State Polymerization

In the topochemical polymerization, the reactivity is essentially dependent on the monomer crystal structure. A guide from an organic synthetic aspect has been demanded for the design of monomers with a crystal structure suitable for topochemical polymerization. In the course of the study on the topochemical polymerization of the diene monomers, we have newly found that the 1-naphthyl-methylammonium salts of (E,E)-muconic acid (32) and sorbic acid (33) polymerized in the crystalline state and provided stereoregular polymers, as well as the (Z,Z)-muconic acid derivatives.

Table II summarizes the results of the crystalline-state photoreaction of several diene mono- and dicarboxylic acid derivatives as shown in Scheme 3. ZZ-1-NaphM provided the polymer in a high yield, similarly to the results with the alkylammonium and benzylammonium salts previously reported (27, 30). EE-1-NaphM also provided a polymer, in contrast to all the other (E,E)-derivatives inert during the photoirradiation. The unreacted monomer examined by NMR spectroscopy after the polymerizations of ZZ-1-NaphM and EE-1-NaphM confirmed that both monomers polymerized without any isomerization. The photoreaction of EZ-1-X yielded the corresponding (E,E)-isomers in low yields. In the powder X-ray diffraction profiles of the crystals of ZZ-1-NaphM and EE-1-NaphM as well as the resulting polymer crystals, it was found that sharp and intense diffraction was retained even after the polymerization and that the diffraction patterns of the polymers closely resembled those of the respective monomers (32). This supports the belief that these monomers polymerized in the crystalline state to yield the polymer crystals via a topochemical polymerization mechanism. The polymers obtained were insoluble in any solvent, including conventional organic solvents, fluoro-containing polar solvents, acids, and alkaline solvents. In order to characterize the stereochemistry of the polymers in solution by ^{13}C NMR spectroscopy, they were transformed to the triethylammonium salts by solid-state polymer reactions. The single peak with a narrow linewidth for each carbon indicates the formation of a stereoregular polymer during the crystalline-state polymerization. The spectrum of the polymer derived from EE-1-NaphM was identical to that from **ZZ-1-NaphM**. From the chemical

Table II. Photoreaction of Various Diene Carboxylic Acid Derivatives in the Crystalline State[a]

Monomer	Photo-product	Yield (%)	Monomer	Photo-product	Yield (%)
ZZ-1-DD	Polymer	80	EZ-1-DD	*EE*-Isomer	2
ZZ-1-OD	Polymer	95	EZ-1-OD	*EE*-Isomer	3
ZZ-1-Bn	Polymer	21	EZ-1-Bn	*EE*-Isomer	2
ZZ-1-NaphM	Polymer	90	EZ-1-NaphM	*EE*-Isomer	9
EE-1-DD	No reaction	-	EE-2-Bn	No reaction	-
EE-1-OD	No reaction	-	EE-2-NaphM	Polymer	98
EE-1-Bn	No reaction	-			
EE-1-NaphM	Polymer	71			

a) Irradiation with high-pressure Hg lamp in air at 30 °C for 8 h.

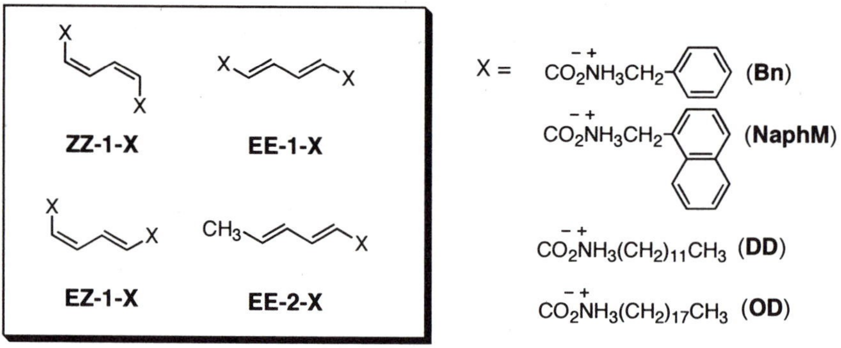

Scheme 3.

shifts and the previous results for the other related polymers, it has been revealed that the polymers obtained from both isomers have a *meso*-diisotactic-*trans*-2,5-structure. We have confirmed from IR, NMR, and X-ray diffraction results that both **ZZ-1-NaphM** and **EE-1-NaphM** polymerize in the crystalline state to give polymer crystals without any isomerization.

The crystalline-state polymerization of the *(E,E)*-muconic acid derivative inspired us to test other types of diene monomers. We selected the *(E,E)*-sorbic acid derivative as the next candidate as the monomer for topochemical polymerization because sorbic acid is popular and inexpensive compared with muconic acid (*33*). We have found that **EE-2-NaphM** polymerized in the crystalline state, while the *n*-butyl, isopropyl, *tert*-butyl, and benzylammonium salts had no reactions. Thus, the introduction of the naphthylmethylammonium moiety is very effective in inducing topochemical polymerization. The polymer structure was determined by NMR spectroscopy after the polymer transformation as the soluble polymer, similarly to the muconate polymers. In this case, it has also been confirmed that a stereoregular polymer is produced.

Crystal Structure and Polymer Stereochemistry

In this section, we discuss the stereochemical structure of the polymers produced during topochemical polymerization on the basis of the *EZ*-configuration of the monomers and the stacking manner of the monomers in a column formed in the crystals. There are four types of possible stereoregular structures for the *trans*-1,4-polymers of 1,4-disubstituted butadiene monomers such as **ZZ-1-X** and **EE-1-X**, as illustrated in Figure 5. The stereochemistry of the polymer was represented by two kinds of relationships as follows; one of which is a relative configuration between two repeating monomer units and another is a relative configuration between the vicinal carbon centers. For the topochemical polymerization of the diene monomers examined in this work, all the polymers obtained have the same stereoregularity, i.e., a *meso*-diisotactic structure. When the monomer molecules stack translationally in a columnar structure of the crystals, the *(Z, Z)*-isomer and the *(E, E)*-isomer provide the same polymer, a *meso*-diisotactic polymer, irrespective of the monomer configurations, as shown in Figure 6. Here, the dotted lines tie the carbons that form a new bond during the polymerization. Actually, we have determined the crystal structures of EMU (*26*) and **ZZ-1-Bn** (*31*) and have demonstrated that the monomer molecules are packed in each column in the crystals. Because we can obtain the polymer single crystals by the topochemical

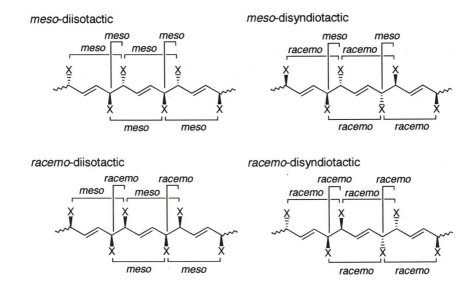

Figure 5. Four types of possible stereoregular structures for *trans*-1,4-polymer obtained from 1,4-disubstituted butadienes.

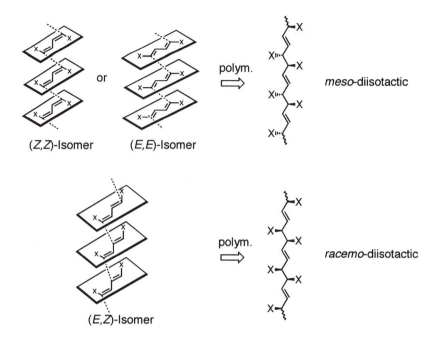

Figure 6. Relationship between the crystal packing of monomer molecules with different configurations and the stereochemical structure of polymers produced during topochemical polymerization of 1,4-disubstituted butadienes.

polymerization, we determine the polymer structure directly by crystallographic study (26,34). The structural parameters determined for the polymer crystals were very similar to those of the monomer crystals, strongly indicating that the polymerization step is an accurate topochemical reaction. It also directly proved the stereochemical structure of the polymer, a *meso*-diiso-tactic-*trans*-2,5-structure, as the repeating units.

The other type of stereoregular polymer such as a *racemo*-diisotactic polymer could be obtained if the topochemical polymerization of an *(E,Z)*-diene isomer proceeds under a similar mechanism. Unfortunately, we have found no polymerization of any *(E,Z)*-muconate derivatives at the present time. On the other hand, if alternating molecular packing is possible, it could result in the formation of disyndiotactic polymers. Such an alternating molecular packing seems to be unfavorable in actual crystals unless some special and intensive interaction between substituents is present.

Conclusions

We have demonstrated that the crystalline-state photopolymerization of EMU proceeds in a radical chain mechanism to yield a high molecular weight and tritactic polymer and that this polymerization process is the first evident topochemical polymerization, which is useful for control of the polymerization reactivity and the structure of the resulting polymers. The photoreactivity of the crystals of a series of benzylammonium salts of muconic acid has been related to the monomer crystal structures, i.e., the molecular arrangement in the crystals, as well as the EMU crystal. It has been revealed that the stacking structure of the diene moiety in a column is suitable for topochemical polymerization. We have also found that the 1-naphthylmethylammonium salts of *(E,E)*-muconic acid and sorbic acid polymerized in the crystalline state and provided stereoregular polymers, as well as the *(Z,Z)*-muconic acid derivatives. The *meso*-diisotactic polymers are produced from both isomers, irrespective of the configurations. This finding of the topochemical polymerization of the *(E,E)*-diene carboxylic acid derivatives is a great step in the progress of topochemical polymerization of diene compounds, because the mode and rate of the topochemical reactions are determined by the molecular packing of the substrates and the geometrical isomers should be regarded as different compounds rather than homologs. In the near future, we hope that structural limitation for the topochemical polymerization of diene monomers will be further expanded and that the crystalline-state polymerization can be widely used for control of the polymer chain structure as well as the polymer crystal structure.

Acknowledgments

The author acknowledges the members of the research group of Osaka City University for their devoted work and Prof. K. Tashiro, Prof. M. Miyata, and Dr. K. Sada of Osaka University for their collaboration and invaluable discussions. This work was partly supported by Grant-in-Aids from the Ministry of Education, Science, Sports, and Culture of Japan.

References

1. Moad, G.; Solomon, D. H. *The Chemistry of Free Radical Polymerization* Pergamon: Oxford, 1995, p. 315-351.
2. Matyjaszewski, K. ed. *Controlled Radical Polymerization: ACS Symposium Series 685* American Chemical Society: Washington, DC, 1998.
3. Otsu, T.; Matsumoto, A. *Adv. Polym. Sci.* **1998**, *136*, 75-137.
4. Sawamoto, M.; Kamigaito, M. *Synthesis of Polymers* Schlüter, A.-Dieter ed. Wiley-VCH: Weinheim, 1999, p. 163-194.
5. Okamoto, Y.; Nakano, T. *Chem. Rev.* **1994**, *94*, 349-372.
6. Nakano, T.; Okamoto, Y. *Controlled Radical Polymerization: ACS Symposium Series 685* Matyjaszewski, K. ed., American Chemical Society: Washington, DC, 1998, p. 451-462.
7. Porter, N. A.; Allen, T. R.; Breyer, R. A. *J. Am. Chem. Soc.* **1992**, *114*, 7676-7683.
8. Curran, D. P.; Porter, N. A.; Giese, B. *Stereochemistry of Radical Reactions* VCH: Weinheim, 1996.

9. Miyata, M. *Comprehensive Supramolecular Chemistry* Vol. 10, Reinhoudt, N. D. ed., Pergamon: Oxford, 1996, p. 557-582.

10. Toda, F. *Acc. Chem. Res.* **1995**, *26*, 480-486.

11. Ohashi, Y. ed. *Reactivity in Molecular Crystals* VCH-Kodansha: Tokyo, 1993.

12. Ramamurthy, V. ed. *Photochemistry in Organized and Constrained Media* VCH: New York, 1991.

13. Paleos, M. ed. *Polymerization in Organized Media* Gordon and Breach: Philadelphia, 1992.

14. Stupp, S. I.; Osenar, P. *Synthesis of Polymers* Schlüter A.-Dieter ed., Wiley-VCH: Weinheim, 1999, p. 513-547.

15. Schmidt, G. M. J. *Pure Appl. Chem.* **1971**, *27*, 647-678.

16. Wegner, G. *Pure Appl. Chem.*, **1977**, *47*, 443-454.

17. Hasegawa, M. *Adv. Phys. Org. Chem.* **1995**, *30*, 117-171.

18. Enkelmann, V. *Adv. Polym. Sci.* **1984**, *63*, 91-136.

19. Matsumoto, A.; Matsumura, T.; Aoki, S. *J. Chem. Soc., Chem. Commun.* **1994**, 1389-1390.

20. Matsumoto, A.; Matsumura, T.; Aoki, S. *Macromolecules* **1996**, *29*, 423-432.

21. Matsumoto, A.; Yokoi, K.; Aoki, S.; Tashiro, K.; Kamae, T.; Kobayashi, M. *Macromolecules* **1998**, *31*, 2129-2136.

22. Tashiro, K.; Kamae, T.; Kobayashi, M.; Matsumoto, A.; Yokoi, K.; Aoki, S. *Macromolecules* **1999**, *32*, 2449-2454.

23. Matsumoto, A.; Yokoi, K.; Aoki, S. *Polym. J.* **1998**, *30*, 361-363.

24. Matsumoto, A.; Yokoi, K. *J. Polym. Sci., Part A, Polym. Chem.* **1998**, *36*, 3147-3155.

25. Matsumoto, A.; Tsubouchi, T. *The 5th International Symposium on Polymers for Advanced Technologies* August 31-September 5, 1999, Tokyo; Abstract, p. 261.

26. Tashiro, K.; Zadorin, A. N.; Saragai, S.; Kamae, T.; Matsumoto, A.; Yokoi, K.; Aoki, S. *Macromolecules* **1999**, *32*, 7946-7950.

27. Matsumoto, A.; Odani, T.; Yokoi, K. *Proc. Japan Acad. Series B* **1998**, *74*, 110-115.

28. Desiraju, G. R. ed. *The Crystals as a Supramolecular Entity: Perspectives in Supramolecular Chemistry Vol. 2* Wiley: Chichester, 1996.

29. Lehn, J. -M. *Supramolecular Chemistry* VCH: Weinheim, 1995.

30. Matsumoto, A.; Odani, T.; Aoki, S. *Polym. J.* **1998**, *30*, 358-360.

31. Matsumoto, A.; Odani, T.; Chikada, M.; Sada, K.; Miyata, M. *J. Am. Chem. Soc.* **1999**, *121*, 11122-11129.

32. Odani, T.; Matsumoto, A. *Macromol. Rapid Commun.* **2000**, *21*, 40-44.

33. Matsumoto, A.; Odani, T. *Polym. J.* **1999**, *31*, 717-719.

34. Matsumoto, A.; Katayama, K.; Odani, T.; Oka, K.; Tashiro, K.; Saragai, S.; Nakamoto, S. in preparation.

NITROXIDE-MEDIATED POLYMERIZATION

Chapter 8

Use of Phosphonylated Nitroxides and Alkoxyamines in Controlled/"Living" Radical Polymerization

C. Le Mercier[1], J.-F. Lutz[2], S. Marque[3], F. Le Moigne[1], P. Tordo[1,5],
P. Lacroix-Desmazes[2], B. Boutevin[2], J.-L. Couturier[4], O. Guerret[4],
R. Martschke[3], J. Sobek[3] , and H. Fischer[3]

[1]Laboratoire de Structure et Reáctivité des Espèces Paramagnétiques,
UMR 6517, CNRS et Universités d'Aix-Marseille 1 et 3,
Av. Esc. Normandie Niemen, 13397 Marseille Cedex 20, France
[2]UMR-CNRS 5076, ENSCM, 8 rue de l'Ecole Normale,
34296 Montpellier Cedex 5, France
[3]Physikalisch-Chemisches Institut der Universität Zürich,
Winterthurerstrasse 190, CH–8057 Zürich, Switzerland
[4]Elf-Atochem, CRRA, Rue Moissan, B.P. 63, 69310 Pierre-Bénite, France

A new series of stable β-phosphonylated nitroxides bearing a β-hydrogen and different corresponding alkoxyamines were prepared in good yields using commercially available chemicals. The N-*tert*-butyl-N-(1-diethylphosphono-2,2-dimethylpropyl) nitroxide **3** (SG1), and the N-*tert*-butyl-N-(1-diethylphosphono-2,2-dimethylpropyl)-N-(1-phenylethoxy) amine **10**, were used in controlled / "living" polymerization of styrene. The use of either the bicomponent system (**SG1** / **AIBN**) or the monocomponent system (**10**) resulted in reasonably fast and well controlled polymerizations. The equilibrium constant K for the reversible homolysis of **10** was shown to be much larger (450 times) than for its TEMPO analog **16**. This difference accounts for the fast kinetic and the negligible kinetic contribution of the thermal self initiation during the polymerization of styrene in the presence of **10**. The X-ray structures of **10** and **16** were determined and compared.

[5]Corresponding author (e-mail address: tordo@srepir1.univ-mrs.fr)

Introduction

A lot of studies have followed the first report of Otsu *et al*[1] on living radical polymerization and nowadays living / controlled radical polymerization is a major topics in free radical polymerization research. Several procedures have been detailed such as stable free radical polymerization (SFRP) using stable nitroxyl radicals[2], atom transfer radical polymerization (ATRP) using transition metal complexes[3] and radical addition fragmentation chain transfer (RAFT) using dithioester derivatives.[4] In the case of a free radical polymerization mediated by a stable nitroxide, the control relies on the reversible trapping of the growing polymer radical by a stable nitroxide to form the corresponding N-alkoxyamine (Scheme 1). The living / controlled character was shown to be closely related to a general phenomenon which appears in reactions where transient and persistent radicals are formed simultaneously, the Persistent Radical Effect.[5] The control and the rate of the polymerization depend on the value of K ($K = k_d / k_c$) and particularly on the value of the dissociation constant k_d. Hence, the bond dissociation energy (BDE) of the NO-C bond appears as a key parameter in these nitroxide controlled free radical polymerizations.

dormant species $K = k_d / k_c$ polymer radical

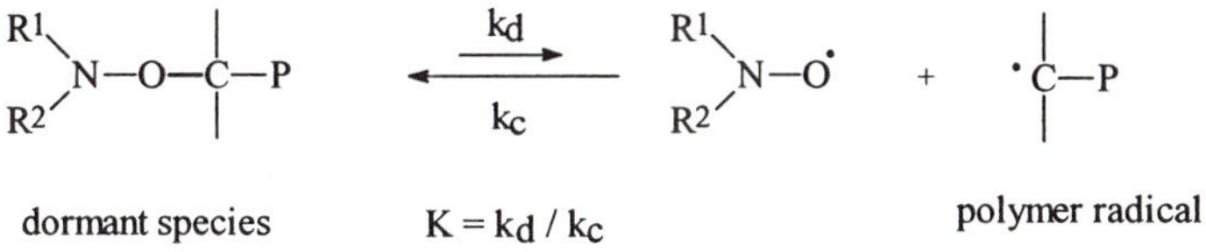

Scheme 1

Among the few commercially available nitroxides, 2,2,6,6-tetramethyl piperidinoxyl (TEMPO) was shown to be the most convenient for SFRP and it has been widely used. However, the use of TEMPO suffers from two main limitations : (i) the rates of polymerization in the presence of TEMPO are dramatically decreased ; (ii) the use of TEMPO is almost limited to styrenic monomers. In order to overcome these limitations, research has focused on the search of additives[6] which could change the equilibrium constant K and on the design of new effective stable nitroxides. As shown by Moad and Rizzardo[7] both the steric size and the electronic effects of the R^1 and R^2 groups of the nitroxide moiety influence the BDE of the NO-C bond in the corresponding alkoxyamines. Puts and Sogah[8] reported that 2,5-dimethyl-2,5-diarylpyrrolidin-1-oxyl provides a significantly faster polymerization reaction, compared to the TEMPO system. Georges *et al.*[9] observed that the polymerization of styrene in the presence of di-*tert*-butylnitroxide is faster than the TEMPO-mediated polymerization reaction. More recently, it has been shown that morpholone and piperazinone based nitroxides and related alkoxyamines also allowed faster polymerizations than TEMPO.[10] The increase of the bulkiness of R^1 or (and) R^2 results on the weakening of the BDE of the NO-C bond. However, a too large steric repulsion between R^1 and R^2 could result in the rapid decomposition of the nitroxide.[11] The preferred conformation adopted by sterically crowded aliphatic *tert*-butyl alkyl nitroxides is shown in scheme 2 (S = small, L = large).

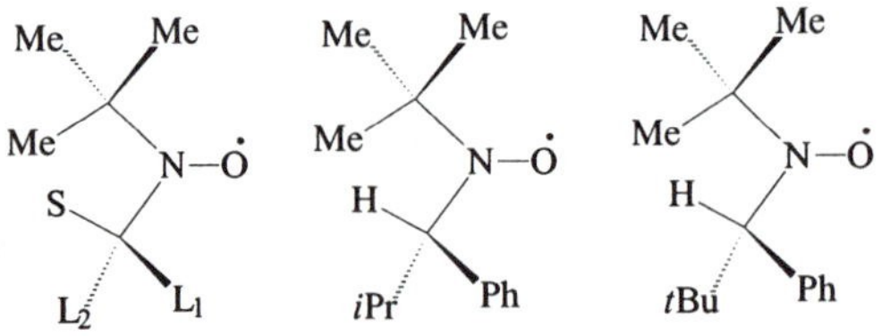

1 *(stable)*   **2** *(persistent)*

Scheme 2

Usually, nitroxides with hydrogen atoms attached to an α-carbon are not stable and cannot be isolated.[12] However, Volodarsky[13] reported that nitroxides bearing one hydrogen atom attached to an α-carbon (referred as β-hydrogen according to the nomenclature of ESR couplings), such as **1** (S = H, scheme 2), can be stable on condition that the hydrogen is locked close to the nodal plane of the nitroxyl function. However, even for this kind of nitroxides the increase of the steric strain is limited and if the bulkiness of the L groups is too large the nitroxides undergo unimolecular decay and their half-life can be very short (nitroxide **2**, Scheme 2). In preliminary communications[14] we reported that nitroxide **1** was more efficient than TEMPO and di-*tert*-butyl nitroxide (DTBN) for the SFRP of styrene. Recently, these preliminary results have been confirmed.[15] Moreover, **1** was shown to be also efficient in the control of the free radical polymerization of various acrylic monomers.[15]

In the course of our program on the search of stable nitroxides, we have found a new series of stable β-phosphonylated nitroxides bearing a β-hydrogen (nitroxides **3** – **9**, figure 1) which are able to efficiently control the free radical polymerization of different monomers.[14, 16] This paper describes briefly the synthesis of these nitroxides and their corresponding alkoxyamines. Then we compare the kinetic parameters of the reversible homolysis of model alkoxyamines with those of the TEMPO analogs, and finally we report on the use of these compounds to control the free radical polymerization of styrene.

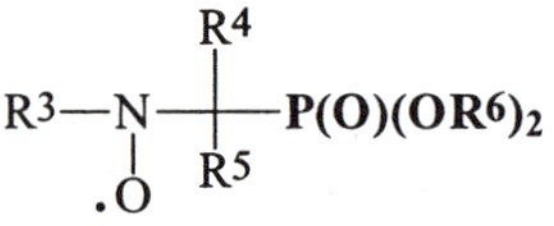

3 (SG1) : $R^3 = R^4 = t$-Bu, $R^5 = H$, $R^6 = Et$
4 : $R^3 = R^4 = t$-Bu, $R^5 = H$, $R^6 = CH_2Ph$
5 : $R^3 = t$-Bu, $R^4 = i$-Pr, $R^5 = H$, $R^6 = Et$
6 : $R^3 = t$-Bu, $R^4 = cy$-Hex, $R^5 = H$, $R^6 = Et$
7 : $R^3 = PhCH(Me)$, $R^4, R^5 = (CH_2)_5$, $R^6 = Et$
8 : $R^3 = i$-Pr, $R^4, R^5 = (CH_2)_5$, $R^6 = Et$
9 : $R^3 = cy$-Hex, $R^4, R^5 = (CH_2)_5$, $R^6 = Et$

Figure 1 : Stable β-Phosphonylated Nitroxides 3 – 9.

Experimental section

Reagents for polymerizations. Dicumyl peroxide, DCP, (98%, Aldrich), benzoic anhydride (98%, Lancaster) were used as received. AIBN (98%, Fluka) was recrystallized in ethanol and styrene (Aldrich) was distilled over CaH₂.

Polymerizations. Typically, a mixture of alkoxyamine and styrene in a Schlenk flask was thoroughly purged with argon. Then, bulk polymerization was conducted at 123°C and samples were withdrawn under positive argon purge and analyzed by Size Exclusion Chromatography (SEC) and ^{1}H NMR.

Analyses. Molecular weights and polydispersities were determined by SEC calibrated with polystyrene standards. Monomer conversion was determined by ^{1}H NMR analysis on crude samples.

General procedure for the preparation of α-aminophosphonates 3' – 9'. A mixture of the amine (50 mmol) and the aldehyde or ketone (50 mmol) was stirred at 40°C for 1 h under nitrogen atmosphere. After addition of diethyl phosphite (75 mmol) at room temperature, the solution was stirred at 40°C for 24 h. The reaction mixture was then diluted with diethyl ether (100 ml) and washed with 5 % aqueous HCl until pH 3. The aqueous phase was extracted with diethyl ether. Sodium hydrogencarbonate was added to the aqueous phase until pH 8, and the aqueous phase was then extracted with diethyl ether (2 × 30 mL). The combined organic extracts were dried over anhydrous Na_2SO_4. Removal of the solvent under reduced pressure afforded the α-aminophosphonates.

General procedure for the oxidation of α-aminophosphonates 3', 4', 7', 8' and 9' into the β-phosphonylated nitroxides 3, 4, 7, 8 and 9. A solution of *m*-chloroperbenzoic acid (8 mmol) in dichloromethane (20mL) was added at 0°C to a solution of the α-aminophosphonate (8 mmol) in dichloromethane (10 mL). The mixture was stirred for 6 h at room temperature. Then a saturated aqueous solution of sodium hydrogencarbonate was added until neutral pH. The organic phase was successively washed with water, 1 M aqueous sulfuric acid, water, saturated aqueous sodium hydrogencarbonate and water. After drying, over anhydrous Na_2SO_4, removal of the solvent under reduced pressure afforded an oil which was purified by silica gel chromatography (pentane / ethyl acetate).

General procedure for the oxidation of α-aminophosphonates 5', 6' into the β-phosphonylated nitroxides 5 and 6. Oxone (40 mmol) was added at room temperature to a solution of the α-aminophosphonate (10 mmol), sodium carbonate (60 mmol) in water (20 mL) and ethanol (70 mL). The mixture was stirred for 24 h at room temperature. The reaction mixture was then filtered and the solvent was evaporated under reduced pressure. The residue was extracted with dichloromethane and the organic phase was dried over anhydrous Na_2SO_4. Removal of the solvent under reduced pressure afforded an oil which was purified by silica gel chromatography (pentane / ethyl acetate).

General procedure for the synthesis of alkoxyamines 10 – 15. Under inert atmosphere, a solution of the nitroxide (5 mmol) and the alkyl bromide (10 mmol) in benzene (8 mL) was transferred to a mixture of CuBr (10 mmol), bipyridine (20 mmol) and Cu(0) (5 mmol if necessary, see Table I) in benzene (8 mL). After 2 days stirring at room temperature, the mixture was filtered and washed with a 5 % w/v aqueous solution of $CuSO_4$. The organic phase was dried over anhydrous Na_2SO_4 and removal of the solvent under reduced pressure afforded an oil which was purified by silica gel chromatography (pentane / diethyl ether).

Results and discussion

Synthesis of the nitroxides 3 – 9 and alkoxyamines 10 - 15

Synthesis of the nitroxides 3 – 9. The β-phosphonylated nitroxides **3 – 9** were prepared by oxidation of the corresponding α-aminophosphonates **3' – 9'**[14] (Figure 2). The compounds **3' – 9'** were obtained in reasonable to good yields (30 - 90 %) through two major approaches : a - the one pot reaction of equimolar amounts of carbonyl compound and amine with a slight excess of dialkylphosphite (**3', 7'**) ; b - the addition of a dialkylphosphite to an imine, either generated in situ (**4', 5', 6', 8'**), or isolated (**9'**). Oxidation of the α-aminophosphonates **3', 4', 7', 8'** and **9'** with *m*-chloroperbenzoic acid (*m*-CPBA) in dichloromethane gave after purification the corresponding nitroxides **SG1, 4, 7, 8** and **9** in 48, 25, 37, 41 and 46 % yields respectively. The improvement of the oxidation of **3'** was investigated and allowed the preparation of **SG1** on a large scale in 80 % yield with a purity superior to 90 % and without any purification. Oxidation of the α-aminophosphonates **5'** and **6'** with *m*-CPBA gave very poor yields and the nitroxides **5** and **6** were isolated in respectively 26 and 18 % yields through oxidation of **5'** and **6'** by Oxone (2 $KHSO_5$ - $KHSO_4$ - K_2SO_4) in a mixture of ethanol, water and Na_2CO_3[17]. Only the nitroxide **4** was obtained as a solid.

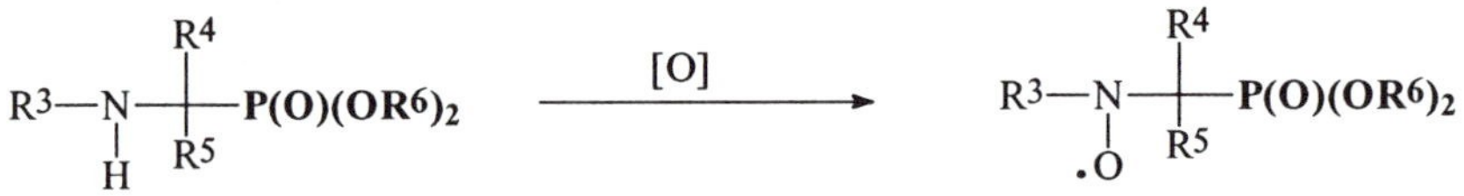

R3—N(H)(R5)—P(O)(OR6)2 →[O]→ R3—N(•O)(R5)—P(O)(OR6)2

3' : R³ = R⁴ = *t*-Bu, R⁵ = H, R⁶ = Et   **3 (SG1)**
4' : R³ = R⁴ = *t*-Bu, R⁵ = H, R⁶ = CH₂Ph   **4**
5' : R³ = *t*-Bu, R⁴ = *i*-Pr, R⁵ = H, R⁶ = Et   **5**
6' : R³ = *t*-Bu, R⁴ = *cy*-Hex, R⁵ = H, R⁶ = Et   **6**
7' : R³ = PhCH(Me), R⁴, R⁵ = (CH₂)₅, R⁶ = Et   **7**
8' : R³ = *i*-Pr, R⁴, R⁵ = (CH₂)₅, R⁶ = Et   **8**
9' : R³ = *cy*-Hex, R⁴, R⁵ = (CH₂)₅, R⁶ = Et   **9**

Figure 2 : Oxidation of the α-Aminophosphonates 3' – 9' to the β-Phosphonylated Nitroxides 3 – 9.

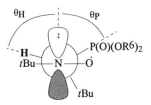

Figure 3 : Predominant conformation of SG1, 4, 8 and 9.

All these nitroxides contain a β-hydrogen and are stable compounds which can be isolated and stored either neat or in solution. Their Electron Spin Resonance (ESR) study in solution[18] showed that the $a_{Hβ}$ coupling was either very small ($a_{Hβ}$ < 0.01 mT) or unresolved (for **SG1, 4, 8** and **9**). This observation allowed us[19] to assume for these compounds the existence at ambient temperature of a largely predominant conformer in which the

H_β atom is eclipsed by a bulky alkyl group and lies close to the nodal plane of the nitroxyl function ($\theta_H \approx 90\,°$, Figure 3). In this preferred conformation, the H_β atom is sterically masked, thus impeding the decay of the nitroxide through disproportionation[12]. The influence of the steric hindrance due to the two *tert*-butyl groups of the nitroxide **4** was also reflected in its X-ray structure[20].

Synthesis of the alkoxyamines 10 – 15. The synthesis of alkoxyamines through the trapping of free radicals with nitroxides has been largely described. Matyjaszewski *et al.*[21] developed a versatile and efficient method based on the ATRP system for synthesizing several alkoxyamines derived from TEMPO. Hawker *et al.*[22] coupled TEMPO and its derivatives with 1-phenyl-ethyl radical generated through addition of styrene to the Jacobsen's Manganese (III) catalyst. Low temperature multi-step syntheses of a variety of alkoxyamine initiators were also reported by Braslau *et al.*[23,15] In order to evaluate their initiator efficiency in the polymerization of styrene and acrylates, and to measure the Bond Dissociation Energy (BDE) of their NO-C bond, the alkoxyamines **10 – 15** derived from the nitroxides **SG1, 8, 9** and **1** were synthesized (Figure 4). They were all prepared following the method derived from ATRP and first reported by Matyjaszewski. In this method, the radicals R[•] are generated via copper (I) reduction of the corresponding organic halides RX. Owing to the facile cleavage of most of the targeted alkoxyamines, their synthesis was carried out at room temperature and excess of copper (I) (2 eq.) and alkyl halides (2 eq.) relative to nitroxides was necessary in order to obtain the complete conversion of nitroxides. In our experimental conditions, we found that the addition of 1 equivalent of copper (0) slightly improved the yield and the reaction time[24] (Table I). Even though they were not optimized, good yields of alkoxyamines were obtained. The compound **12** was characterized only in solution and could not be isolated since the cleavage of its NO-C bond occurs below room temperature.

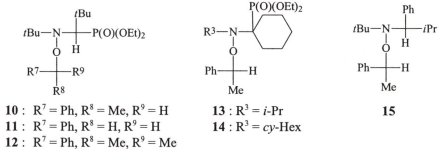

10 : R⁷ = Ph, R⁸ = Me, R⁹ = H   13 : R³ = *i*-Pr
11 : R⁷ = Ph, R⁸ = H, R⁹ = H   14 : R³ = *cy*-Hex   15
12 : R⁷ = Ph, R⁸ = Me, R⁹ = Me

Figure 4 : Alkoxyamines 10 – 15.

In order to roughly appreciate the thermal stability of the prepared alkoxyamines, we measured by ESR their cleavage temperature *ie* the temperature at which a significant signal of the nitroxide moiety is observed (Table I). It is interesting to point out that for the alkoxyamines **10 - 12** derived from the nitroxide **SG1**, the cleavage temperature was at least 35 °C lower than for the TEMPO analogs (data not reported). Moreover, in the series **11, 10, 12**, there is a significant decrease of the cleavage temperature. This trend can be explained by an increase of the steric strain in the alkoxyamine and by an increase of the stability of the released radical.

Table I. Synthesis and cleavage temperature of the alkoxyamines 10 - 15.

Alkoxyamine	Amount of Cu(0) (eq.)	Yield (%)	Cleavage Temperature (°C)
10	0	*95*	*60*
11	0	*40*	*90*
12	0	*Not isolated*	*< rt*
13	1	*79*	*80*
14	1	*88*	*70*
15	1	*90*	*60*

Rate constants for the reversible homolysis of the alkoxyamines 10, 15 and TEMPO-CH(Me)Ph, 16

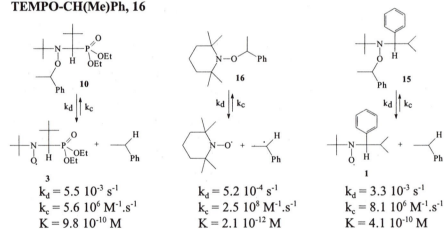

$k_d = 5.5 \ 10^{-3} \ s^{-1}$ $k_d = 5.2 \ 10^{-4} \ s^{-1}$ $k_d = 3.3 \ 10^{-3} \ s^{-1}$

$k_c = 5.6 \ 10^{6} \ M^{-1}.s^{-1}$ $k_c = 2.5 \ 10^{8} \ M^{-1}.s^{-1}$ $k_c = 8.1 \ 10^{6} \ M^{-1}.s^{-1}$

$K = 9.8 \ 10^{-10} \ M$ $K = 2.1 \ 10^{-12} \ M$ $K = 4.1 \ 10^{-10} \ M$

Figure 5. Rate Constants at 123° C in tert-butylbenzene for the reversible cleavage of the alkoxyamines 10, 15 and 16.

The values of k_d, k_c and their temperature dependence were measured according to the procedures previously described for TEMPO-C(Me)$_2$Ph.[25] The individual rate constants, k_d and k_c, yield the equilibrium constant (K) of the cleavage and the bond dissociation energy (BDE) of the NO-C bond. The values obtained are listed in Figure 5. The rate constant of the cleavage of **10** is about 10 times larger than for its TEMPO analog. Furthermore, the rate constant of trapping of the 1-phenyl-ethyl radical is about 45 times larger with TEMPO than with nitroxide **SG1**, and this difference is certainly a consequence of a greater steric hindrance for **SG1**. The value of K for **10** is about 450 times larger than for **16**, then, in nitroxide mediated radical polymerization of styrene the concentration of active species and the rate of polymerization are expected to be much larger with nitroxide **SG1** than with TEMPO. The K value for **10** is only about 2 times higher than for **15**, nevertheless, the rate of the "living" / controlled polymerization of styrene in the presence of **10** is significantly higher (2-3 times, see below) than in the presence of **15**.

With the aim to get information on the influence of structural parameters on the homolysis of alkoxyamines we determined the X-ray geometry of **10** and **16**. These geometries are shown in Figure 6 and some selected bond lengths and angles are listed in Table II.

Table II. Selected X-ray Bond Lengths (Å) and Angles (deg) for the alkoxyamines 10 and 16.

	d_{CO}	d_{NO}	CON	Σ angles(N)	d_{CN}(in NOC)
10	1.454	1.453	113.2	329.2	2.42
16	1.452	1.458	112.4	330.6	2.41

The values of the angles indicate that for **10** and **16** the geometry is pyramidal around the nitrogen. The C-N distance in the C-O-N fragment is a good measure of the influence of steric factors. This distance is the same for **10** and **16** thus indicating that the steric strain in this fragment is the same for the two alkoxyamines. However, some bond length and angle values observed for **10** are significantly larger than their mean value ($d_{C7C11} = 1.57$ Å, $N_4C_7C_{11} = 112.8°$) and suggest that the total steric strain for **10** is higher than for **16**.

For the nitroxide **2** (Scheme 2) the steric hindrance between the phenyl and *tert*-butyl groups is too large and the nitroxide undergoes unimolecular decay. For our series of phosphonylated nitroxides the large value of the C-P bond length ($d_{C-P} = 1.831$ Å for **4**[20]) reduces the steric hindrance between L_1 and L_2 (Scheme 2) and accounts for the stability of these highly constrained molecules. However, the phosphonyl group contributes largely to mask the nitroxide oxygen and to make difficult the coupling with the 2-phenyl-ethyl radical.

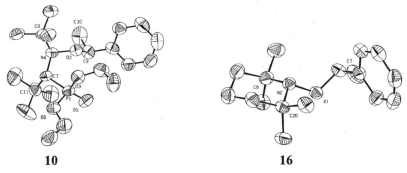

10   **16**

Figure 6 : X-ray structure of 10 and 16

Polymerization results

Polymerization of styrene in the presence of alkoxyamine 10. We have studied the bulk polymerization of styrene in the presence of the alkoxyamine **10**. Four different molecular weights were targeted from 10000 up to 100000 g / mol. First, as shown on Figure 7, the linear character of the plots of M_n versus conversion reflects the fast initiation by the alkoxyamine **10**. This result is in accordance with the rather high value of the dissociation rate constant $k_d = 5.5 \; 10^{-3} \; s^{-1}$ determined for this alkoxyamine at 123°C, corresponding to a short half-life $t_{1/2} = 126$ s. In their work on alkoxyamines based on TEMPO, Hawker et al.[26] had already noticed that the success of a controlled polymerization strongly depends on the ability of the alkoxyamine to dissociate. Moreover, Figure 7 also shows that the agreement between experimental and theoretical values of M_n is very good in all cases and that the polydispersity index remains low, in the range of 1.15-1.25.

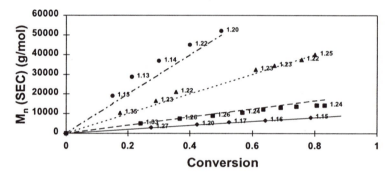

Figure 7. M_n versus conversion for the bulk polymerization of styrene at 123°C with [alkoxyamine 10] = 0.00909 M (●), 0.0175 M (▲), 0.044 M (■), 0.0874 M (◆). Theoretical M_n = [styrene]$_0$×conversion×104.15/[10]$_0$ (lines).

The effect of the initial concentration of alkoxyamine **10** on the kinetics of the polymerization is shown in Figure 8. In all cases, the polymerization is fast : high conversions are obtained within 4 and 6 hours instead of more than 24 hours in the case of TEMPO. The plots of $Ln([M]_0/[M])$ versus time clearly indicate that the polymerization rate increases with the initial concentration of **10**. The lowest trace shows the negative deviation from a linear behavior which is expected from theoretical considerations for low conversion.[5, 27] The others are more or less linear, and this may be caused by the known decrease of the termination constant with increasing conversion. Plots of $Ln([M]_0/[M])$ versus $t^{2/3}$ allowed us to determine a dependence on [alkoxyamine]$_0^\alpha$ with $\alpha = 0.34$ and an equilibrium constant $K = 1.04 \; 10^{-8} \; M$. The value of α is in excellent agreement with the theoretical value of 1/3 while K is in the range of values determined in previous studies.[28] This non zero order behavior is strikingly different from the results obtained with TEMPO. In the classic case of TEMPO mediated polymerization of styrene, it is well established that $R_p/[M]$ is only proportional to $(R_{th})^{0.5}$, where R_p is the rate of the polymerization and

R_{th} is the rate of thermal self initiation of styrene.[29] Thus, the slope of the curve Ln([M]$_0$/[M]) does not depend on the initial concentration of the alkoxyamine, and this behavior is a consequence of the very low value of the equilibrium constant between dormant and active species (K = 2.1 10^{-12} M). On the contrary, with nitroxide **SG1**, the contribution of the thermal self initiation is very low with respect to that of the equilibrium between active and dormant species.[28] Indeed, the equilibrium constant is about two orders of magnitude higher for nitroxide **SG1** in comparison with TEMPO.

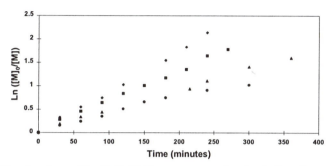

Figure 8. Kinetic plots of Ln([M]$_0$/[M]) versus time for the bulk polymerization of styrene at 123°C with [alkoxyamine 10] = 0.00909 M (●), 0.0175 M (▲), 0.044 M (■), 0.0874 M (◆).

Comparison of monocomponent (alkoxyamine) and bicomponent (Initiator/Nitroxide) systems. Some studies have already been successfully conducted with bicomponent systems (Initiator / Nitroxide **SG1**) involving AIBN as the initiator[16]. We have compared the kinetic behaviors for three systems involving : alkoxyamine **10**, (system A) ; AIBN / Nitroxide **SG1**, (system B) ; and DCP / Nitroxide **SG1** (system C). In this respect, in the case of the bicomponent systems, we have used a stoechiometric ratio between initiator and nitroxide whereas previous studies often used a substantial excess of nitroxide versus initiator. DCP was chosen in reference to promising results obtained with TEMPO[30, 31]. It must be noticed that at the temperature of the polymerization (123°C), the decomposition of AIBN is almost instantaneous ($t_{1/2}$ = 1 min) while the decomposition of DCP is much slower ($t_{1/2}$ = 172 min).

As shown in Figure 9, the plots of Ln([M]$_0$/[M]) versus time in the case of systems A and B indicate that the polymerization is faster for the monocomponent system A. This kinetic difference can be explained by the occurrence of a cage effect for the decomposition of AIBN[32]. Thus, the efficiency of AIBN is lower than one and the concentration of alkoxyamine formed in situ is lower than predicted. This is in accordance with the upper deviation of the molecular weights versus conversion as shown on Figure 10. On the contrary, in the case of the monocomponent system, the cage effect does not occur[25,33]. It results that the concentration of the initiated polymer chains is almost equal to the theoretical one (efficiency very close to unity) as demonstrated by the very good agreement between experimental and theoretical

118

molecular weight (Figure 10). This kinetic difference between the alkoxyamine and the system AIBN / Nitroxide also illustrates the non zero order kinetic behavior.

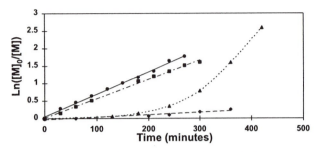

Figure 9. Kinetic plots of Ln([M]$_0$/[M]) versus time for the bulk polymerization of styrene at 123°C with [alkoxyamine 10]=0.044 M (●) (system A), [AIBN]/[SG1] = 0.022M/0.044M (■) (system B), [DCP]/[SG1] = 0.022M/0.044M (▲) (system C) and [SG1] = 0.044 M (thermal initiation) (◆).

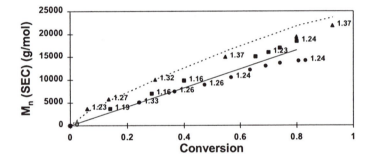

Figure 10. M$_n$ versus conversion for the bulk polymerization of styrene at 123°C with [alkoxyamine 10] = 0.044 M (●) (system A), [AIBN]/[SG1] = 0.022M/0.044M (■) (system B) and [DCP]/[SG1]=0.022M/0.044M (▲) (system C). Theoretical M$_n$ = [styrene]$_0$×conversion×104.15/[Initiator]$_0$ (—); for DCP as initiator, considering the slow initiation rate, theoretical M$_n$ = [styrene]$_0$×conversion×104.15/(2f×[Initiator]$_0$(1-exp(-k$_d$t))) with a value of k$_d$=6.71×10^{-5} s^{-1} (- - - -).

Considering the slow decomposing initiator DCP, the rate of polymerization of styrene in the presence of system C starts slowly and accelerates to become faster than previous systems (Figure 9). In a first phase, the decomposition of the initiator mainly serves to build up the concentration of the dormant species. However, thanks to the rather high value of the equilibrium constant between dormant and active species, the polymerization is not completely inhibited as in the case of TEMPO. Then, due to the non zero order of the kinetics versus the concentration of alkoxyamine, the polymerization rate gradually increases with the concentration of initiated polymer chains. For system C, a non linear evolution of M$_n$ versus conversion is expected because of the low initiation rate constant k$_d$ of DCP (Figure

10). Initiation takes place during a longer period of time leading to higher polydispersity indexes (still in the range 1.2-1.5)[34].

A recent work of Malmström *et al.*[6] reported the efficiency of acylating agents as accelerating additives of TEMPO mediated polymerization of styrene. In this respect, we have tested the effect of benzoic anhydride on our system. For the three systems (A, B, C), the anhydride does not induce any significant acceleration of the polymerization. Moreover, we could see only minor changes of the evolution of molecular weights versus conversion upon the addition of the anhydride[35]. Similar results were recently obtained by Hawker *et al.*[15] for the bulk polymerization of styrene in the presence of alkoxyamine **15** with acetic anhydride as additive.

Comparison of alkoxyamines 10 and 15. The nitroxide **1** has been previously successfully tested with AIBN as initiator.[14a, 15] We have compared the polymerization of styrene in the presence of alkoxyamines **10** and **15**. In both cases,

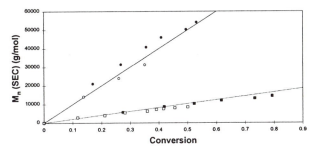

Figure 11. M_n versus conversion for the bulk polymerization of styrene at 123°C with [alkoxyamine 10] = 0.044 M (■) , 0.00909 M (●) and [alkoxyamine 15] = 0.044 M (□), 0.00909 M (O).

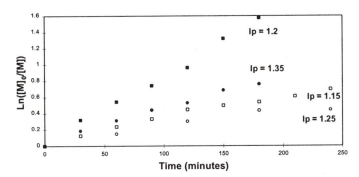

Figure 12. Kinetic plots of Ln([M]$_0$/[M]) versus time for the bulk polymerization of styrene at 123°C with [alkoxyamine 10] = 0.044 M (■) , 0.00909 M (●)and [alkoxyamine 15] = 0.044 M (□), 0.00909 M (O).

120

we observed a good control of the molecular weight together with narrow polydispersity indexes (between 1.1 and 1.3) (Figure 11). However, the kinetic plot of $Ln([M]_0/[M])$ versus time clearly shows that the polymerization is much faster in the case of the phosphonylated alkoxyamine 10 (Figure 12). This behavior is in agreement with the larger value of the equilibrium constant for the phosphonylated alkoxyamine (K = 9.8 10^{-10} M at 123°C) in comparison with the alkoxyamine 15 (K = 4.7 10^{-10} M at 123°C). Again with alkoxyamine 15 and for low conversions, the data show the expected deviations from linearity.[5, 27]

Controlled polymerization of acrylate monomers. Promising results were previously obtained for the polymerization of n-butyl acrylate mediated by nitroxide **SG1** in the presence of AIBN as the initiator[16]. We have confirmed these preliminary results with the alkoxyamine 10, using very small amounts of the free nitroxide to moderate the polymerization rate. The control of the molecular weight was shown to be as good as for styrene. A detailed study of the kinetic behavior of this system will be published elsewhere.

Conclusions

Owing to the large value of the C-P bond length a new series of sterically constrained β-phosphonylated nitroxides bearing a β-hydrogen and different corresponding alkoxyamines were prepared in good yields. A very good control of the polymerization of styrene can be obtained using the alkoxyamine 10. The polymerization rate is not zero order in alkoxyamine, indicating that the kinetic contribution of the thermal self initiation of styrene is low. This behavior is predicted from the rather large value of the equilibrium constant between dormant and active species. Moreover, as anticipated by its structural study, this phosphonylated alkoxyamine turned out to be well suited for the control of the polymerization of styrene and acrylates. Especially, it was shown to provide a faster polymerization than alkoxyamine 15 which was another good candidate derived from the series of α-hydrogen bearing nitroxides studied in our laboratories. Otherwise, the alkoxyamine 10 offered a better control of the polymerization in comparison with bicomponent systems. In the case of the bicomponent system AIBN / **SG1**, the efficiency of initiation is a little bit lower than for alkoxyamine 10, leading to slightly slower polymerization and to upper deviation of the molecular weight.

The design of new effective nitroxides and alkoxyamines has permitted to overcome the limitations inherent to TEMPO mediated polymerization : a wide range of monomers can be controlled, polymerizations proceed at faster rate and lower reaction temperature are required opening the door to controlled polymerization in emulsion. Considering their efficacy in controlled / "living" polymerization of different monomers and their facile synthesis, these new nitroxides and alkoxyamines appear as particularly appropriate for industrial uses.

Acknowledgment. Elf-Atochem is acknowledged for financial and scientific support.

References

1. Otsu, T.; Yoshida, M.; Tazaki, T. *Makromol. Chem., Rapid. Commun.* **1982**, *3*, 133.
2. (a) Solomon, D. H.; Waverly, G.; Rizzardo, E.; Hill, W.; Cacioli, P. *US Patent* **1986**, US 4,581,429. (b) Georges, M. K.; Veregin, R. P. N.; Kazmaier, P. M.; Hamer, G. K. *Macromolecules.* **1993**, *26*, 2987. (c) Devonport, W.; Michalak, L.; Malmström, E.; Mate, M.; Kurdi, B.; Hawker, C. J.; Barclay, G.G.; Sinta, R. *Macromolecules.* **1997**, *30*, 1929. (d) Goto, A.; Fukuda, T. *Macromolecules.* **1997**, *30*, 4272. (e) Greszta, D.; Matyjaszewski, K. *Macromolecules.* **1996**, *29*, 7661. (f) Bertin, D.; Destarac, M.; Boutevin, B in *Polymers and surfaces. A versatile combination,* Hommel H. (Ed.) **1998**, 47.
3. (a) Wang, J. S.; Matyjaszewski, K. *Macromolecules* **1995**, *28*, 7901. (b) Uegaki, H.; Kotani, Y.; Kamigaito, M; Sawamoto, M. *Macromolecules* **1997**, *30*, 2249. (c) Percec, V.; Kim, H. J.; Barboiu, B. *Macromolecules* **1997**, *30*, 6702. (d) Haddleton, D. M.; Jasieczek, C. B.; Hannon, M. J.; Shooter A. J. *Macromolecules* **1997**, *30*, 2190.
4. (a) Chiefari, J.; Chong, Y. L. ; Ercole, F. ; Krstina, J. ; Jeffery, J. ; Le, T. P. T. ; Mayadunne, R. T. A. ; Meijs, G. F. ; Moad, C. L.; Rizzardo, E. ; Thang, S. H. *Macromolecules* **1998**, *31*, 5559. (b) Chang, Y. K.; Le, T. P. T.; Moad, G.; Rizzardo, E.; Thang, S. H. *Macromolecules* **1999**, *32*, 2071. (c) Hawthorne, D. G.; Moad, G.; Rizzardo, E.; Thang, S. H. *Macromolecules* **1999**, *32*, 5457.
5. (a) Fischer, H. *J. Am. Chem. Soc.* **1986**, *108*, 3925. (b) Fischer, H. *Macromolecules* **1997**, *30*, 5666. (c) Fischer, H. *J. Polym. Sci., Part A: Polym Chem,* **1999**, *37*, 1885.
6. (a) Georges, M. K.; Kee, R. A.; Veregin, R. P. N.; Hamer, G. K.; Kazmaier, P. M. *J. Phys. Org. Chem.* **1995**, *8*, 301. (b) Malmström, E.; Miller, R. D.; Hawker, C. J. *Tetrahedron* **1997**, *53*, 15225.
7. Moad, G.; Rizzardo, E. *Macromolecules* **1995**, *28*, 8722.
8. (a) Puts, R. D.; Sogah, D. Y. *Macromolecules* **1996**, *29*, 3323. (b) Sogah, D. Y.; Puts, R. D.; Trimble, A.; Sherman, O. *Polym.Prepr. (Am. Chem. Soc., Div. Polym. Chem.)* **1997**, *38(1)*, 731.
9. Kazmaier, P. M.; Moffat, K. A.; Georges, M. K.; Veregin, R. P. N.; Hamer, G. K. *Macromolecules* **1995**, *28*, 1841.
10. Puts, R.; Lai, J.; Nicholas, P.; Milam, J.; Tahilliani, S.; Masler, W.; Pourahmady, N. *Polym.Prepr. (Am. Chem. Soc., Div. Polym. Chem.)* **1999**, *40(2)*, 323.
11. Reznikov, V. A.; Gutorov, I. A.; Gatlov, Y. V.; Rybalova, T. V.; Volodarsky, L. B. *Russ. Chem. Bull.* **1996**, *45(2)*, 384.
12. Bowman, D. F.; Gillan, T.; Ingold, K. U. *J. Am. Chem. Soc.* **1971**, *93*, 6555.
13. Reznikov, V. A.; Volodarsky, L. B. *Tetrahedron Lett.* **1994**, *35*, 2239.

14. (a) Grimaldi, S.; Finet, J.-P.; Zeghdaoui, A.; Tordo, P.; Benoit, D.; Gnanou, Y.; Fontanille, M.; Nicol, P.; Pierson, J.-F. *Polym. Prepr.* **1997**, *38(1)*, 651. (b) Benoit, D.; Grimaldi, S.; Finet, J.-P.; Tordo, P.; Fontanille, M.; Gnanou, Y. *Polym. Prepr.* **1997**, *38(1)*, 729-730. (c) Grimaldi, S.; Le Moigne, F.; Finet, J.-P.; Tordo, P.; Nicol, P.; Plechot, M. *International Patent* WO 96/24620 **1996**.
15. Benoit, D.; Chaplinski, V.; Braslau, R.; Hawker, C. J. *J. Am. Chem. Soc.* **1999**, *121(16)*, 3904.
16. Benoit, D.; Grimaldi, S.; Finet, J.-P.; Tordo, P.; Fontanille, M.; Gnanou, Y. *ACS Symp.Ser.* (Controlled Radical Polymerization) Matyjaszewski, K., Ed.; **1998**, *685*, 225.
17. Ratajczak, F. *Thesis*, University of Grenoble, France, **1997**.
18. Le Mercier, C.; Bernard-Henriet, C.; de Sainte Claire, V.; Le Moigne, F.; Tordo, P.; Couturier, J.-L.; Gillet, J.-Ph.; Guerret, O. *Polym. Prepr. (Am. Chem. Soc., Div. Polym. Chem.)* **1999**, *40(2)*, 403.
19. Tordo, P.; Boyer, M.; Friedmann, A.; Santero, O. *J. Phys. Chem.* **1978**, *82*, 1742.
20. Grimaldi, S.; Siri, D.; Finet, J.-P.; Tordo, P. *Acta Cryst* **1998**, *C54*, 1712.
21. Matyjaszewski, K.; Woodworth, B.E.; Zhang, X.; Gaynor, S.G.; Metzner, Z. *Macromolecules* **1998**, *31*, 5955.
22. Dao, J.; Benoit, D.; Hawker, C.J. *J. Polym. Sci. A* **1998**, *36*, 2161.
23. Braslau, R.; Burrill, L.C.II; Sciano, M.; Naik, N.; Howden, R.K.; Mahal, L.K. *Macromolecules* **1997**, *30*, 6445.
24. Wang, J.-L.; Grimaud, T.; Matyjaszewski, K. *Macromolecules* **1997**, *30*, 6507.
25. (a) Kothe, T. ; Marque, S. ; Martschke, R. ; Popov, M. ; Fischer, H. *J. Chem. Soc., Perkin Trans. 2* **1998**, 1553. (b) Skene, W. G. ; Belt, S. T. ; Connolly, T. J. ; Hahn, P. ; Scaiano, J. C. *Macromolecules* **1998**, *31*, 9103.
26. Hawker, C. J.; Barclay, G. G.; Orellana, A.; Dao, J.; Devonport, W. *Macromolecules* **1996**, *29*, 5245.
27. Ohno, K.; Yoshinobu, T.; Miyamoto, T.; Fukuda, T.; Goto, M.; Kobayashi, K.; Akaike, T. *Macromolecules* **1998**, *31*, 1064.
28. (a) Benoit, D.; Robin, S.; Grimaldi, S.; Tordo P.; Gnanou, Y. *J. Am. Chem. Soc.*, *submitted*. (b) Lacroix-Desmazes, P.; Lutz, J.-F.; Boutevin, B. *Macromol. Chem. Phys., in press*.
29. Fukuda, T.; Terauchi, T.; Goto, A.; Ohno, K.; Tsujii, Y.; Miyamoto, T.; Kobatake, S.; Yamada, B. *Macromolecules* **1996**, *29*, 6393.
30. Bertin, D. *Thesis*, University of Montpellier II, France, **1997**.
31. Greszta, D.; Matyjaszewski, K. *J. Polym. Sci., Part A : Polym. Chem.* **1997**, *35*, 1857.
32. Odian, G. in *La Polymérisation – principes et applications,* Third Edition, *Polytechnica* **1994** , 259
33. Marque, S.; Le Mercier, C.; Tordo, P.; Fischer, H. *Macromolecules, submitted*.
34. Lutz, J.-F.; Lacroix-Desmazes, P.; Boutevin, B. *in preparation*.
35. Lutz, J.-F.; Lacroix-Desmazes, P.; Boutevin, B. *Polym. Prepr. (Am. Chem. Soc., Polym. Div.)* **1999**, *40(2)*, 319.

Chapter 9

Progress Toward a Universal Initiator for Nitroxide Mediated 'Living' Free Radical Procedures

Didier Benoit, Eva Harth, Brett Helms, Ian Rees, Robert Vestberg, Marlene Rodlert, and Craig J. Hawker[1]

Center for Polymeric Interfaces and Macromolecular Assemblies, IBM Almaden Research Center, 650 Harry Road, San Jose, CA 95120–6099

Examination of novel alkoxyamines has demonstrated the pivotal role that the structure of the nitroxide plays in mediating the 'living', or controlled polymerization of a wide range of vinyl monomers. Surveying a variety of different alkoxyamine structures led to α-hydrido derivatives based on a 2,2,5-trimethyl-4-phenyl-3-azahexane-3-oxy, **1**, skeleton which were able to control the polymerization of numerous monomer families, such as acrylates, acrylamides, etc. In comparison with 2,2,6,6-tetramethyl-1-piperidinoxy (TEMPO), these new systems represent a dramatic improvement and overcome many of the limitations traditionally associated with nitroxide mediated 'living' free radical procedures.

The development of a free radical polymerization, which displays many of the characteristics of a living polymerization, has long been a goal of synthetic polymer chemists' *(1)*. The driving force for this interest is the desire to couple the non-demanding reactions conditions, such as bulk or emulsion conditions, associated with radical polymerizations with the ability to polymerize essentially any vinyl monomer, irrespective of the presence of functional groups. The development of a 'living' free radical procedure which combines the level of control associated with traditional living systems such as anionic polymerization, with the functional group tolerance of free radical chemistry, would therefore be an extremely attractive addition to synthetic polymer chemistry.

Early attempts to realize a "living", or controlled free radical process involved the concept of reversible termination of growing polymer chains by iniferters *(2,3)*. This concept was however plagued by high polydispersities and

---

[1]Corresponding author.

poor control over molecular weight and chain ends. Following this approach, Moad and Rizzardo introduced the use of stable nitroxide free radicals, such as 2,2,6,6-tetramethylpiperidinyloxy (TEMPO), as reversible terminating agents to "cap" the growing polymer chain *(4)*. In a seminal contribution, the use of TEMPO in 'living' free radical polymerizations was refined by Georges et al who demonstrated that at elevated temperatures narrow molecular weight distribution polymers (PD = 1.1-1.3) could be prepared using bulk polymerization conditions *(5)*. In 1995, a new approach to living free radical polymerization, termed Atom Transfer Radical Polymerization or ATRP, was independently reported by Matyjaszewski *(6)* and Sawamoto *(7)*. While similar to nitroxide mediated systems, the primary difference between these procedures is the mode of generation of the propagating radical center. In ATRP, a metal catalyst, typically a copper (I), ruthenium (II), nickel (II), or iron (II) species reversibly abstracts a chlorine or bromine atom from the dormant polymer chain end to give the propagating radical. This is in direct contrast to nitroxide mediated systems in which no catalyst is required and the reaction is thermally driven. One consequence of this subtly different mechanism is that the range of monomers that can be polymerized in a living fashion by ATRP is considerably greater than TEMPO mediated nitroxide systems. For example, the polymerization of functionalized acrylates and methacrylates can be controlled to give low polydispersity macromolecules and block copolymers using ATRP. Similarly, the recently introduced Radical Addition Fragmentation Transfer (RAFT) process *(8)*, which utilizes thiocarbamate and dithioesters transfer agents, has also been demonstrated to have applicability to a diverse family of monomers.

The use of TEMPO mediated living free radical procedures for the controlled polymerization of vinyl monomers therefore suffers from two main challenges; 1) essentially only styrenic homopolymers or styrene-containing random copolymers can be prepared under living conditions; and 2) the rate of polymerization is slow compared to ATRP. This is unfortunate since the absence of transition metal catalysts, or odorous sulfur compounds, is a major benefit and attraction of nitroxide mediated systems. The development of nitroxide mediated systems which can overcome both of these challenges would be of major interest and significantly advance the rapidly expanding field of living free radical chemistry.

Experimental

General. All reactions were run under N_2 unless noted. Solvents were dried as follows: THF and toluene were distilled under N_2 from sodium-benzophenone, and CH_2Cl_2 was distilled from calcium hydride. Analytical thin layer chromatography (TLC) was performed on commercial Merck plates coated with silica gel GF254 (0.25 mm thick). Silica gel for flash chromatography was either Merck Kieselgel 600 (230-400 mesh) or Universal Scientific Inc. Silica Gel 63-200. Nuclear magnetic resonance (NMR) spectroscopy was performed on Bruker ACF 250, AM 500 MHz NMR spectrometers using deuterated chloroform ($CDCl_3$) as solvent and the internal solvent peak as reference. Gel permeation chromatography (GPC) was carried out on a Waters chromatograph (four Waters Styragel HR columns HR1, HR2, HR4 and HR5E in series) connected to a Waters 410 differential refractometer with THF as the carrier solvent. Molecular weight standards used for calibration of the GPC system was either narrow polydispersity polystyrene or poly(t-butyl acrylate). IR spectra were recorded in $CDCl_3$ solution.

2,2,5-Trimethyl-4-phenyl-3-azahexane-3-nitroxide (1). *N-tert*-butyl-α-*iso*-propylnitrone (66.0 g, 461 mmol) was dissolved in 500 ml of THF and the solution cooled to 0°C. A 3.0 M solution of phenylmagnesium bromide (310 ml, 920 mmol) in diethyl ether was added by cannula at this temperature over 5 min. During the addition some precipitate formed. The mixture was allowed to warm to room temperature. After 12 h, excess Grignard reagent was decomposed by the addition of 100 ml of concentrated ammonium chloride solution followed by 300 ml of water until all solids had dissolved. The organic layer was separated and the aqueous layer was extracted with 500 ml of diethyl ether. The organic layers were combined and dried over magnesium sulfate, filtered, concentrated, and the residue was treated with a mixture of 2000 ml of methanol, 150 ml of concentrated NH_4OH, and 4.59 mg (23 mmol) of $Cu(OAc)_2$ to give a pale yellow solution. A stream of air was bubbled through the yellow solution until it became dark blue (5-10 min). This was concentrated and the residue dissolved in a mixture of 2000 ml of chloroform, 500 ml of concentrated $NaHSO_4$ solution and 2000 ml of water. The organic layer was separated and the aqueous layer was extracted with 500 ml of chloroform. The organic layers were combined and washed with 600 ml of saturated sodium bicarbonate solution, dried over magnesium sulfate and concentrated *in vacuo* to give 101.6 g of crude nitroxide. The nitroxide was then purified by flash column chromatography (20:1 hexane:ethyl acetate) to afford 72.6 g (71% yield) of pure **1** as an orange oil, which crystallized at temperatures below 4°C. TLC: 16:1 hexane:ethyl acetate, molybdenum stain, R_f = 0.49; ^{1}H-NMR (250 MHz, $CDCl_3$) in the presence of pentafluorophenyl hydrazine: δ 7.60-7.25 (m, 5H, Ph), 3.41 (d, 1H, J = 6.5 Hz), 2.28 (m, 1H), 1.44 and 0.97 (s, 9H), 1.20 and 0.58 (d, 6H, J = 6.8 Hz); ^{13}C-NMR (62.5 MHz, $CDCl_3$) in the presence of pentafluorophenyl hydrazine: δ 154.26, 142.06, 141.20, 136.02, 129.50, 128.77, 128.43, 127.82, 127.25, 126.61, 73.37, 71.31, 63.30, 59.10, 31.51, 31.23, 30.19, 26.85, 21.54, 20.55, and 18.48.

2,2,5-Trimethyl-3-(1'-phenylethoxy)-4-phenyl-3-azahexane (6). To a solution of styrene (4.16 g, 40.0 mmol) and 2,2,5-trimethyl-4-phenyl-3-azahexane-3-nitroxide, **1**, (4.40 g, 20.0 mmol) in 1:1 toluene/ethanol (150 ml) was added [N,N'-bis(3,5-di-t-butylsalicylidene)-1,2-cyclohexanediaminato] manganese(III) chloride (2.80 g, 4.0 mmol) followed by di-t-butyl peroxide (4.30 g, 30.0 mmol) and sodium borohydride (2.28 g, 60.0 mmol). The reaction mixture was then stirred at room temperature for 12 hours, evaporated to dryness, partitioned between dichloromethane (150 ml) and water (200 ml), and the aqueous layer further extracted with dichloromethane (3 x 100 ml). The combined organic layers were then dried, evaporated to dryness, and the crude product purified by flash chromatography eluting with 1:9 dichloromethane/hexane gradually increasing to 1:3 dichloromethane/hexane. The desired alkoxyamine, **6**, was obtained as a colorless oil, (5.26 g, 81%); IR(CDCl$_3$) 2950, 1490, 1450, 1390, 1210, 1065 cm^{-1}; ^{1}H-NMR (250 MHz, CDCl$_3$, both diastereomers) δ 7.5-7.1 (m, 20H), 4.90 (q+q, 2H, J = 6.5 Hz, both diastereomers), 3.41 (d, 1H, J = 10.8 Hz, major diastereomer), 3.29 (d, 1H, J = 10.8 Hz, minor diastereomer), 2.35 (two m, 2H, both diastereomers), 1.62 (d, 3H, J = 6.8 Hz, major diastereomer), 1.54 (d, 3H, J = 7.0 Hz, minor diastereomer), 1.31 (d, 3H, J = 6.3 Hz, major diastereomer), 1.04 (s, 9H, minor diastereomer), 0.92 (d, 3H, minor diastereomer), 0.77 (s, 9H, major diastereomer), 0.54 (d, 3H, J = 6.5 Hz, major diastereomer), 0.22 (d, 3H, J = 6.5 Hz, minor diastereomer); ^{13}C-NMR (APT) (63 MHz, CDCl$_3$, both diastereomers) δ 145.87 (s), 145.08 (s), 142.56 (s), 142.35 (s), 131.03 (d), 128.15 (d), 127.46 (d), 127.37 (d), 127.28 (d), 127.09 (d), 126.70 (d), 126.43 (d), 126.25 (d), 83.58 (d), 82.91 (d), 72.30 (d), 72.21 (d), 60.59 (s), 60.45 (s), 32.12 (d), 31.69 (d), 28.48 (q), 28.3 (q), 24.81 (q), 23.26 (q), 23.23 (q), 22.05 (q), 21.26 (q), 21.14 (q). MS (FAB) m/z 326 ([M+1]$^+$, 3), 221 (11), 178 (31), 148 (15), 133 (75), 122 (21), 106 (17), 105 ([C$_8$H$_9$]$^+$, 100); HRMS exact mass calcd for [M+1]$^+$ C$_{22}$H$_{32}$NO 326.2484, found 326.2485.

General Procedure for Isoprene Polymerization from 6. A mixture of the α-hydrido alkoxyamine, **6**, (32.5 mg, 0.1 mmol), and isoprene (1.70 g, 25.0 mmol) were degassed by 3 freeze/thaw cycles, sealed under argon and heated at 125°C under nitrogen for 36 hours. The viscous reaction mixture was then dissolved in dichloromethane (5 ml) and precipitated (2x) into methanol (200 ml). The gum was then collected and dried to give the desired polyisoprene, **8**, as a colorless gum (1.35 g, 78%), M$_n$ = 20 000 a.m.u., PD. = 1.08; ^{1}H-NMR (250 MHz, CDCl$_3$) δ 5.6-5.5, 5.1-5.0, and 4.9-4.6 (m, olefinic H), 2.1-1.9, and 1.7-1.3 (m, aliphatic H); ^{13}C-NMR (63 MHz, CDCl$_3$) δ 15.97, 23.38, 26.71, 28.25, 28.47, 30.84, 31.97, 38.49, 39.72, 40.0, 110.5, 124.23, 125.02, and 134.88.

General Procedure for N-Methyl Maleimide Copolymerization from 6 and 1. A mixture of the α-hydrido alkoxyamine, **6**, (130 mg, 0.4 mmol), nitroxide, **1**, (5 mg, 0.02 mmol), styrene (9.36 g, 90.0 mmol), N-Methyl maleimide, **9**, (1.11 g, 10.0 mmol) and dimethyl formamide (2.0 ml) were degassed by 3 freeze/thaw cycles, sealed under argon and heated at 125°C under nitrogen for 10 hours. The viscous reaction mixture was then dissolved in dichloromethane (25 ml) and precipitated (2x) into methanol (500 ml). The white solid was then collected and dried to give the desired block copolymer, **10**, as a colorless solid (8.95 g, 85%), M$_n$ = 21 000 a.m.u., PD. = 1.15; IR 2950, 1850, 1780, 1490,

1450, and 700 cm^{-1}; ^{1}H-NMR (250 MHz, CDCl$_3$) δ 1.2-2.0 (m, backbone CH and CH$_2$), 2.65 (br s, N-Me) and 6.3-7.2 (aromatic H).

Results and Discussion

In order to overcome the limitations typically associated with TEMPO mediated processes, research has focused on two major areas; the use of additives *(9)* and the design of new nitroxides *(10)*. While some success has been observed for additives such as acetic anhydride *(11)* for increasing the rate of polymerization of styrene, or acetol for controlling the polymerization of acrylates *(12)*, the degree of improvement is still not comparable to that obtained in ATRP systems. A much more promising approach has been the design of new nitroxides, which are structurally different to TEMPO. Puts and Sogah *(13)* have reported that 2,5-dimethyl-2,5-diarylpyrrolidin-1-oxyl based nitroxides, **2**, lead to a doubling in the rate of polymerization for styrene when compared to TEMPO. Similarly, Georges and others *(14)* have shown that di-t-butyl nitroxide, morpholone, and piperazinone derivatives display similar polymerization rate enhancements when compared to the traditional TEMPO systems. While these studies point towards the crucial role that nitroxide structure plays in mediating these polymerizations, the most instructive report was from the groups of Gnanou and Tordo *(15)*. In this seminal contribution, an initiating system consisting of a 2.5:1 mixture of a β-phosphonylated nitroxide, **3,** and AIBN gave poly(acrylates) with polydispersities of between 1.1 and 1.3. Interestingly, the structure of **3** is significantly different to that of other

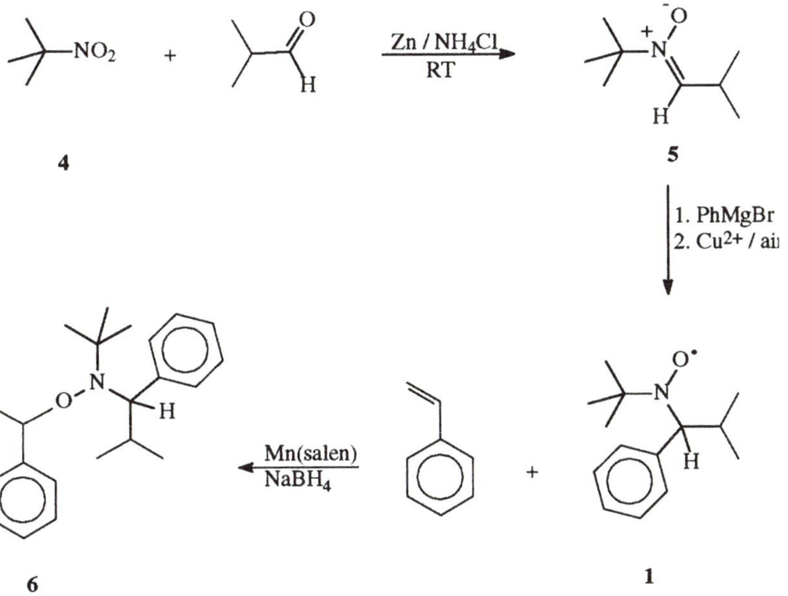

Scheme 1

nitroxides that have been examined and is a member of a unique family of nitroxides originally introduced by Volodarsky *(16)* which have a hydrogen in a α-position. Normally nitroxides with α-hydrogens are unstable and rapidly undergo disproportionation to a hydroxyamine and a nitrone, however as Volodarsky was able to demonstrate, if the steric size of the groups attached to the α-carbon bearing the hydrogen are of the appropriate size the hydrogen can be locked close to the nodal plane of the nitroxy function which leads to a dramatic increase in stability.

Based upon these pioneering results, we have explored a range of structurally diverse alkoxyamines in an effort to develop a universal system suitable for the polymerization of a significantly wider selection of monomers than is currently available *(17)*. Interestingly, the results from this diverse set of structures revealed that the most promising candidates were all based on α-hydrogen nitroxides. In our hands the superior set of materials, when features such as molecular weight control, polydispersity, and rate of polymerization are considered collectively, were based on the 2,2,5-trimethyl-4-phenyl-3-azahexane-3-oxy, **1**, skeleton. While changes in the nature of the aryl group, isopropyl and t-butyl substituents were shown to have an influence on the effectiveness of these initiators, the changes were not as significant as for that observed on going from TEMPO to **1**. For this reason all of the subsequent discussion will be directed to the parent nitroxide, **1**, and the associated alkoxyamine.

A synthetic strategy was developed for the synthesis of **1** from the corresponding tertiary nitro and aldehydes according to Scheme 1. Reductive condensation of 2-methyl-2-nitropropane, **4**, with isobutyryl aldehyde gave the desired nitrone, **5**, in a single high yielding step, which obviates the need to isolate the easily oxidizable alkyl hydroxylamine intermediates. Addition of a phenyl magnesium bromide followed by copper (II) catalyzed oxidation under ambient atmosphere resulted in the formation of the desired nitroxide **1**. Synthesis of the alkoxyamine initiators from these nitroxides was primarily accomplished by a manganese catalyzed procedure based on Jacobsen's reagent *(18)*, addition of the nitroxide, **1**, to a variety of styrene derivatives under very mild conditions gave the desired alkoxyamine, **6**, in ca.70-80% yield after purification.

Polymerization of 1,3-dienes

The controlled polymerization of 1,3-dienes by living free radical procedures in the presence of **6** is another excellent example of the advantages of these second generation alkoxyamines over the earlier TEMPO based systems. An additional feature of this system that deserves comment is based on the realization that polymerization of 1,3-dienes by ATRP systems is problematic due to their ability to act as ligands with the transition metals leading to partial deactivation of the catalyst *(20)*. The development of a viable nitroxide based system for these 'problematic' monomers would therefore be an valuable complement to ATRP procedures.

Initially a similar procedure to that used for the successful polymerization of acrylates using **6** was examined. Therefore 100 equivalents of isoprene was heated at 120°C in the presence of 1 equivalent of the alkoxyamine, **6**, and 0.05 equivalents of free nitroxide, **1**. The resulting polymerization was observed to be extremely sluggish and after heating for 72 hours, poly(isoprene), **8**, was isolated in only 25% yield. While this was considered unsatisfactory, **8** was found to have a molecular weight, M_n, of 3 500 and a polydispersity of 1.08, indicating a controlled, living process. A possible explanation for this low conversion (ca. 30%) is that the rationale for adding excess free nitroxide to slow down and control the polymerization of acrylates, i.e. very high k_p, does not apply in this case due to the significantly reduced k_p for isoprene when compared to acrylate monomers. On the basis of this hypothesis, repetition of the above experiment in the absence of free nitroxide lead to a 75% conversion after 36 hours while still maintaining accurate control over polydispersity and molecular weight, M_n = 9 800; PD. = 1.07 (Scheme 2).

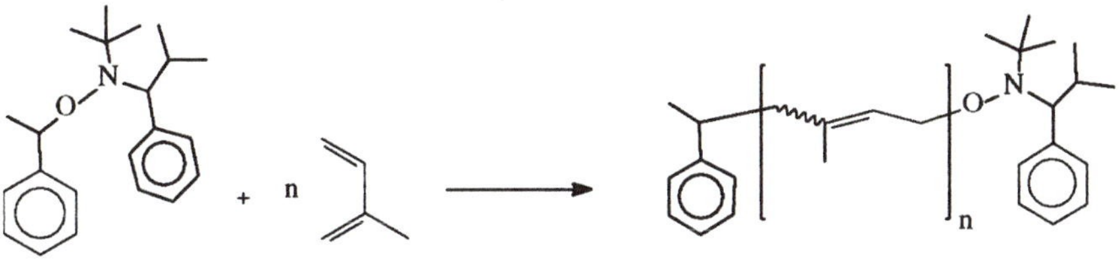

6   **8**

Scheme 2

The living nature of the polymerization was further demonstrated by examining the evolution of molecular weight with conversion. As shown in Figure 1, the relationship between molecular weight, M_n, and percent conversion is linear with low polydispersities (1.07-1.14) being obtained for all the samples. Examination of the chemical shifts and peak intensities of the olefinic carbons in the ^{13}C NMR spectrum and vinyl hydrogens (ca. 5.05 ppm) in the ^1H NMR spectrum of **8** showed a microstructure of predominately 1,4-cis and trans repeat units which is essentially the same as that obtained in a conventional free radical isoprene polymerization *(21)*.

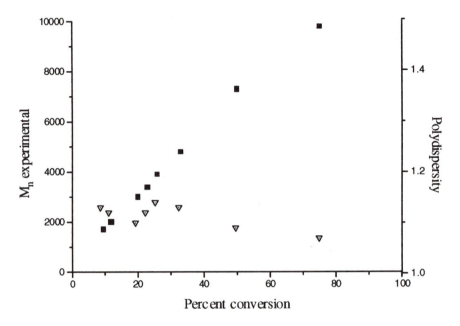

Figure 1. *Evolution of experimental molecular weight, M_n, (■) and polydispersity (▽) with conversion for the polymerization of isoprene and 6 at 120°C for 36 hours.*

The living nature and high fidelity of the polymerization suggested that the molecular weight of the poly(isoprenes) could be simply controlled by varying the ratio of monomer to initiator. A series of polymerizations were then conducted at 130°C for 48 hours with initiator:isoprene ratios of 25:1 to 1500:1. For each example, the polymerization was stopped at 80% conversion, or less, due to the observation of a minor higher molecular weight shoulder at conversions greater than 85-90%. Presumably this high molecular weight shoulder is due to chain-chain coupling which is favored at high conversion and extended reaction times. The low glass transition temperature of poly(isoprene), coupled with the extremely low viscosity of the bulk polymerization mixtures, even at 90% conversion, would also be expected to facilitate chain migration and hence coupling. As can be seen in Figure 2, molecular weight can be controlled up to 100 000 amu with polydispersities typically being less than 1.20. This level of structural control is comparable to the polymerization of styrenic and acrylate monomers by either nitroxide mediated or ATRP processes and at low molecular weights (<25,000) approaches anionic procedures for isoprene.

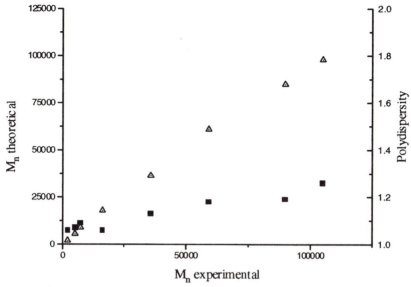

Figure 2. *Relationship between theoretical molecular weight ($\triangle$); polydispersity ($\blacksquare$) and experimental molecular weight, M_n, for the polymerization of isoprene in the presence of 6 at 120°C for 36 hours.*

While the use of **1** as an initiator for isoprene polymerization may approximate living anionic procedures in terms of polydispersity and molecular weight control, one of the major advantages of living free radical techniques is their compatibility with functional groups and the ability to readily prepare a variety of random copolymers *(22)*. Both of these issues were addressed by the synthesis of a wide variety of well-defined random copolymers of isoprene with styrenic and acrylate monomers under standard conditions. For styrene and acrylate monomers, well-defined random copolymers could be prepared over the whole comonomer range from 90% isoprene to 10% isoprene with no change in molecular weight control or polydispersity (PD. = 1.1 – 1.2). Of particular note is the ability to readily include functionalized monomers at high levels of incorporation with little, if any, change in the degree of control. From these results it can be concluded that the living free radical polymerization of isoprene, mediated by the α-hydrido nitroxide, **1**, is a controlled process and random copolymers can be prepared with a wide range of functionalized monomer units, i.e. styrenics, acrylates, methacrylates. Interestingly, the addition of free nitroxide was unnecessary in the preparation of acrylate or methacrylate random copolymers with isoprene, which is analogous to styrene-based systems. This observation, coupled with the findings above, suggest that 1,3-dienes such as isoprene resemble styrenics in their polymerization behavior rather than acrylates.

Polymerization of Maleimides

Due to their high degree of functionality and ability to undergo alternating polymerization with electron deficient monomers such as styrene, the controlled polymerization of maleimide derivatives and maleic anhydride is a desirable target. Unfortunately the polymerization of these monomers under ATRP conditions is problematic and their high reactivity precludes their polymerization under anionic conditions. As for the case of 1,3-dienes, the development of a nitroxide mediated living free radical procedure for the controlled polymerization of maleimides and maleic anhydride would be a significant addition to the arsenal of living free radical techniques.

The ability to control the copolymerization of these monomers would also represents a unique opportunity to design a novel one-step procedure for the preparation of functionalized block copolymers. Such an approach would take advantage of the alternating copolymerization nature of these monomers under radical conditions. For example, during normal free radical polymerization, a 9:1 mixture of styrene and N-methyl maleimide will lead to a heterogeneous mixture of homopolystyrene and various styrene/N-methyl maleimide copolymers. Significantly different behavior may be expected for the corresponding living free radical copolymerization. In this case, all chains will be homogeneous due to the rapid initiation and living nature of the

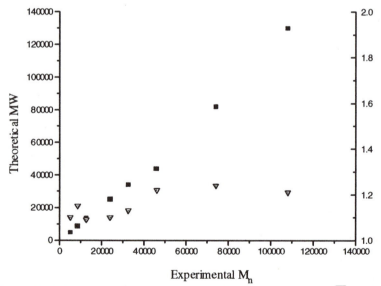

*Figure 3. Relationship between theoretical molecular weight (■); polydispersity (▽) and experimental molecular weight, M_n, for the polymerization of a 9:1 mixture of styrene and N-methyl maleimide, **9**, in the presence of **6** and **1** at 120°C for 12 hours.*

polymerization with only the monomer composition along these homogeneous chains varying. During the early stages of polymerization, preferential incorporation of N-methyl maleimide will occur and upon its depletion the monomer feed will consist of essentially pure styrene. In contrast to the normal free radical case in which pure homopolystyrene chains are produced at this stage, chain growth of a polystyrene block occurs in the living free radical case to give a poly[(styrene-r-N-methyl maleimide)-b-polystyrene] block copolymer.

As was observed in both the acrylate and 1,3-diene cases, initial polymerizations in the presence of 2,2,6,6-tetramethylpiperidinyloxy (TEMPO) as the mediating nitroxide were unsuccessful. Non-living behavior was observed under a variety of different conditions with little or no control over molecular weights and polydispersity. The substitution of TEMPO with alkoxyamines such as 6 resulted in extremely high rates of polymerization at 120°C with >90% conversion being obtained after 1 hour. While the material obtained was polydisperse (PD. = 1.4-1.7), the living character was enhanced when compared to TEMPO and suggests that the system is behaving similar to the homopolymerization of acrylates and additional nitroxide mediating agent may be required to give true living character. A small excess (5 mole%) of free nitroxide, 1, was therefore added to the reaction mixture. This produced a linear relationship between molecular weight and conversion with the molecular weights being controlled up to 100,000 a.m.u. while maintaining low polydispersities (Figure 3). Examining the monomer conversion as a function of time by ^{1}H NMR spectroscopy provided insight into the compositional nature of the chains. As can be seen in Figure 4, the polymerization is initially fast with preferential consumption of N-methyl maleimide, 9. After 1-1.5 hours, no detectable amounts of 9 could be observed in the reaction mixture, while the conversion of styrene was ca 25-30%. At this point, the monomer feed is pure styrene and so further polymerization involves growth of a pure polystyrene block leading to a block copolymer, 10, consisting of an initial ca 1:3 copolymer of N-methyl maleimide and styrene, respectively, followed by a block of polystyrene which is roughly twice the molecular weight of the initial anhydride functionalized block (Scheme 3).

Examining the monomer conversion as a function of time by ^{1}H NMR spectroscopy provided insight into the compositional nature of the chains. As can be seen in Figure 4, the polymerization is initially fast with preferential consumption of N-methyl maleimide, 9. After 1-1.5 hours, no detectable amounts of 9 could be observed in the reaction mixture, while the conversion of styrene was ca 25-30%. At this point, the monomer feed is pure styrene and so further polymerization involves growth of a pure polystyrene block leading to a block copolymer, 10, consisting of an initial ca 1:3 copolymer of N-methyl maleimide and styrene, respectively, followed by a block of polystyrene which is roughly twice the molecular weight of the initial anhydride functionalized block (Scheme 3).

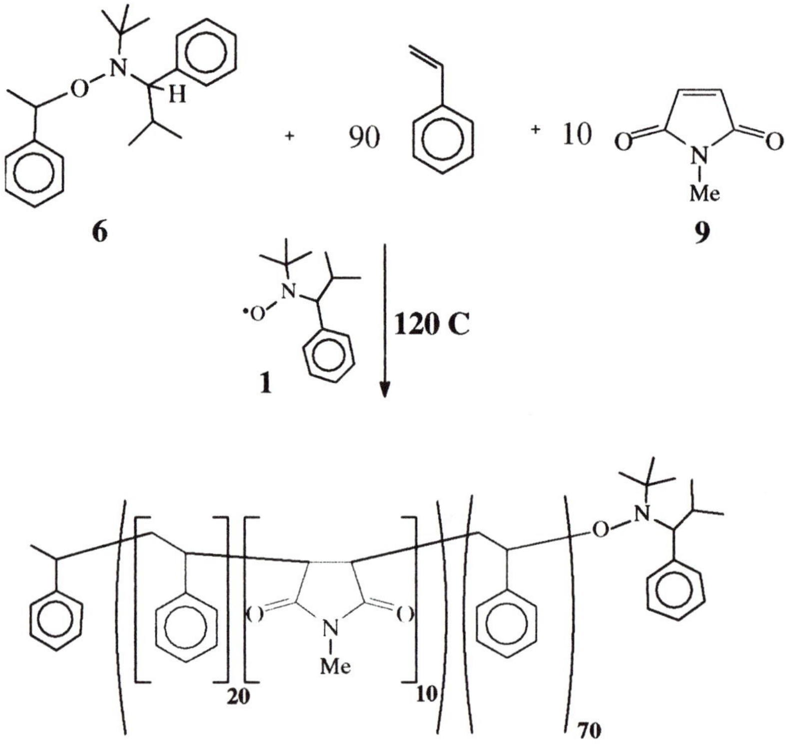

10

Scheme 3

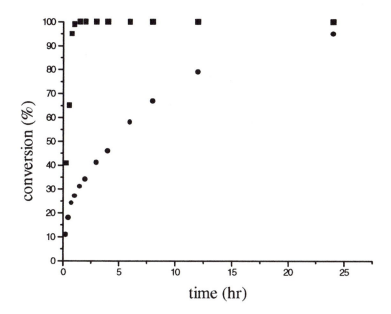

Figure 4. *^{1}H NMR determined monomer conversions for styrene (●) and N-methyl maleimide (■), 9, as a function of time during polymerization at 120°C in the presence of 6 and 1.*

The extension of nitroxide mediated living free radical procedures to reactive monomers such as N-methyl maleimide, **9**, by the use of a mixture of α-hydrido alkoxyamine, **6**, and nitroxide **1**., further demonstrates the utility of nitroxide mediated systems in general and these new α-hydrogen initiators in particular. In copolymerizations with styrene, the living nature of the polymerization is preserved which leads to preferential consumption of N-methyl maleimide and the one-step production of functionalized block copolymers, whose molecular weights can be readily controlled up to 100 000 a.m.u. while retaining low polydispersities. These materials can be considered to be a limiting example of gradient copolymers.

Conclusion

The development of second generation nitroxides and alkoxyamines, such as **6**, which overcomes many of the problems typically associated with TEMPO based systems represents a significant advancement in the general field of living free radical procedures. The ability to control the polymerization of functionalized monomers such as acrylates, 1,3-dienes, maleimides, etc. offers a viable alternative to ATRP and RAFT procedures while retaining the advantages of a simplified, thermally driven reaction. As we progress towards the development of a universal initiator for nitroxide mediated living free radical procedures, a number of important issues still need to be addressed. The polymerization of methacrylates is still a challenge, while reduction in the time to reach high conversions from 8-12 hours to less than 2-4 hours and decreases in the polymerization temperature from 100-125°C to 60-80°C would dramatically increase the viability and versatility of these procedures. A more complete understanding of the role of nitroxide structure, influence of reaction conditions, and the development of 3rd generation initiators may all play important roles in achieving these goals.

Acknowledgments: The authors gratefully acknowledge financial support from the NSF funded Center for Polymeric Interfaces and Macromolecular Assemblies and the IBM Corporation.

References

(1)   Webster, O. W. *Science* **1994**, *251*, 887; Fréchet, J.M.J. *Science* **1994**, *263*, 1710.

(2)   Otsu, T.; Matsunaga, T.; Kuriyama, A.; Yoshida, M. *Eur. Polym. J.* **1994**, *25*, 643.

(3)   Turner, S.R.; Blevins, R.W. *Macromolecules* **1990**, *23*, 1856.

(4)   Moad, G.; Solomon, D.H.; Johns, S.R.; Willing, R.I. *Macromolecules* **1982**, *15*, 1188.

(5)   Georges, M.K.; Veregin, R.P.N.; Kazmaier, P.M.; Hamer, G.K. *Macromolecules* **1993**, *26*, 2987.

(6)   Wang, J. S.; Matyjaszewski, K. *Macromolecules* **1995**, *28*, 7901; Wang, J. S.; Matyjaszewski, K. *J. Am. Chem. Soc.* **1995**, *117*, 5614.

(7)   Kato, M.; Kamigaito, M.; Sawamoto, M.; Higashimura, T. *Macromolecules* **1995**, *28*, 1721.

(8)   Chiefari, J.; Chong, Y.K.; Ercole, F.; Krstina, J.; Jeffery, J.; Le, T.P.T.; Mayadunne, R.T.A.; Meijs, G.F.; Moad, C.L.; Moad, G.; Rizzardo, E.; Thang, S.H. *Macromolecules*, **1998**, *31*, 5559.

(9)   Georges, M. K.; Veregin, R. P. N.; Kazmaier, P. M.; Hamer, G. K.; Saban, M. *Macromolecules* **1994**, *27*, 7228.

(10)   Kazmaier, P.M.; Moffat, K.A.; Georges, M.K.; Veregin, R.P.N.; Hamer, G.K. *Macromolecules* **1995**, *28*, 1841.

(11)   Malmstrom, E.E.; Miller, R.D.; Hawker, C.J. *Tetrahedron* **1997**, *53*, 15225.

(12)   Keoshkerian, B.; Georges, M.K.; Quinlan, M.; Veregin, R.; Goodbrand B. *Macromolecules*, **1998**, *31*, 7559.

(13)   Puts, R. D.; Sogah, D. Y. *Macromolecules* **1996**, *29*, 3323.

(14)   Puts, R.; Lai, J.; Nicholas, P.; Milam, J.; Tahilliani, S.; Masler, W.; Pourahmady, N. Polym. Prep., **1999**, *40(2)*, 323; Hawker, C.J. *Acc. Chem. Res.*, **1997**, *30*, 373; Colombani, D. *Prog. Polym. Sci.* **1997**, *22*, 1649;.

(15)   Benoit, D.; Grimaldi, S.; Finet, J. P.; Tordo, P.; Fontanille, M.; Gnanou, Y. *ACS Symp. Ser.* Matyjaszewski, K. Ed.; **1998**, *685*, 225.

(16)   Reznikov, V.A.; Gutorov, I.A.; Gatlov, Y.V.; Rybalova, T.V.; Volodarsky, L.B. *Russ. Chem. Bull.* **1996**, *45*, 384.

(17)   Benoit, D,; Chaplinski, V.; Braslau, R.; Hawker, C.J. *J. Am. Chem. Soc.*, **1999**, *121*, 3904.

(18)   Dao, J.; Benoit, D,; Hawker, C.J. *J. Polym. Sci., Part A: Polym. Chem.*, **1998**, *36*, 2161.

(19)   Patten, T.E.; Matyjaszewski, K. *Adv. Mater.* **1998**, *10*, 901.

(20)   Matyjaszewski, K., personal communication.

(21)   Sato, H.; Tanaka, Y. *J. Polym. Sci., Polym. Chem.*, **1979**, *17*, 3551; Sato, H.; Ono, A.; Tanaka, Y. *Polymer*, **1977**, *18*, 580.

(22)   Hawker, C.J.; Elce, E.; Dao, J.; Volksen, W.; Russell, T.P.; Barclay, G.G. *Macromolecules*, **1996**, *29*, 4167; Kotani, Y.; Kamigaito, M.; Sawamoto, M. *Macromolecules* **1998**, *31*, 5582; Haddleton, D.M.; Crossman, M.C.; Hunt, K.H.; Topping, C.; Waterson, C.; Suddaby, K.G. *Macromolecules* **1997**, *30*, 3992; Greszta, D.; Matyjaszewski, K. *Polym. Prepr. (Am. Chem. Soc., Div. Polym. Chem.)* **1996**, *37(1)*, 569; Listigovers, N.A.; Georges, M.K.; Odell, P.G.; Keoshkerian, B. *Macromolecules*, **1996**, *29*, 8992.

Chapter 10

Nitroxide-Mediated Controlled Free-Radical Emulsion and Miniemulsion Polymerizations of Styrene

M. Lansalot[1], C. Farcet[1], B. Charleux[1, 4], J.-P. Vairon[1], R. Pirri[2], and P. Tordo[3]

[1]Laboratoire de Chimie Macromoléculaire, Université Pierre et Marie Curie, T44, E1 4, Place Jussieu, 75252 Paris Cedex 05, France
[2]Elf-Atochem, GRL, 64170 LACQ, France
[3]Laboratoire de Structure et Réactivité des Espèces Paramagnétiques, Université d'Aix-Marseille, 13397 Marseille Cedex, France

Nitroxide-mediated controlled free-radical polymerization of styrene was studied in miniemulsion and emulsion systems. The use of an acyclic β-phosphonylated nitroxide enabled work to be performed at a temperature below 100 °C, typically 90 °C. A bicomponent initiating system was chosen, i.e. a radical initiator in conjunction with added free nitroxide. Kinetics of polymerization, characteristics of the polymer and particles diameter were determined. Effects of the initial concentrations of initiator and nitroxide, and of the water phase pH, were examined in detail.

Controlled free-radical polymerization (CRP) enables one to synthesize (co)polymers with controlled molar masses, narrow molar mass distributions and well-defined architectures (1). In nitroxide-mediated CRP, the nitroxide stable radical acts as a reversible terminating agent and control is based on an equilibrium between the active macromolecular radical and a dormant covalent counterpart which is an alkoxyamine. Until now, TEMPO (2,2,6,6-tetramethyl piperidinyl-1-oxy) has been the most widely used nitroxide and was mainly applied for the controlled polymerization of styrenic monomers at T > 120 °C (1-4). CRP has been predominantly studied in homogeneous

[4]Corresponding author (e-mail:charleux@ccr.jussieu.fr)

systems, i.e. bulk or solution polymerizations. However, in industry, radical polymerization is widely performed in aqueous dispersed systems and particularly as emulsion polymerization (5). This technique has many advantages over homogeneous polymerizations such as the absence of organic volatile compounds and the possibility to reach high molar mass polymers with high conversion and with larger rates of polymerization than in bulk or solution. The final aqueous suspension of stable polymer particles (also called latex) is easy to handle owing to a generally low viscosity, even at high solid content and can be used directly for coating applications or as a dried polymer after removal of water.

In the case of styrene, nitroxide-mediated CRP in aqueous dispersed media was previously reported using various conditions: suspension (2), dispersion (6), seeded emulsion (7), batch emulsion (8) and miniemulsion (9-11) polymerizations. In all cases, TEMPO or a derivative was used as a mediator and polymerization was performed at a temperature above 120 °C, which implied to work under pressure. Furthermore, at such high temperatures, stabilization of the latexes becomes a challenge and thermal self-initiation of styrene cannot be avoided.

The objective of the present work was twofold. The first goal was to carry out nitroxide-mediated CRP of styrene in aqueous dispersed systems at a temperature below 100 °C. For that purpose an acyclic β-phosphonylated nitroxide stable radical, the N-tert-butyl-N-(1-diethylphosphono-2,2-dimethylpropyl) nitroxide (SG1, Scheme 1) was used (12). It was previously reported to give excellent results in the controlled radical polymerization of styrene and acrylic esters (13-16). Owing to a significantly higher equilibrium constant of activation-deactivation ($K = 1.9.10^{-8}$ mol.L^{-1} at 125 °C) (17) than that of TEMPO for styrene polymerization ($K = 2.10^{-11}$ mol.L^{-1} at 125 °C) (18) this stable radical can be used as a mediator at a lower temperature, i.e. from 90 °C to 130 °C.

Scheme 1 :
Structure of SG1

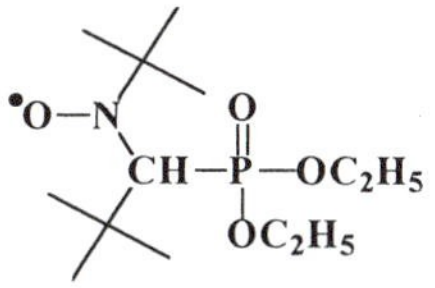

The second objective was to obtain stable latexes with particles diameter in the range of 50 to 500 nm, which is that of emulsion polymerization. This technique usually employs hydrophobic monomers and the system requires the use of a water soluble radical initiator together with a surfactant. The initial system is an unstable oil-in-water emulsion containing large monomer droplets (diameter > 1 μm) and small surfactant micelles swollen by monomer. Polymerization does not directly take place in the monomer droplets but in particles which are created in the early stage of the polymerization from water-phase generated oligoradicals. The mechanism of particles nucleation is very complex and more information can be found in Ref. 5. Once formed, the stabilized growing polymer particles are swollen by monomer and the consumption of monomer owing to propagation is continuously compensated by its diffusion from the droplets, via the water phase. Therefore, the droplets only act as monomer reservoirs. However, at elevated temperature, thermal self-initiation of styrene cannot be neglected and might lead to significant initiation in them, resulting in unstable latex or at least in broad particle size distribution. In order to overcome this problem and also to avoid the very complicated nucleation step, a miniemulsion process can be applied (19,20). This polymerization technique leads to the same type of final latex but differs from emulsion polymerization by the initial stage. Actually, the initial monomer in water emulsion is strongly sheared in order to divide the organic phase into very small droplets (diameter < 1 μm). In addition of a classical surfactant, the use of a cosurfactant (e.g. hexadecanol) or of a hydrophobe (e.g. hexadecane) is required to stabilize the droplets from Ostwald ripening (19,20). The advantage of the system is that, owing to their large surface area, the droplets can be directly nucleated and become polymer particles, providing that no micelles are present. Moreover, to ensure nucleation in all the droplets, it was shown that a high molecular weight polymer can be initially dissolved in the monomer (19).

Thus, in this work, batch miniemulsion and emulsion polymerization processes were applied for the SG1-mediated CRP of styrene in aqueous dispersed system at 90 °C. In order to keep experimental conditions close to the usual ones, a bicomponent initiating system was chosen, i.e. a conventional radical initiator together with added free nitroxide. In the case of emulsion polymerization a water-soluble radical initiator ($K_2S_2O_8/Na_2S_2O_5$) was used. In the case of miniemulsion polymerization, the same initiating system was used and an oil-soluble radical initiator (azobisisobutyronitrile, AIBN) was also tested which enabled to compare the results with bulk polymerization. All the latexes were stabilized by sodium dodecyl sulfate (SDS). In this presentation, we will particularly focus on the kinetics of polymerization and on the evolution of molar masses versus monomer conversion. We have more carefully examined the parameters which affect the rate of polymerization and the polymer molar mass distribution.

Experimental

Styrene (St) was distilled under vacuum before use and all the other reagents were used without further purification. SG1 was provided by Elf-Atochem. Bulk polymerizations were carried out in sealed tubes after degasing by freeze-pump-thaw cycles. Batch emulsion and miniemulsion polymerizations were performed in a conventional 250 mL thermostated reactor, under nitrogen atmosphere. For emulsion polymerizations, a conventional process was applied. For miniemulsions, hexadecane was used as a hydrophobe (5 wt% with respect to styrene) and a high molar mass polystyrene (M_w = 330 000, 1 wt% with respect to styrene) was dissolved in the monomer to play the role of nucleation enhancer ; afterwards, the monomer in water initial emulsion was sheared by ultrasonification (Branson 450 Sonifier). Samples were withdrawn at various times in order to follow the evolution of monomer conversion and of polymer molar mass. Monomer conversion was determined by gravimetry. After water evaporation, molar masses were measured by size exclusion chromatography (SEC) using a Waters apparatus equipped with three columns

(Shodex KF 802.5, KF 804L, KF 805L) and working at room temperature with THF eluent at a flow rate of 1 mL.min⁻¹. A differential refractive index detector was used and molar masses were derived from a calibration curve based on polystyrene standards. Particle diameters (D) were measured by dynamic light scattering using a Zetasizer4 from Malvern. The ESR measurements were performed using a Bruker ESP 300 equipment.

Results and Discussion

Miniemulsion Polymerization using AIBN as an Oil-Soluble Initiator

Comparison with Bulk Polymerization
Before starting the polymerizations in aqueous dispersions, it was necessary to choose the best concentrations of initiator and nitroxide in order to obtain a controlled polymerization in bulk at 90 °C. As it was described in previous studies (14,15), AIBN was chosen as a radical initiator and various [nitroxide]/[AIBN] initial ratios were tested. It appeared that a value of 1.2 corresponds to the best compromise leading to an acceptable control together with a relatively fast polymerization (Figure 1). When a 1.6 ratio was used, the final polydispersity index was close to 1.1 but 40 % conversion was only obtained after 16 hours of polymerization.

Miniemulsion polymerizations using the same oil-soluble initiator were performed in order to directly compare the results with bulk polymerization. Experimental conditions are reported in Table I. Two initial concentrations of SDS were used (0.6 and 2 times the critical micelle concentration) and two different weight percents of styrene (10 wt% and 20 wt%). A blank experiment (ME0) was carried out in the absence of added nitroxide. In all the other cases except ME4, an initial [nitroxide]/[AIBN] ratio of 1.2 was used. In Figures 2 and 3, monomer conversions are plotted versus time and molar masses are plotted versus conversion.

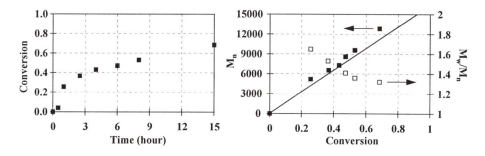

Figure 1. Bulk CRP of styrene at 90 °C with SG1 as a mediator (See Table I) ; (In all figures, straight line = theoretical Mₙ).

Table I. Bulk and Miniemulsion Polymerizations of Styrene at 90 °C using AIBN as a Radical Initiator and SG1 as a mediator.

Expt.	Symbol	St/water (wt/wt)	$[AIBN]_0$ mol/L_{St}	r	[SDS] mol/L_{water}	Final D (nm)
Bulk	■	-	0.037	1.2	-	-
ME0 [a]	▲	1/9	0.037	0	$1.5\ 10^{-2}$	114
ME1	○	1/9	0.037	1.2	$5.2\ 10^{-3}$	220
ME2	◇	1/9	0.037	1.2	$1.5\ 10^{-2}$	180
ME3	✳	2/8	0.018	1.2	$1.5\ 10^{-2}$	207
ME4	◆	1/9	0.037	1.6	$1.5\ 10^{-2}$	157

$r = [SG1]_0/[AIBN]_0$; $[NaHCO_3]_0 = 9.10^{-3}$ mol/L_{water} ; [a] Miniemulsion with $[SG1]_0 = 0$

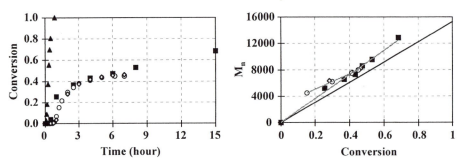

Figure 2. CRP of styrene in bulk (■) and in miniemulsion (ME0 : ▲ ; ME1 : ○ ; ME2 : ◇ ; St/water = 1/9 wt/wt) using AIBN as a radical initiator at 90 °C.

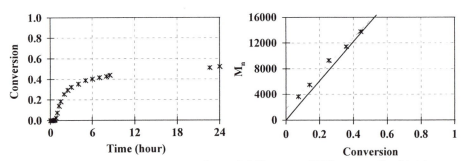

Figure 3. CRP of styrene in miniemulsion (ME3) using AIBN as a radical initiator at 90 °C with r = 1.2.

In all cases, stable latexes without coagulum were obtained ; they had relatively broad particle size distributions and mean diameters were between 114 nm and 220 nm, depending on the experimental conditions. Concerning the kinetics, the very first observation illustrated in Figure 2 is that addition of the nitroxide in the polymerization medium leads to a drastic decrease of the polymerization rate. In the presence of SG1, the general shapes of the conversion versus time plots are very similar for homogeneous and heterogeneous polymerizations. A slight difference may be observed in the last period where rate of polymerization in bulk appears to be higher than in miniemulsion. No significant effect of the initial concentration of SDS can be evidenced. An induction period is first observed during which no polymer can be recovered. This corresponds to the decomposition of the radical initiator (AIBN half lifetime $\approx$ 23 min at 90 °C ($k_d = 5.10^{-4}$ s^{-1}) (21)) followed by fast trapping by SG1 of the formed oligoradicals. Polymerization can start only when the concentration of free SG1 is small enough to allow the propagation step. Since the ratio $[SG1]_0/[AIBN]_0 = 1.2$, this means that potentially initiating radicals are in excess with respect to the scavenger, providing that AIBN efficiency is close to 1. In that case, radicals are still supplied by decomposition of the initiator while free SG1 concentration is low. This explains the large polymerization rate observed at the beginning. However, it slows down significantly after about 40 % conversion to reach a plateau with only slow propagation. At that stage, a polymer sample from experiment ME2 (7 hours reaction time, 46 % conversion, M_n = 7 010 g.mol^{-1}, M_w/M_n = 1.6) was isolated and used as a macroinitiator for the bulk polymerization of styrene at 90 °C. The expected chain extension was observed (M_n = 27 300 g.mol^{-1}, M_w/M_n = 1.4) indicating that polymer chains were still living, i.e. a majority of them still had the alkoxyamine end-functionality. In the size exclusion chromatograms, about 20 mol% of dead chains could be detected. Therefore, the final decrease of the polymerization rate can be assigned to the continuous increase of SG1 concentration owing to the unavoidable radical-radical terminations (persistent radical effect (22)). The consequence is a shift of the activation / deactivation equilibrium towards the favored formation of dormant species until the system reaches a pseudo-equilibrium state. Since thermal self initiation of styrene is slow at 90 °C, the only source of radicals is the reversible decomposition reaction of the alkoxyamines. This leads to incomplete conversion after 24 hours (see Figure 3).

For the bulk polymerization, a linear increase of M_n with monomer conversion can be seen with the expected intercept at $M_n = 0$. The experimental values follow the theoretical line. For the miniemulsion polymerizations, an increase of M_n with conversion can also be observed but, in that case, molar masses are slightly higher than expected at low conversion (the experimental line does not pass through the origin). This situation is however improved when monomer content is increased (see experiment ME3, Figure 3). In all cases, M_w/M_n values decrease when monomer conversion increases but remain systematically larger when the polymerization is carried out in a dispersed medium (1.6 to 1.7 at final stage instead of 1.3 for bulk polymerization). The size exclusion chromatograms are displayed in Figure 4 for the experiment ME3. In contrast to the homogeneous system, the chromatograms obtained at low conversion in the heterogeneous system are broad and display a maximum on the high molar mass side together with a significant tailing on the low molar mass side which disappears when conversion increases. The broad molar mass distribution obtained at low conversion is the consequence of a poor control due to the low concentration of nitroxide in the early stage of the polymerization, which leads also to

144

a larger rate of polymerization as noted above. Moreover, molar masses higher than the theoretical ones at low conversion are generally assigned to a slow initiation process indicating that the number of chains is still increasing. However, as the same initiator was used in bulk and in miniemulsion polymerizations this can only be a consequence of the heterogeneity of the system since it is not observed in bulk polymerization. Actually, AIBN is partitioned between the organic and the aqueous phases and partly decomposes in water where low molar mass alkoxyamines can be formed owing to the fast trapping by SG1. Those very short alkoxyamines propagate at a slower rate compared to their longer counterparts because of their possible partitioning between the two phases (the concentration of styrene in the aqueous phase is 4.3 mmol.L^{-1} at 50 °C (23)). They have to reach a sufficient length to irreversibly enter a particle.

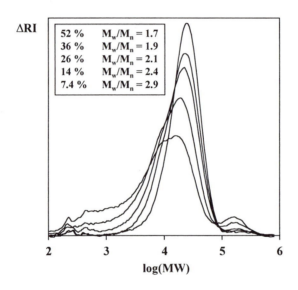

Figure 4. Size exclusion chromatograms for the CRP of styrene in miniemulsion ME3 using AIBN as an initiator at 90 °C with r = 1.2. (The high MW peak corresponds to the added polystyrene)

Effect of the [Nitroxide]/[Initiator] Initial Ratio
As it was previously mentioned for bulk polymerization, an increase of the initial [nitroxide]/[initiator] ratio from 1.2 to 1.6 leads to a significant decrease of the rate of polymerization, the general shape of the conversion versus time plot being unchanged (Experiment ME4, Figure 5). This effect is accompanied by a narrowing of the molar mass distribution, since polydispersity indexes close to 1.4 were obtained.

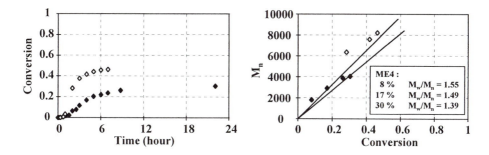

Figure 5. CRP of styrene in miniemulsion (ME2 : ◇ and ME4 : ◆) using AIBN as a radical initiator at 90 °C: effect of the ratio r = [SG1]₀/[AIBN]₀.

Mn/Mn values shown in figure: $M_w/M_n = 1.55$ (8 %), $M_w/M_n = 1.49$ (17 %), $M_w/M_n = 1.39$ (30 %).

Miniemulsion Polymerization using a Water-Soluble Radical Initiator.

Comparison with the Miniemulsion Polymerizations Initiated by AIBN

A miniemulsion polymerization (ME5) was carried out using a water soluble redox initiator $K_2S_2O_8/Na_2S_2O_5$ under the same experimental conditions as the miniemulsion ME3 (Table II). Conversion-time and M_n-conversion plots are displayed in Figure 6 ; size exclusion chromatograms are shown in Figure 7. It appears that polymerization is faster for experiment ME5, specially in the pseudo-equilibrium regime. The final latex is stable and average particle diameter is 212 nm. There is a linear increase of M_n with monomer conversion but molar mass distributions are broader than for the polymers obtained with AIBN initiator. As previously, the chromatograms show the existence of a tailing towards the lower molar masses which disappears as polymerization progresses.

Another example of miniemulsion polymerization with the same water-soluble radical initiator is displayed in Figure 8. The initial concentrations of initiator and SDS were respectively decreased and slighly increased. It appears in that case that conversion reaches completion within approximately 10 hours. The final latex is stable and average particle diameter is 100 nm. There is a linear increase of M_n with monomer conversion but molar mass distribution was found to be quite broad ($M_w/M_n > 2$). Average molar masses as high as 90 000 g.mol^{-1} could be obtained.

Table II. Miniemulsion Polymerizations of Styrene using $K_2S_2O_8/Na_2S_2O_5$ as a Radical Initiator at 90 °C and SG1 as a mediator.

Expt.	Symbol	St/water (wt/wt)	$[K_2S_2O_8]_0$ mol/L_{water}	r	$[SDS]_0$ mol/L_{water}	Buffer	Final D (nm)
ME5	□	2/8	$4.5 \ 10^{-3}$	1.21	$1.5 \ 10^{-2}$	NaHCO$_3$	212
ME6	●	2/8	$2.1 \ 10^{-3}$	1.20	$2.3 \ 10^{-2}$	NaHCO$_3$	100
ME7	○	2/8	$4.5 \ 10^{-3}$	1.25	$1.5 \ 10^{-2}$	NaHCO$_3$	170
ME8	◉	1/9	$4.5 \ 10^{-3}$	1.20	$1.5 \ 10^{-2}$	NaHCO$_3$	145
ME9	★	1/9	$4.5 \ 10^{-3}$	1.20	$1.5 \ 10^{-2}$	K$_2$CO$_3$	130

[Buffer] = $9.3 \ 10^{-3}$ mol/L_{water} ; r = [SG1]₀/[K_2S_2O_8]₀ ; [Na_2S_2O_5]₀ = [K_2S_2O_8]₀

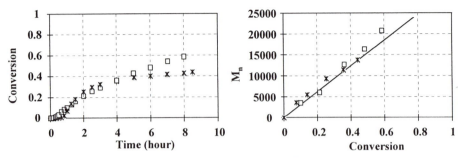

Figure 6. CRP of styrene in miniemulsion (ME5 : □) using $K_2S_2O_8/Na_2S_2O_5$ as a water-soluble radical initiator at 90 °C. Comparison with miniemulsion ME3 (✱) initiated with AIBN (see experimental conditions in Tables I and II).

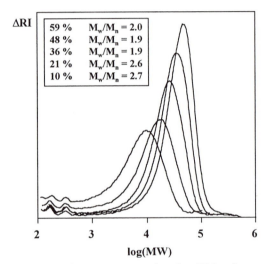

Figure 7. Size exclusion chromatograms for the CRP of styrene in miniemulsion ME5 using $K_2S_2O_8/Na_2S_2O_5$ as a water-soluble radical initiator at 90 °C with $[SG1]_0/[K_2S_2O_8]_0 = 1.21$.

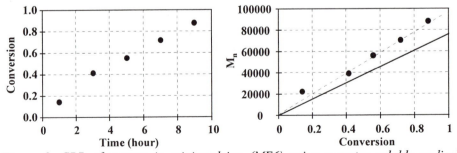

Figure 8. CRP of styrene in miniemulsion (ME6) using a water soluble radical initiator at 90 °C. (see experimental conditions in Table II).

Effect of the [Nitroxide]/[Initiator] Initial Ratio

As with AIBN initiator, the effect of the initial concentration of nitroxide was examined. Similarly, when a ratio of 1.6 was applied, polymerization rate decreased drastically and polydispersity was much narrower. Thus, we have tested the effect of a slight increase of the initial concentration of the nitroxide from r = 1.21 (ME5) to r = 1.25 (ME7). Results are displayed in Figure 9. On the conversion-time plot, one can see that the induction period is longer when the concentration of nitroxide is increased. Afterwards, the rate of polymerization is not significantly modified. A similar evolution of the average molar masses with monomer conversion can be seen but the molar mass distribution is improved when increasing r. Indeed, M_w/M_n decreases below 2. For instance, for ME7, after 8 hours of reaction, 47 % of polymer was recovered and it had the following characteristics: M_n = 11 300 g.mol^{-1}, M_w/M_n = 1.55. For the miniemulsion ME5 with r = 1.21, after 6 hours, 48 % conversion was reached and the polymer had M_n = 16 400 g.mol^{-1} with M_w/M_n = 1.93.

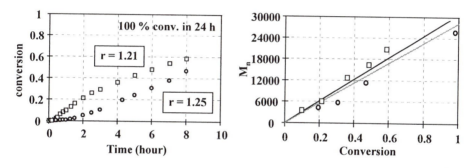

Figure 9. CRP of styrene in miniemulsion using $K_2S_2O_8/Na_2S_2O_5$ as a water-soluble radical initiator at 90 °C: effect of the initial ratio r. Comparison of experiments ME5 (□ ; r = 1.21) and ME7 (O ; r = 1.25) (see experimental conditions in Table II).

Effect of the Water-Phase pH

Another important parameter which was shown to affect the rate of the polymerizations initiated with the $K_2S_2O_8/Na_2S_2O_5$ system is the pH of the water phase. Actually, an important feature of this type of water-soluble radical initiator is that it leads to an acidification of the latex serum (5). Usually, sodium hydrogencarbonate was used as a buffer. However, in all the previous experiments, its concentration was insufficient since the pH decreased from about 7 to 3 throughout the polymerization. When shifting to potassium carbonate buffer, the initial pH value of 8 remained stable. As illustrated in Figure 10, this change in experimental conditions has a significant effect on the rate of polymerization: the more acidic the medium, the faster the polymerization. It has been shown just above that the main factor which may influence the rate of polymerization is the concentration of free nitroxide during the polymerization. Therefore, it is likely that an acidic pH might affect the concentration of free nitroxide. Two explanations can be proposed. The first one is that the SG1 partition coefficient can be dependent upon pH: it would be more water-soluble at low pH and therefore less able to control the polymerization in the polymer particles. The second one is that side reactions occur at acidic pH leading to a slow decomposition of the nitroxide. To check those assumptions, the stability of SG1 at 90 °C in the presence of the various components of a miniemulsion polymerization was evaluated by ESR. It appeared that this nitroxide is stable over 7 hours in distilled water, in the presence of the surfactant and in the presence of the two buffers. Moreover, pH has no significant effect on its stability (pH = 2 and 4 were tested over 90 min at 90 °C). However, its concentration decreased very quickly when $Na_2S_2O_5$, $K_2S_2O_8$, or both are respectively added in the aqueous solution at 90 °C. In Table III, the half life-times of SG1 are reported according to the various conditions. This nitroxide was found to be very unstable in the presence of the two components of the initiating system but it can be seen that stability is improved when pH is increased from 6 to 8. Some redox reactions might be responsible for the decomposition of the nitroxide and they are most probably pH-dependent. Therefore, in a polymerization system, one can assume that there is initially a competition between the recombination of SG1 with the active chain end and its decomposition induced by the initiating system. This situation leads to a decrease of the free nitroxide concentration. When working at slightly alkaline pH, the polymerization is slower but the molar mass distribution becomes much narrower. For experiment ME9, all the polymers had M_w/M_n comprised between 1.4 and 1.6 whereas values were in the range of 2 to 3 for experiment ME8.

Table III. Half-lifetime of SG1 under various experimental conditions as measured by ESR

	Buffer	pH	t1/2 (min)
$Na_2S_2O_5$	No	< 4	5[a]
	$NaHCO_3$	6	153
	K_2CO_3	8	stable
$K_2S_2O_8$	No	< 4	1[a]
	$NaHCO_3$	6	6
	K_2CO_3	8	15
$Na_2S_2O_5 / K_2S_2O_8$	No	< 4	too fast
	$NaHCO_3$	6	1.7
	K_2CO_3	8	2.5

T = 90 °C ; [Buffer] = 9.10^{-3} mol/L ; $[Na_2S_2O_5] = [K_2S_2O_8] = 4.5.10^{-3}$ mol/L ; [SG1] = 5.10^{-3} mol/L ; [a][SG1] = 10^{-4} mol/L

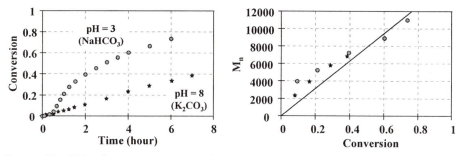

Figure 10. CRP of styrene in miniemulsion using K₂S₂O₈/Na₂S₂O₅ as a water-soluble radical initiator at 90 °C: effect of the pH (ME8 : pH = 3 : ● ; ME9 : pH = 8 : ★) (see Table II).

Emulsion Polymerization using a Water-Soluble Radical Initiator.

Although miniemulsion polymerization process has the great advantage of simplifying the nucleation and particle growth mechanisms, it is essentially only used in laboratories. Indeed, for industrial production, emulsion polymerization is preferred because of a much simpler experimental procedure. Therefore, SG1-mediated controlled radical polymerization of styrene has been performed at 90 °C using a very conventional emulsion polymerization process with $K_2S_2O_8/Na_2S_2O_5$ as a water-soluble radical initiator. The initial monomer in water emulsion was stabilized only by the addition of SDS and by mechanical stirring. The effect of different parameters was studied. The initial [nitroxide]/[initiator] ratio and the pH were found to affect the polymerization rate and the molar mass distribution in the same way as in miniemulsion polymerization. Nevertheless, when comparing the kinetics of a miniemulsion and an emulsion polymerizations carried out with the same concentration of reagents, an important difference can be highlighted. This concerns the initial stage of the polymerization: the induction period is much longer in emulsion than in miniemulsion polymerizations as illustrated in Figure 11. This can be assigned to the nucleation step (i.e. particle formation) which does not exist in miniemulsion. During that period, kinetics of polymerization in the aqueous phase is of major importance (5). In the case of SG1-mediated controlled radical polymerization, it can be supposed that the oligoradicals generated in the water-phase are trapped by SG1 in this phase before reaching a sufficient degree of polymerization to be totally oil-soluble and to contribute to particle formation. This leads to an initially slow polymerization until particles are formed wherein propagation can take place at a normal rate. The evolution of molar masses with monomer conversion was not found to be significantly different in both processes, but as observed in this experiment and in many other cases, molar mass distribution was broader in emulsion. For instance in the experiment shown in Figure 11, M_w/M_n was 2.3 at 45 % conversion for the emulsion polymerization while it was 1.9 at the 48 % conversion for the miniemulsion polymerization.

150

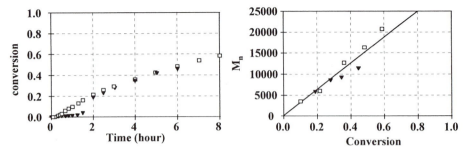

Figure 11. CRP of styrene in emulsion (▼) using $K_2S_2O_8/Na_2S_2O_5$ as a water-soluble radical initiator at 90 °C ; comparison with miniemulsion ME5 (□). St/water = 2/8 (wt/wt) ; $[K_2S_2O_8]_0 = [Na_2S_2O_5]_0 = 4.5.10^{-3}$ mol/L_{water} ; $r = [SG1]_0/[K_2S_2O_8]_0 = 1.2$; $[NaHCO_3]_0 = 9.10^{-3}$ mol/L_{water} ; $[SDS]_0 = 1.5.10^{-2}$ mol/L_{water} ; Final D = 120 nm (no coagulum).

Another important feature of emulsion polymerization is that a few percent of coagulum are usually obtained at the end of the polymerization which was not the case in miniemulsion. This is not yet fully understood but might be related to the long period of nucleation resulting in broad particle size distribution. To better understand the real mechanism of a controlled radical polymerization in emulsion and to obtain well-defined latex particles with narrow particle size distribution, it is obvious that a better understanding of the nucleation step is required.

Conclusion

For the first time, nitroxide-mediated controlled free-radical polymerization of styrene was studied in miniemulsion and emulsion systems at temperature below 100 °C. The use of an acyclic β-phosphonylated nitroxide, the N-tert-butyl-N-(1-diethylphosphono-2,2-dimethylpropyl) nitroxide (SG1) enabled polymerization to be performed at 90 °C. A bicomponent initiating system was chosen, i.e. a radical initiator in conjunction with added free nitroxide. The initiators were conventional ones: AIBN for bulk and miniemulsion polymerizations and the $K_2S_2O_8/Na_2S_2O_5$ redox system for the miniemulsion and the emulsion polymerizations. The polymer particles were stabilized by the widely used sodium dodecyl sulfate, which proved to be an efficient stabilizer, although hydrolysis occurs at such elevated temperature. As in bulk, the rate of polymerization was shown to strongly depend on the [nitroxide]/[initiator] ratio ; the larger the ratio, the slower the polymerization. When using the $K_2S_2O_8/Na_2S_2O_5$ initiating system, the effect of pH was also demonstrated. The molar mass distribution was broader when polymerization was faster and was systematically broader in dispersed medium than in bulk. Miniemulsion polymerization proved to be an efficient process to obtain stable latex particles.

Acknowledgments: Elf-Atochem is acknowledged for financial support. M. Culcasi and R. Lauricella are particularly thanked for the ESR analyses.

References

1. "Controlled Radical Polymerization", K. Matyjaszewski Ed., *ACS Symp. Series.,* **1998**, 685
2. Georges, M.K.; Veregin, R.P.N.; Kazmaier, P.M.; Hamer, G.K. *Macromolecules* **1993**, 26, 2987
3. Solomon, D.H.; Rizzardo, E.; Cacioli, P; *U.S. Patent 4,581,429*, March 27, **1985**
4. Hawker, C.J. *J. Am. Chem. Soc.* **1994**, 116, 11185
5. Gilbert, R. G. in *"Emulsion Polymerization: a Mechanistic Approach",* Academic Press, **1995**
6. Gabaston, L.I.; Jackson, R. A.; Armes, S.P. *Macromolecules* **1998**, 31, 2883
7. Bon, S.A.F.; Bosveld, M.; Klumperman, B.; German, A.L. *Macromolecules* **1997**, 30, 324
8. Marestin, C.; Noël, C.; Guyot, A.; Claverie, J. *Macromolecules* **1998**, 31, 4041
9. Prodpran, T.; Dimonie, V.L.; Sudol, E.D.; El-Aasser, M.S. *PMSE* **1999**, 80, 534
10. MacLeod, P.J.; Keoshkerian, B.; Odell, P.; Georges, M.K. *PMSE* **1999**, 80, 539
11. Durant, Y.G. *PMSE* **1999**, 80, 538
12 Lansalot, M.; Charleux, B.; Vairon, J.P.; Pirri, R.; Tordo, P. *Polym. Prepr. (Am. Chem. Soc., Div. Polym. Chem.)* **1999**, 40(2), 317
13. Grimaldi, S.; Finet, J.-P.; Zeghdaoui, A.; Tordo, P.; Benoit, D.; Gnanou, Y.; Fontanille, M.; Nicol, P.; Pierson, J.-F. *Polym. Prepr. (Am. Chem. Soc., Div. Polym. Chem.)* **1997**, 38(1), 651
14. Benoit, D.; Grimaldi, S.; Finet, J.; Tordo, P.; Fontanille, M.; Gnanou, Y. *Polym. Prepr. (Am. Chem. Soc., Div. Polym. Chem.)* **1997**, 38(1), 729
15. Benoit, D.; Grimaldi, S.; Finet, J.; Tordo, P.; Fontanille, M.; Gnanou, Y. in "Controlled Radical Polymerization", K. Matyjaszewski Ed., *ACS Symp. Ser.* **1998**, 685, 225
16. Le Mercier, C.; Gaudel, A.; Siri, D.; Tordo, P.; Marque, S.; Martschke, R.; Fischer, H. *Polym. Prepr. (Am. Chem. Soc., Div. Polym. Chem.)* **1999**, 40(2), 313
17. Lacroix-Desmazes, P.; Lutz, J.-F.; Boutevin, B. *Macromol. Chem. Phys.* (in press)
18. Goto, A.; Terauchi, T.; Fukuda, T.; Miyamoto, T. *Macromol. Rapid Commun.* **1997**, 18, 673
19. Miller, C. M.; Sudol, E. D.; Silebi, C. A.; El-Aasser, M. S. *Macromolecules* **1995**, 28, 2754 ; 2765 ; 2772
20. Landfester, K.; Bechthold, N.; Tiarks, F.; Antonietti, M. *Macromolecules* **1999**, 32, 5222
21. *Polymer Handbook*, 4th Edition, **1999**; Brandrup, J.; Immergut, E.H.; Grulke, E.A. Eds, John Wiley & Sons
22. Fischer, H. *J. Polym. Sci., Part A, Polym. Chem.* **1999**, 37, 1885
23. Lane, W.H. *Ind. Eng. Chem.* **1946**, 18, 295

Chapter 11

Triazolinyl Radicals: Toward a New Mechanism in Controlled Radical Polymerization

Markus Klapper, Thorsten Brand, Marco Steenbock, and Klaus Müllen[1]

MPI für Polymerforschung, Ackermannweg 10,
D–55128 Mainz, Germany

Abstract:
We describe the use of triazolinyl counter radicals in controlled radical polymerization as an alternative to the nitroxides. This class of radicals allows for the controlled polymerization of monomers other than styrene, such as acrylates, methacrylates and vinylacetate, and blockcopolymers of various composition are accessible by this metal-free process. It is demonstrated that due to a well controlled decomposition of the counter radical the concentration of growing macroradicals is adjusted. The specific behavior of the triazolinyl radicals is explained by a self-regulation process.

1. Introduction

One way of achieving control over radical polymerization is to minimize the amount of side reactions of the growing free macroradical by reducing its concentration in the equilibrium with a counter radical (Scheme 1).[1-4]

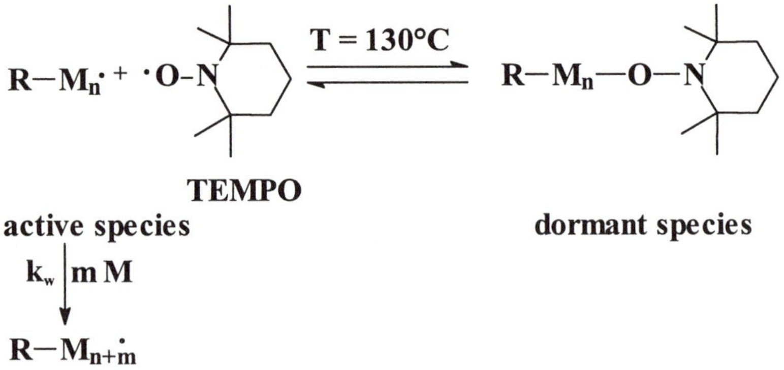

$$R-M_n^{\cdot} + \cdot O-N \quad \underset{}{\overset{T = 130°C}{\rightleftharpoons}} \quad R-M_n-O-N$$

TEMPO

active species   dormant species

$k_w \bigg| m\,M$

$R-M_{n+m}^{\cdot}$

Scheme 1: The function of TEMPO as controlling agent in radical polymerization.

[1]Corresponding author.

The use of 2,2,6,6-tetramethylpiperidine-1-oxide (TEMPO) as counterradical leads to a series of restrictions which prompted us to search for a new mechanism of control in radical polymerization.[2-5] Since unwanted side reactions of the free macroradical cannot rigorously be excluded, at one point in time the concentration of the free counter radical increases, the equilibrium is further shifted to the dormant side, and the polymerization stops. With that in mind, our approach is intended to include some sort of self-regulation. A crucial aspect thereby is the stability of the counterradical or, to be more specific, its controlled decomposition.

While in most cases the counterradicals used have their unpaired electron confined, e.g., to a nitrogen-oxygen bond, we introduce here the triazolinyl radical 1[6] as a counterradical which has its spin density delocalized in a more extended π-system. It should be noted, however, that 1 is structurally related to the verdazyl radical, which has not been particularly useful as counterradical in controlled radical polymerization.[7]

2. The triazolinyl moiety as stable free radical

According to Neugebauer[6] the triazolinyl radicals 1 and 2 are available by condensation of the „chlorohydrazone" 3 and a diarylmethanamine (4 or 5) and subsequent oxidation of the resulting triazoline with hexacyanoferrate (Scheme 2). It is important that triazolinyl 1 can be structurally modified by substituting the phenyl substituents at the positions 2, 3 and 5 (Table 1).

	R_1	R_2	R_3
1	H—⬡—	H—⬡—	H—⬡—
2	H—⬡—	H—⬡—	fluorenyl (spiro)
8a	naphthyl	H—⬡—	H—⬡—
8b	MeO—⬡—	H—⬡—	H—⬡—
8c	Cl—⬡—	H—⬡—	H—⬡—
8d	NO_2—⬡—	H—⬡—	H—⬡—
8e	H—⬡—	Cl—⬡—	H—⬡—
8F	H—⬡—	NO_2—⬡—	H—⬡—

Table 1: Some examples of triazolinyl radicals.

This approach allows us to affect the steric and electronic conditions for the interaction between 1 and the free macroradical. Another modification implies transition from the diphenyl-substituted case 1 to the corresponding spirofluorenyl compound **2**. While **1** and **2** are closely related, their thermal stability appears to be significantly different which is highly relevant for their function as counterradicals.

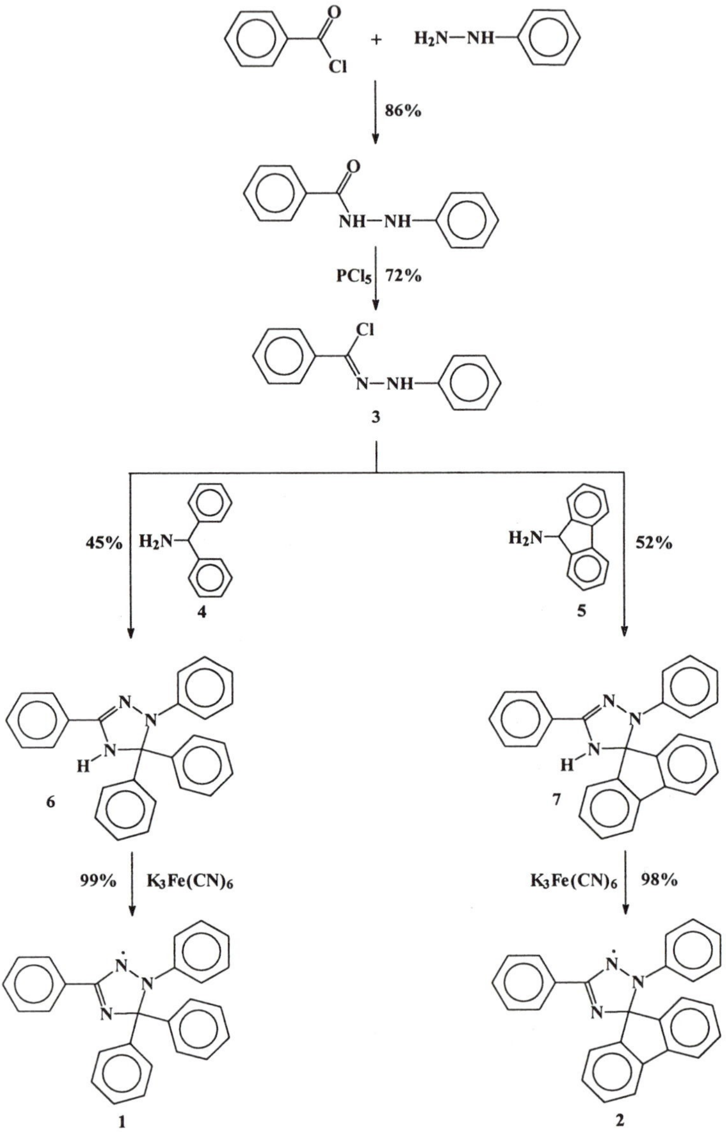

Scheme 2: *Syntheses of triazolinyl 1 and spirotriazolinyl 2.*

3. Controlled polymerization of different monomers

In a typical experiment, the polymerization of styrene is performed in bulk at 120 °C with dibenzoylperoxide (BPO) as initiator and triazolinyl **1** as additive in a slight excess compared to the initiator. GPC curves clearly show a systematic increase of the molecular weight of the resulting polystyrene with conversion which is an important requirement for controlled radical polymerization. The most convincing evidence for controlled polymerization of styrene in the presence of the triazolinyl radical is the linear plot of the molecular weight versus conversion and the decrease of the molecular weight distribution (Figure 1). [8-11]

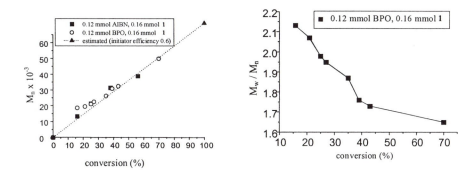

Figure 1: *Controlled radical polymerization of styrene in presence of triazolinyl 1 (T = 120 °C, bulk); M_n and polydispersity index versus conversion.*

Figure 1 also shows that the observed development of molecular weights can be compared with the calculated values resulting from monomer/initiator ratio and conversion for an estimated initiator effecency of 0.6. In accordance with the linear M_n vs. conversion plot the polydispersity decreases with conversion although the limiting value of about 1.6 is relatively high. This aspect will be referred to below.

Figure 2 demonstrates the first order kinetics for styrene polymerization in the presence of different triazolinyl radicals; in all cases, however, the M_n versus conversion plot remains linear. The same criteria convincingly reveal controlled radical polymerization for a series of acrylates such as butylacrylate and even methylmethacrylate (MMA). [8-11] The latter is known to be a particularly critical case in the use of TEMPO as counter radical. [2-3] In this context, it should be noted that the formation of unsaturated end groups in PMMA by hydrogen abstraction is widely surpressed. This might be due to the conformational situation in the end group of the dormant species or due to the reactivity of the counter radical, but this has not been systematically investigated so far.

Figure 3 exhibits the first-order kinetics plot for the thermal and the BPO initiated polymerization of MMA in presence of triazolinyl 1. It should be noted that the temperatures at which controlled polymerization is possible depends sensitively upon the nature of the monomer. Thus, while controlled polymerization of styrene occurs in the temperature range between 120 and 140 °C, the analogous process for methylmethacrylate requires a much lower temperature, i.e. between 70 and 90 °C. What comes somewhat as a surprise is that with **1** as counterradical the polymerization of MMA can also be achieved in a thermally initiated process. This important finding will be referred to in section 5.

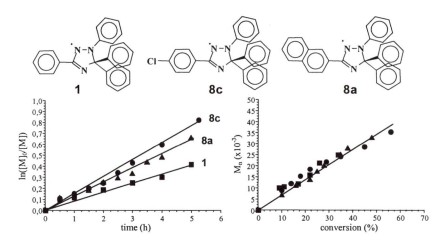

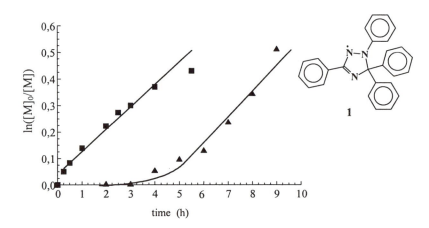

Figure 2: *Influence of differently substituted triazolinyls on the kinetics of controlled styrene polymerization (1: ■ ;8a ▲;8c: •)*

Figure 3: *First order kinetic plot for the controlled polymerization of MMA in bulk at 70 °C:■ = BPO-initiated: MMA / 1 / BPO = 1000 / 1.6 / 1.2;▲ = thermally initiated: MMA / 1 = 1000 / 3.0.*

4. Blockcopolymer synthesis

The key question toward blockcopolymer synthesis is whether the growing free macroradical can be reversibly functionalized by the counterradical and unwanted termination reactions supressed.

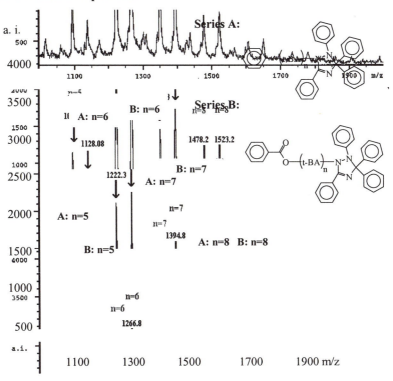

Figure 4: *MALDI-TOF spectrum of an oligo(t-butylacrylate) obtained by controlled polymerization with BPO and triazolinyl 1.*

MALDI-TOF spectra of both, *t*-butylacrylate oligomers (Figure 4) and of polystyrene samples obtained with BPO and triazolinyl **1** support the essentially quantitative endfunctionalization with a triazolinyl group.

As expected, the triazolinyl termination of the poly(methylmethacrylate) (PMMA) brings about a reduced thermal stability compared to a sample obtained by free radical polymerization (Figure 5). It is possible, however, to reductively remove the triazolinyl group upon treatment with hydroquinone.

Furthermore, the endfunctionalization of similarly obtained polystyrene samples could be carefully analyzed by UV-spectroscopy, which is an important experimental advantage of triazolinyl controlled polymerization.[12] It must be concluded from such experiments that the degree of endfunctionalization is continuously reduced upon increasing the conversion or the molecular weight M_n (Figure 6). One has clearly to distinguish, how far one would like to go with the conversion to keep a still high degree of endfunctionalization.

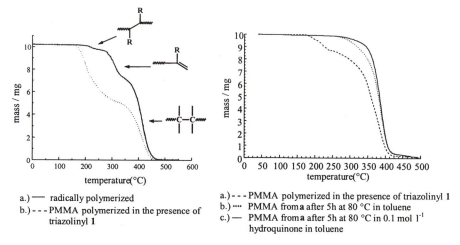

a.) —— radically polymerized
b.) - - - PMMA polymerized in the presence of triazolinyl **1**

a.) - - - PMMA polymerized in the presence of triazolinyl **1**
b.) ···· PMMA from **a** after 5h at 80 °C in toluene
c.) —— PMMA from **a** after 5h at 80 °C in 0.1 mol l^{-1} hydroquinone in toluene

Figure 5: *Thermal stability of PMMA samples with different endgroups in comparison.*

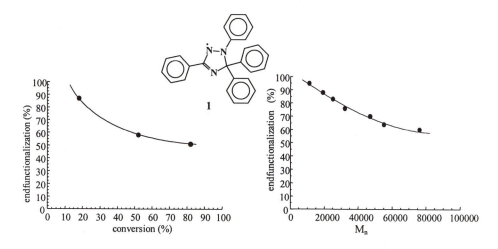

Figure 6: *UV spectroscopic determination of the degree of triazolin endfunctionalization of polystyrenes after different conversions and at different molecular weights.*

A triazolin endfunctionalized polystyrene obtained by precipitation after about 50 % conversion can be used as a new macroinitiator. The GPC curves shown in Figure 7 leave no doubt as to the chain extension of the polystyrene.

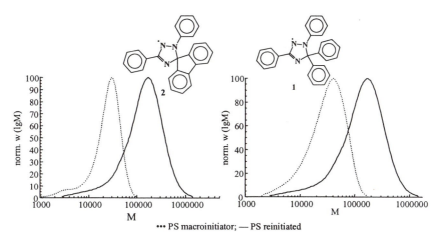

Figure 7: *Polystyrene chain extension with triazolin functionalized samples: GPC traces of the starting block versus the resulting longer chains.*

Various blockcopolymers could successfully be made using this polystyrene macroinitiator: We obtained for example polystyrene-b-MMA, polystyrene-b-butylacrylate and, remarkably, polystyrene-b-vinylacetate. The polystyrene chains were reinitiated with the second monomer at 120°C.

One possible complication should be mentioned, however, which follows from the different reaction temperatures required for different monomers. Reinitiation of a polystyrene block requires a relatively high temperature (see above); at that temperature a methylmethacrylate or vinylacetate polymerization is no longer controlled so that only partial control over blockcopolymer formation is obtained, even if the temperature is lowered rapidly.

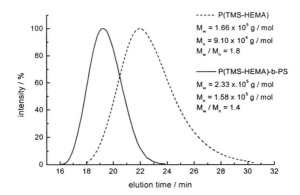

Figure 8: *GPC traces of poly(TMS-HEMA) macroinitiator versus the resulting poly(TMS-HEMA)-b-polystyrene.*

The other possibility is to work vice versa, that is to polymerize the (meth)acrylic monomer first at low temperature and to use the resulting endfunctionalized poly(meth)acrylate for initiation of styrene at increased temperature. We did this for example in the case of trimethylsiloxyethyl methacrylate (TMS-HEMA) monomer at 70 °C to obtain the first block and applied it as macroinitiator for styrene polymerization at 100 °C. Figure 8 shows the GPC traces of the first block versus the resulting poly(TMS-HEMA)-b-polystyrene.[13]

5. Concept of self-regulation

It has been noted in section 2 that modifying the substitution type of triazolinyl radicals not only affects the electronic structure, but also the thermal stability. A good example can be made when comparing **1** with the corresponding spirofluorenyl case **2**. Figure 9 convincingly demonstrates the controlled polymerization of MMA in the presence of **1**. It should be noted for the M_n versus conversion plot that initiation requires a temperature of about 90 °C, after which the temperature is lowerd to 70 °C in order to achieve controlled polymerization.

- Spiro-triazolinyl *2*: ➲ Polymerization inhibited

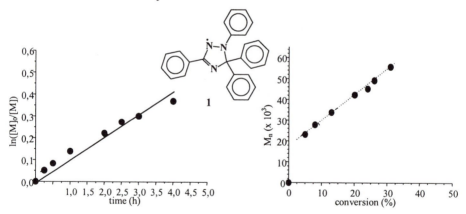

Figure 9: *Polymerization of MMA in the presence of 1 in bulk at 70 °C; [triazolinyl]₀ / [BPO]₀ = 1.5; a) conversion versus time b) M_n versus conversion.*

Remarkably enough, under similar conditions the corresponding spiro-substituted triazolinyl **2** acts as an inhibitor. A comparison of the thermal stabilities of **1** and **2** is crucial. Figure 10, in which the intensity of the ESR signal is used as a measure of the amount of the counterradicals, shows that while **2** is more or less stable at 95°C, **1** undergoes a rapid decomposition ($t_{1/2} \sim 20$ min.).[11]

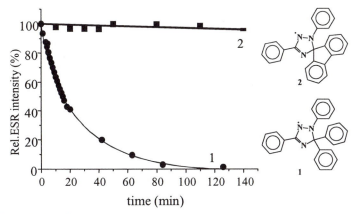

Figure 10: *ESR-spectroscopic investigation of the thermal stability of triazolinyl radicals 1 and 2 in toluene at 95 °C.*

Upon cleavage of the carbon-carbon bond between the heterocycle and one of the phenyl groups in 3-position, a phenyl radical and a stable aromatic triazene is formed. (Scheme 3). It will be useful for the discussion below to note that the thermal decomposition occurs for the free radical **1**, not, however, for the triazolin **6** (Scheme 2) or the endfunctionalized polymer where **1** forms a covalent bond with the macroradical.

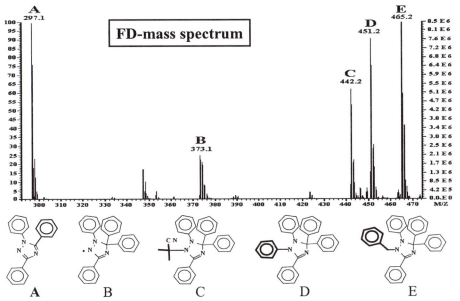

Figure 11: *Thermal decomposition of AIBN in the presence of triazolinyl 1 in toluene (95 °C).*

162

A mass spectrometric analysis of the products obtained during the thermal decomposition of **1** in the presence of AIBN supports this conclusion, in particular, by the detection of species A and D (Figure 11).

Further insight into the different behavior of **1** and **2** can be derived from the thermal polymerization of styrene. The polymerization of styrene in the presence of **1** proceeds in a controlled fashion, even in the absence of an initiator (Figure 12).

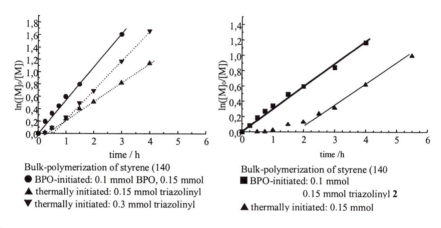

Bulk-polymerization of styrene (140
● BPO-initiated: 0.1 mmol BPO, 0.15 mmol
▲ thermally initiated: 0.15 mmol triazolinyl
▼ thermally initiated: 0.3 mmol triazolinyl

Bulk-polymerization of styrene (140
■ BPO-initiated: 0.1 mmol
  0.15 mmol triazolinyl **2**
▲ thermally initiated: 0.15 mmol

Figure 12: Thermal polymerization of styrene in the presence of triazolinyl 1 and 2.

The same holds true for the polymerization in the presence of **2**. Without an initiator, however, this reaction requires an induction period. Thus, while the thermal decomposition of **1** provides phenyl radicals which can act as initiator, the corresponding process with **2** rests upon the formation of free radicals by the autopolymerization of styrene. This finding readily explains the occurrence of an induction period.

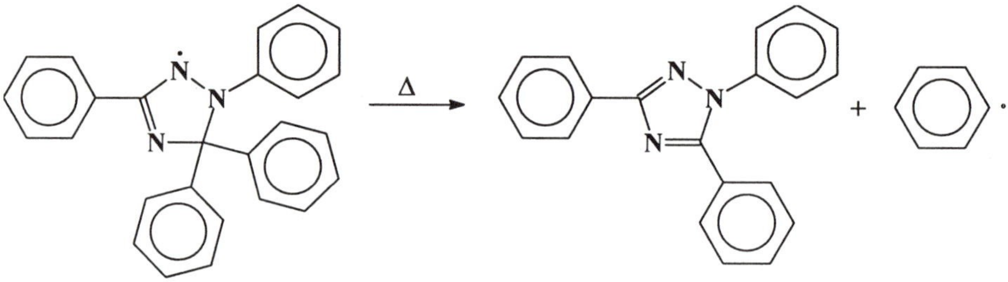

1   **9**

Scheme 3: Thermal decomposition of triazolinyl 1.

It is also in line with this outcome that the propagation rate during the triazolinyl controlled polymerization of styrene is higher in the case of **1**, because radicals are formed, on the one hand, by autopolymerization and, on the other hand, by the decomposition of **1**. Thus, while the M_n versus conversion plot is linear for **2** up to relatively high conversion, the deviation from the linear behavior detected for **1** points toward a reduced degree of control (Figure 13). Not surprisingly, therefore, the achievable polydispersity (ca. 1.2) is significantly lower for **2** than for **1**.

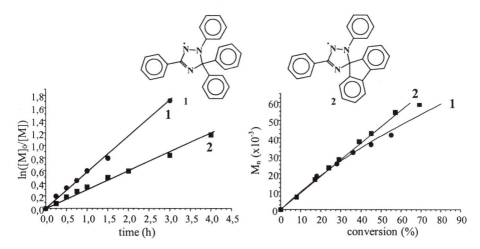

Figure 13: *Controlled radical polymerization of styrene in bulk with both triazolinyl-radicals **1** and **2**. T = 140 °C, [triazolinyl]$_0$ / [BPO]$_0$ = 1.5 for both cases.*

It follows that the description of the equilibrium between active and dormant species has to be modified to account for the remarkable properties of counterradical **1** (Figure 14). If the growing macroradical escapes the equilibrium by unavoidable termination reactions (Figure 14 path **c**), the free counterradical, T•, reduces its concentration by a thermal dissociation process. Thereby, however, a phenyl radical is formed which can now act to initiate a new chain process. This is the essence of our self-regulation process in which the system develops an ability to control the free concentration of T• and thus responds to inevitable side reactions of the macroradical.[10,11] In this case autopolymerization is no longer necessary to keep the concentration of macroradicals constant. For the herein presented concept it is essential to mention that the triazolinyl attached to the polymer chain is stable to this mode of decomposition, as decomposition would not lead to an aromatic heterocycle (e. g. triazole **9** in Scheme 3). Furthermore, the period during the radical polymerization where the macroradical is growing and the counter radical is free is very short, and therefore the decomposition of T• is nearly excluded. Decomposition only occurs if an excess of counter radical, due to side reactions (**c**), is formed.

164

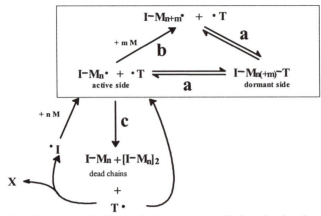

Figure 14: Concept of self-regulation in controlled radical polymerization.

This concept is proven as well by simulation with 'Predici' (CiT GmbH, v. 4.7). From these calculations it was shown that the control not only depends on the nature of the equilibrium but also on the decomposition rate. The influence of various substituents on the thermal stability and the polymerization behavior is under investigation.

It should be mentioned that the problem of controlling the concentrations of counter radical and growing macroradical has been dealt with by other authors as well, since Georges et al. have added acid to decompose excess counterradical, while Fukuda et al.[3] have continuously added further initiator. In our approach, however, an external control of the equilibrium is not necessary, the radical has the ability to control the equilibrium by itself.

6. Conclusion and outlook

We conclude that the controlled radical polymerization is possible not only for styrene, but also for other monomers, such as acrylates, methacrylates and vinylacetates. However, due to the different electronic structure the reaction conditions and temperatures strongly depend on the monomer (e.g styrene ~130°C, acrylates or vinylacetate ~70-90°C) and have to be optimized in the future. High molecular weights (M_n higher than 100.000) are obtained and structurally defined triazolinyl endfunctionalized polymers can be isolated. This allows the synthesis of a broad series of blockcopolymers.

Another practically important result is the fact that the polymerization rates and also the molecular weights are higher than with the corresponding TEMPO process (Table 2). Moreover it should be emphasized that reactivity can be controlled by electronic as well as by steric effects (Figure 15). This should allow for a fine-tuning and a further increase in the polymerization rate.

Electron spin density:	localized	delocalized	delocalized
Radical polymerization of styrene			
T>100°C : $\ln([M]_0/[M])$ versus time	linear increase	linear increase	linear increase
M_n versus conversion	linear increase	linear increase	linear increase
Polydispersity	>1.1	>1.2	>1.5
Polymerization rate 120°C; V_p [mol $l^{-1}s^{-1}$]	$2.8*10^{-4}$	$2.4*10^{-4}$	$9.1*10^{-4}$
Controlled polymerization of non-autopolymerizable monomers(MMA, acrylates, vinylacetate)	-	-	possible
Radical polymerization of MMA			
$\ln([M]_0/[M])$ versus time	-	-	linear increase
M_n-versus conversion-	-	-	linear increase
Polydispersity	-	-	>1.2

Table 2: *Comparison of the polymerization results with TEMPO, 1 and 2.*

On the other hand also the rate of decomposition can be influenced by introducing various substituents into the phenyl rings attached in the 3-position of the heterocycle.

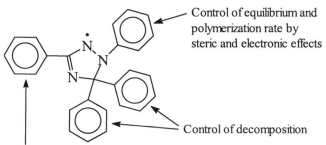

Figure 15: Functionalization of the triazolinyl-radical.

Acknowledgement: Financial support by the BASF AG, Ludwigshafen is gratefully acknowledged.

References

1   Salomon D.H.; Rizzardo E.; Cacioli P., Eur. Pat.Appl. **1985**, 135280.
2   Georges M.K.; Veregin R.P.N.; Kazmaier P.M.; Hamer G.K., Macromolecules **1993**; 26, 2987.
3   Fukuda T.; Terauchi T.; Goto A.; Ohno K.; Tsujii Y.; Miyamoto T.; Kobatake S.; Yamada B., Macromolecules **1996**, 29, 6393.
4   Listigovers N.A.; Georges M.K.; Odell P.G.; Keoshkerian B., Macromolecules **1996**, 29, 8993.
5   Steenbock M.; Klapper M.; Müllen K.; Pinhal N.; Hubrich M., Acta Polymer. **1996**, 47, 276.
6   Neugebauer F.A.; Fischer H., Tetrahedron **1995**, 51,1288.
7   Yamada, , Y. Nobuka, and Y. Miura, Polym. Bull., **1998**, 41, 539.
8   Colombani D.; Steenbock M.; Klapper M.; Müllen K.; Fischer M.; Koch J.; Paulus W., Ger. Pat. Appl. 196 36 8867.
9   Colombani D.; Steenbock M.; Klapper M.; Müllen K., Macromol. Rapid Commun. **1997**, 18, 243.
10 Steenbock M.; Klapper M.; Müllen K., Acta Polymer. **1998**; 49, 376.
11 Steenbock M.; Klapper M.; Müllen K.; Bauer C.; Hubrich M., Macromolecules **1998**, 31, 5223.
12 Steenbock M.; Klapper M.; Müllen K., Macromol. Chem. Phys. **1998**, 199, 763.
13 Dasgupta A.; Brand T.; Klapper M.; Müllen K., in preparation.

ATOM TRANSFER RADICAL POLYMERIZATION

Chapter 12

Living Radical Polymerization of Styrene: RuCl₂(PPh₃)₃ and Alkyl Iodide-Based Initiating Systems

Yuzo Kotani, Masami Kamigaito, and Mitsuo Sawamoto[1]

Department of Polymer Chemistry, Graduate School of Engineering, Kyoto University, Kyoto 606–8501, Japan

A series of 1-phenylethyl halides [CH₃CH(Ph)X: X = I, Br, and Cl] were employed as initiators for living radical polymerization of styrene with ruthenium(II) dichlorotris(triphenylphosphine) [RuCl₂(PPh₃)₃] in the presence of Al(Oi-Pr)₃ in toluene at 100 °C. With X = I, the number-average molecular weights (M_n) were controlled by the feed ratio of the monomer to the initiator, and the molecular weight distributions (MWDs) were relatively narrow (M_w/M_n ~ 1.5). In contrast, the MWDs were broader with X = Br, while the M_n was controlled. With X = Cl, the M_n was much higher than the calculated values. End-group analysis by ¹H NMR supported that the polymerization proceeds via reversible activation of the terminal C–I bond derived from CH₃CH(Ph)I as the initiator. In addition, F(CF₂)₆I and CHI₃ were effective for the styrene living polymerization. Ti(Oi-Pr)₄ was also effective as additives, which induced faster polymerization than that did Al(Oi-Pr)₃ at 60 °C. The resulting polystyrene with Al(Oi-Pr)₃ and Ti(Oi-Pr)₄ showed narrow MWDs (M_w/M_n ~ 1.2).

Introduction

The last several years have witnessed remarkable developments in control of radical polymerization, among which transition metal complexes had led up to novel living radical polymerization systems (*1*). The initiating systems that permit precision control of molecular weights and distributions generally consist of alkyl halides as initiators and group 7–11 transition metal complexes as catalysts. Thus far, various kinds of metal complexes, such as Ru(II) (*2–6*), Cu(I) (*7–12*), Fe(II) (*13, 14*), Ni(II) (*15–17*), Ni(0) (*18, 19*), Rh(I) (*20, 21*), Pd(II) (*22*), and Re(V) (*23*) with suitable ligands, have been employed for expanding the scope of the living radical polymerization. Design and use of a suitable metal complex for a certain monomer seems necessary to achieve the living radical polymerization that involves reversible

---

[1]Corresponding author.

and homolytic cleavage of carbon–halogen (C–X) terminal via single-electron redox reaction of the metal center [~~~C–X + M(n) $\rightleftarrows$ ~~~C• X–M(n+1)] (*1g, 1h*).

We have developed Ru(II)-based initiating systems for methyl methacrylate (MMA) living radical polymerization, which consist of $RuCl_2(PPh_3)_3$, various alkyl chlorides or bromides (R–X), and aluminum alkoxides (*2–5*). As indicated in Scheme 1, for the Ru(II)-mediated system, the homolytic cleavage of the C–X bond in the initiator (R–X) occurs via a single electron oxidation reaction of Ru(II), followed by the addition of R• to the monomer, and the generated Ru(III) species is reduced to the original Ru(II) to give an adduct of R–X and the monomer that possesses a terminal C–X bond. The polymerization proceeds via a similar repetitive addition of monomer to the radical species, reversibly generated from the covalent species with a C–X terminal and Ru(II). The $RuCl_2(PPh_3)_3$-based initiating systems, initially developed for MMA, have been also applied for acrylates (*24*), styrenes (*25*), and acrylamides (*26*). Styrene was indeed polymerized by $CCl_4/RuCl_2(PPh_3)_3/Al(Oi\text{-}Pr)_3$ in toluene at 60 °C but to give polymers with broad molecular weight distributions (MWDs) (M_w/M_n = 3–4), in contrast to the narrow MWDs (M_w/M_n = 1.2–1.3) of PMMA obtained under similar conditions.

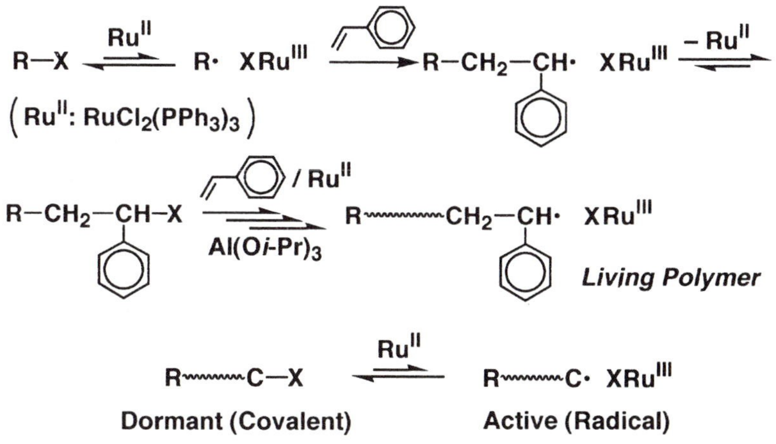

Scheme 1. Living radical polymerization of styrene with Ru(II) complex.

This paper deals with new haloalkyl initiators (R–X; X = Cl, Br, I) for the $RuCl_2(PPh_3)_3$-mediated living radical polymerization of styrene (Scheme 2). The halide initiators should be carefully selected so that the C–X bond is reactive enough to generate radical species [C• XRu(III)] at a considerable rate via interaction with $RuCl_2(PPh_3)_3$. For this, the choice of both the alkyl group (R) and the halogen (X) in R–X is important. Especially, the halogen in R–X, which will be transferred to the growing polymer terminal (R~~~C–X), is crucial for the control of the polymerization, because it affects the rate of interconversion between the dormant and the active species as well as the stability of the growing terminal. The reactivity of C–X bonds in Kharasch addition reactions via peroxy and related free radicals is in the following order; C–I > C–Br > C–Cl (*27*), which agrees with their bond dissociation energy (*28*). Because of this, iodo-compounds have often been employed for free radical addition reactions (*29*). In addition, the C–X bond activation with an iridium(I) complex via single electron transfer is significantly slower for CCl_4 than for CBr_4 or CI_4 (*30*).

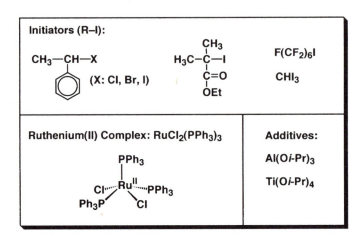

Scheme 2. Ru(II)-Based initiating systems for living radical polymerization of styrene.

The use of alkyl iodides for controlled radical polymerization was first reported by Tatemoto (*31*), where fluorinated monomers such as CF_2CF_2 were polymerized by $K_2S_2O_8$ in the presence of fluoroalkyl iodides such as $F(CF_2)_6I$. Similar systems were also applied for styrene (*32–35*), acrylates (*33, 34*), and vinyl acetate (*36*) to give polymers with controlled molecular weights, although the control seems inferior to that with the metal catalyzed systems. However, alkyl iodides have rarely been employed in metal-mediated radical additions, where usually chlorides or bromides are preferentially employed. There were only a few examples coupled with zero-valent metals like Cu and Ag for the radical addition reactions (*37*) and just a preliminary report for styrene polymerization (*38*).

With these backgrounds, we employed three 1-phenylethyl halides [$CH_3CH(Ph)X$: X = Cl, Br, and I], unimer models of polystyrene with a C–X terminal, as initiators for the Ru(II)-mediated living radical polymerization of styrene. Among the three halogens, iodine proved very effective for this purpose.

Results and Discussion

Effects of Initiator-Halogens

The three phenylethyl halides were employed as initiators for the polymerization of styrene to be coupled with $RuCl_2(PPh_3)_3$ in the presence of $Al(Oi\text{-}Pr)_3$ in toluene at

100 °C. All three systems induced polymerization without an induction phase, and monomer conversion reached over 90% within 10–14 h (Figure 1). The polymerization with CH$_3$CH(Ph)I proceeded faster than that with CH$_3$CH(Ph)Br or CH$_3$CH(Ph)Cl. Without Al(Oi-Pr)$_3$, almost no polymerization proceeded.

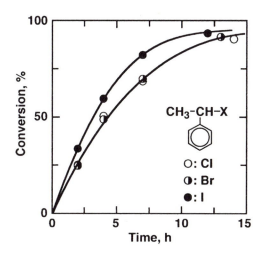

Figure 1. Polymerization of styrene with CH$_3$CH(Ph)X (X = Cl, Br, and I)/ RuCl$_2$(PPh$_3$)$_3$/Al(Oi-Pr)$_3$ in toluene at 100 °C: [styrene]$_0$ = 6.0 M; [CH$_3$CH(Ph)X]$_0$ = 60 mM; [RuCl$_2$(PPh$_3$)$_3$]$_0$ = 30 mM; [Al(Oi-Pr)$_3$]$_0$ = 100 mM.

The M_n and MWD of the obtained polystyrene depended on the initiator halogens, more than polymerization rate did, as shown in Figure 2. With CH$_3$CH(Ph)I or CH$_3$CH(Ph)Br, the M_n increased with monomer conversion and almost agreed with the calculated values based on the assumption that one molecule of the initiator generates one living polymer chain. Furthermore with the iodide, the polydispersity ratios remained narrow ($M_w/M_n \sim 1.5$) throughout the polymerization. CH$_3$CH(Ph)Cl gave polymers with broad MWDs ($M_w/M_n \sim 2$) and much higher M_n ($M_n \sim 2 \times 10^4$), independent of conversion. This is due to the greater bond strength of C–Cl, which in turn leads to a slow initiation from the chloride and to a slow interconversion between the dormant and the active species (cf. Scheme 1). In contrast, with the bromide- and the iodide-based initiating systems, the initiation occurs enough fast to control the molecular weights. However, the MWDs with the bromide were still broad probably due to slow exchange reaction. Among the halides, the CH$_3$CH(Ph)I-based initiating system proved the best for the living radical polymerization of styrene mediated by RuCl$_2$(PPh$_3$)$_3$. Thus, the following section will discuss the details on this CH$_3$CH(Ph)I-based initiating system.

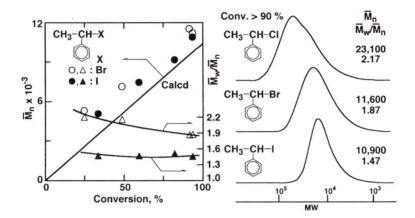

Figure 2. M_n, M_w/M_n, and SEC curves of polystyrene obtained with $CH_3CH(Ph)X$ (X = Cl, Br, and I)/$RuCl_2(PPh_3)_3$/$Al(Oi-Pr)_3$ in toluene at 100 °C: [styrene]$_0$ = 6.0 M; [$CH_3CH(Ph)X]_0$ = 60 mM; [$RuCl_2(PPh_3)_3]_0$ = 30 mM; [$Al(Oi-Pr)_3]_0$ = 100 mM.

$CH_3CH(Ph)I$ Initiating System

Effects of Monomer Concentration

A series of styrene polymerizations were carried out using $CH_3CH(Ph)I$/$RuCl_2(PPh_3)_3$/$Al(Oi-Pr)_3$ system in toluene at 100 °C at varying initial monomer concentrations ([M]$_0$), while the initial ratio of the styrene to the initiator,

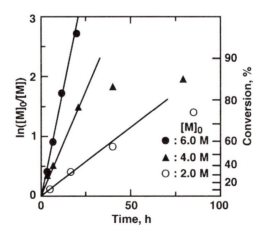

Figure 3. First-order plots for the polymerization of styrene with $CH_3CH(Ph)I$/$RuCl_2(PPh_3)_3$/$Al(Oi-Pr)_3$ in toluene at 100 °C: [styrene]$_0$/[$CH_3CH(Ph)I]_0$/[$RuCl_2(PPh_3)_3]_0$/[$Al(Oi-Pr)_3]_0$ = 6000/60/30/100 mM (●); 4000/40/20/80 mM (▲); 2000/20/10/40 mM (O).

$[M]_0/[CH_3CH(Ph)I]_0$, was fixed at 100 (Figure 3). At any $[M]_0$, styrene polymerization occurred without induction phase. The lower $[M]_0$, the slower the polymerization; for example, time for 50% monomer conversion was 3, 5, and 19 h at $[M]_0$ = 6.0, 4.0, and 2.0 M, respectively. As shown in Figure 3, the logarithmic conversion data, $\ln([M]_0/[M])$ plotted against time t, gave straight line at $[M]_0$ = 6.0 M over 90%, which shows constant concentrations of growing species during the polymerization. However, at lower $[M]_0$ (4.0 M or 2.0 M), the plots shows downward curvatures, and then the polymerization ceased around 80–85% monomer conversion.

Figure 4 shows the M_n, M_w/M_n, and MWDs of polystyrene obtained under the same conditions indicated for Figure 3. The M_n of the polystyrene obtained with $[M]_0$ = 6.0 M increased in direct proportion to monomer conversion over 90%. The M_n also increased even at the lower $[M]_0$, although the polymerization was not quantitative. This showed no chain transfer and no recombination but primary termination reaction like elimination of hydrogen iodide from the polymer terminal during the polymerization at lower $[M]_0$ (4.0 M and 2.0 M). The MWDs obtained at $[M]_0$ = 4.0 and 6.0 M were relatively narrow and almost constant (M_w/M_n = 1.4–1.5) throughout the reaction whereas the MWDs at lower $[M]_0$ (2.0 M) were became broader with conversion. This suggests the existence of dead polymer chain.

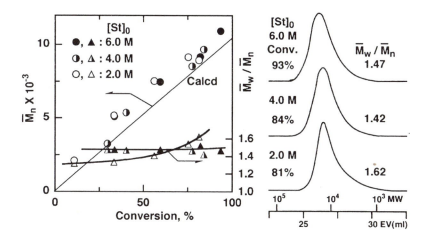

Figure 4. M_n, M_w/M_n, and SEC curves of polystyrene obtained with $CH_3CH(Ph)I/$ $RuCl_2(PPh_3)_3/Al(Oi\text{-}Pr)_3$ in toluene at 100 °C: $[styrene]_0/[CH_3CH(Ph)I]_0/$ $[RuCl_2(PPh_3)_3]_0/[Al(Oi\text{-}Pr)_3]_0$ = 6000/60/30/100 mM (●, ▲); 4000/40/20/80 mM (◑, ▲); 2000/20/10/40 mM (○, △).

End Group Analysis

In order to clarify these results, the terminal structure of the obtained polystyrene was investigated by ^1H NMR spectroscopy. Figure 5 shows the ^1H NMR spectra (4.0–6.4 ppm) of the polystyrenes obtained with $CH_3CH(Ph)I/RuCl_2(PPh_3)_3/Al(Oi\text{-}$ $Pr)_3$ in toluene at 100 °C. The absorption around 4.6 ppm (peak *a*) is attributed to the methine proton adjacent to the terminal iodine, and peak b (the small absorption near peak a) is from the $-C-Cl$ terminal generated through the halogen-exchange reaction between the $-C-I$ ω-end and the $RuCl_2$ complex (4, 39, 40). At an early stage of polymerization (conversion ~ 40%), the peak a could be observed in both spectra for $[M]_0 = 6.0$ M and 2.0 M (A1 and B1 in Figure 5, respectively), and the number-average end functionality (F_n) for the whole halogen end ($a + b$) was 0.78 and 0.62, respectively.

After the polymerization proceeded, however, peak a became small [$F_n(a+b) = 0.36$], then new peaks c and d appeared around 6.1 ppm at $[M]_0 = 2.0$ M (B2). The new peaks c and d are attributed to the olefin proton of the dead polymer chain produced via β-H elimination from the ω-end (denoted in Figure 5). On the other hand, almost no olefin-related peak was observed and peak a still remained in the spectrum (A2) for $[M]_0 = 6.0$ M [$F_n(a+b) = 0.71$]. These results prove that the C–I growing terminal is maintained throughout the polymerization even at 100 °C at high monomer concentrations, though some of them are replaced with chlorine. The propagation reaction is considered to compete with the β-H elimination and the former is faster at a higher monomer concentration. However, at the later stage of the polymerization at $[M]_0 = 2.0$ M, the monomer concentration decreases so that the

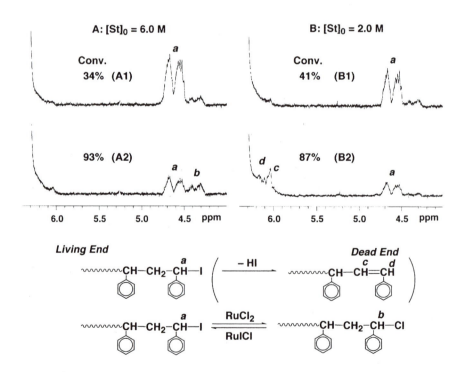

Figure 5. ^{1}H NMR spectra of polymers obtained with $CH_3CH(Ph)I/RuCl_2(PPh_3)_3/ Al(Oi-Pr)_3$ in toluene at 100 °C: $[styrene]_0/[CH_3CH(Ph)I]_0/[RuCl_2(PPh_3)_3]_0/[Al(Oi-Pr)_3]_0 = 6000/60/30/100$ mM (A1 and A2); 2000/20/10/40 mM (B1 and B2).

iodide terminal is converted to the olefinic terminal via β-H elimination. These analytical data are consistent with the results shown in Figures 3 and 4.

Molecular Weight Control with CH₃CH(Ph)I

To confirm the living nature of the polymerization at $[M]_0$ = 6.0 M, the monomer-addition experiment was carried out (Figure 6); thus styrene (100 equiv to the initiator) was polymerized with the $CH_3CH(Ph)I/RuCl_2(PPh_3)_3/Al(Oi\text{-}Pr)_3$ system in toluene at 100 °C, and a fresh feed of styrene (100 eq) was added to the reaction mixture when the initial charge had almost been consumed. The M_n of the polystyrene further increased in direct proportion to monomer conversion after monomer addition and agreed with the calculated values based on the monomer/initiator ratio. Furthermore, the polymer MWDs showed no shoulder and remained narrow in the second-phase. These results demonstrate that living polymerization of styrene was achieved with $CH_3CH(Ph)I$.

High molecular weight polystyrenes were also synthesized with the same initiating system; thus, styrene, 400 eq to the initiator, was polymerized with the same system in bulk at 100 °C (Figure 7). The polymerization proceeded faster than that in toluene to reach 90% in 48 h. The SEC curves were unimodal and shifted to high molecular weight as the polymerization proceeded. The M_n increased linearly up to ~ 5×10^4 with conversion, and the MWDs were narrow (M_w/M_n = 1.3–1.4).

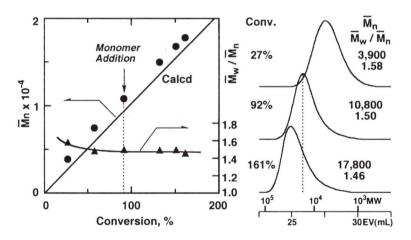

Figure 6. M_n, M_w/M_n, and SEC curves of polystyrene obtained in a monomer-addition experiment with $CH_3CH(Ph)I/RuCl_2(PPh_3)_3/Al(Oi\text{-}Pr)_3$ in toluene at 100 °C: $[styrene]_0 = [styrene]_{add} = 6.0\,M$; $[CH_3CH(Ph)I]_0 = 60$ mM; $[RuCl_2(PPh_3)_3]_0 = 30$ mM; $[Al(Oi\text{-}Pr)_3]_0 = 100$ mM. The diagonal solid line indicates the calculated M_n assuming the formation of one living polymer per $CH_3CH(Ph)I$ molecule.

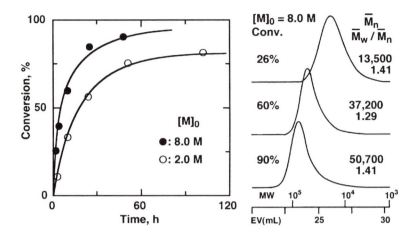

Figure 7. Bulk polymerization of styrene with $CH_3CH(Ph)I/RuCl_2(PPh_3)_3/Al(Oi-Pr)_3$ at 100 °C. [styrene]$_0$ = 8.0 M (●) and 2.0 M (○); [CH_3CH(Ph)I]$_0$ = 20 mM; [RuCl_2(PPh_3)_3]$_0$ = 10 mM; [Al(Oi-Pr)_3]$_0$ = 40 mM.

Iodoalkyl Initiators and Metal Alkoxide Additives

Other than $CH_3CH(Ph)I$, perfluoroalkyl iodide [$F(CF_2)_6I$] and iodoform (CHI_3) were employed with $RuCl_2(PPh_3)_3/Al(Oi-Pr)_3$ in bulk at 100 °C (Figure 8). Although both iodoalkyl compounds give initiating radicals different in structure from the

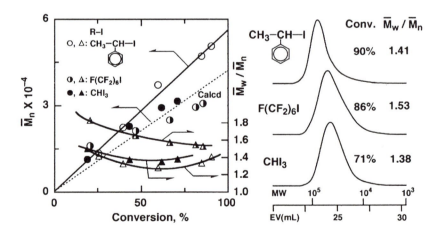

Figure 8. M_n, M_w/M_n, and SEC curves of polystyrene obtained with R–I/ $RuCl_2(PPh_3)_3/Al(Oi-Pr)_3$ at 100 °C: [styrene]$_0$ = 8.0 M; [R–I]$_0$ = 20 mM; [RuCl_2(PPh_3)_3]$_0$ = 10 mM; [Al(Oi-Pr)_3]$_0$ = 40 mM. R–I: $CH_3CH(Ph)I$ (○, △); $F(CF_2)_6I$ (◑, ▲); CHI_3 (●, ▲).

styrene radical, they could initiate styrene polymerization at almost the same rate as with CH$_3$CH(Ph)I. The M_n of the polystyrene increased linearly with conversion and showed good agreement with the calculated values. With F(CF$_2$)$_6$I, the MWD was initially very broad due to its slow initiation but narrowed as the polymerization proceeded. On the other hand, the polystyrene with CHI$_3$ showed a narrow distribution throughout the polymerization, which was incomplete, however.

As an additive, Ti(Oi-Pr)$_4$ was employed in place of Al(Oi-Pr)$_3$. In the presence of these isopropoxides, styrene was polymerized with an α-haloester initiator, (CH$_3$)$_2$C(CO$_2$Et)I, and RuCl$_2$(PPh$_3$)$_3$ in toluene at 60 °C. It took 100 h for 90% conversion with Ti(Oi-Pr)$_4$, whereas 500 h with Al(Oi-Pr)$_3$. The trend is the same as for the Ru(II)-mediated polymerization of MMA (*41*). The M_n of the polystyrene with both additives increased in direct proportion to monomer conversion and well agreed with the calculated values. As shown in Figure 9, the GPC curves exhibited narrow MWDs (M_w/M_n ~ 1.2). Thus, not only Al(Oi-Pr)$_3$ but Ti(Oi-Pr)$_4$ was proved to be efficient additives on the living radical polymerization of styrene.

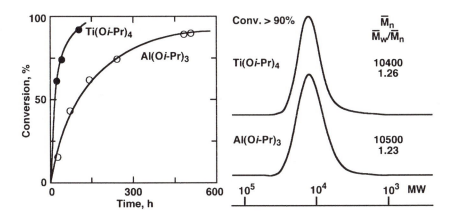

Figure 9. Polymerization of styrene with (CH$_3$)$_2$C(CO$_2$Et)I/RuCl$_2$(PPh$_3$)$_3$/Al(Oi-Pr)$_3$ or Ti(Oi-Pr)$_4$ in toluene at 60 °C: [styrene]$_0$ = 6.0 M; [(CH$_3$)$_2$C(CO$_2$Et)I]$_0$ = 60 mM; [RuCl$_2$(PPh$_3$)$_3$]$_0$ = 10 mM; [Al(Oi-Pr)$_3$]$_0$ = [Ti(Oi-Pr)$_4$]$_0$ =100 mM.

Conclusions

In conclusion, CH$_3$CH(Ph)I has proved to induce living radical polymerization of styrene in conjunction with RuCl$_2$(PPh$_3$)$_3$ in the presence of Al(Oi-Pr)$_3$ in toluene or bulk (*42*). The corresponding bromide [CH$_3$CH(Ph)Br] was also effective but gave broad MWDs. The chloride [CH$_3$CH(Ph)Cl] resulted in much higher molecular weight than the calculated values and broad MWDs. The end-group analysis by ^{1}H NMR shows that the C–I growing terminal remained intact without β-elimination

178

reaction throughout the polymerization even at 100 °C under high monomer concentration. Not only CH$_3$CH(Ph)I but F(CF$_2$)$_6$I, CHI$_3$, and (CH$_3$)$_2$C(CO$_2$Et)I serve as effective initiators. Also, in the presence of Ti(Oi-Pr)$_4$, living polymerization proceeded faster to yield polystyrene with narrow MWDs ($M_w/M_n \sim 1.2$).

Experimental

Materials

Styrene (Wako Chemicals; purity > 99%) was dried overnight over calcium chloride and distilled twice over calcium hydride under reduced pressure before use. RuCl$_2$(PPh$_3$)$_3$ (Merck; purity > 99%), Al(Oi-Pr)$_3$ (Aldrich; purity > 99.99%), and Ti(Oi-Pr)$_4$ (Kanto Chemicals; purity > 97%) were used as received and handled in a glove box (M. Braun) under dry (< 1.0 ppm) and oxygen-free (< 1.0 ppm) argon. Toluene (solvent) and tetralin (internal standards for gas chromatographic analysis of styrene and MMA) were dried overnight over calcium chloride, distilled twice over calcium hydride, and bubbled with dry nitrogen for more than 15 min immediately before use.

Initiators

CH$_3$CH(Ph)Br and CH$_3$CH(Ph)Cl (both Tokyo Kasei; purity > 99%) were dried overnight over calcium chloride and distilled twice over calcium hydride under reduced pressure before use. CH$_3$CH(Ph)I was prepared by adding solution of 1.04 M HI solution (in n-hexane, 0.5 mL) into styrene (250 mM in toluene, 1.58 mL) at –20 °C. CHI$_3$ (Wako; purity > 97%) and F(CF$_2$)$_6$I iodide (Tokyo Kasei; purity > 98%) were used as received. Ethyl 2-iodoisobutyrate (CH$_3$)$_2$C(CO$_2$C$_2$H$_5$)I was prepared by the method of Curran et al. (29); bp 50 °C/9 torr; identified by 500 MHz ^{1}H NMR. Anal. Calcd for C$_6$H$_{11}$O$_2$I: C, 29.8, H, 4.58, I, 52.4. Found: C, 29.7, H 4.59, I, 52.3.

Polymerization Procedures

Polymerization was carried out by the syringe technique under dry nitrogen in sealed glass tubes. A typical example for polymerization of styrene with CH$_3$CH(Ph)I/RuCl$_2$(PPh$_3$)$_3$/Al(Oi-Pr)$_3$ is given below. Al(Oi-Pr)$_3$ (0.1916 g) was dissolved with styrene (6.46 mL) and tetralin (1.35 mL). Then this aluminum solution (6.81 mL) and a toluene solution of CH$_3$CH(Ph)I (0.960 mL) were added into RuCl$_2$(PPh$_3$)$_3$ (0.2301 g), sequentially in this order. Immediately after mixing, the solution was placed in an oil bath at 100 °C. The polymerization was terminated by cooling the reaction mixtures to –78 °C. Monomer conversion was determined from the concentration of residual monomer measured by gas chromatography with tetralin as internal standards for styrene. The quenched reaction solutions were diluted with toluene (~20 mL) and rigorously shaken with an absorbent [Kyowaad-2000G-7 (Mg$_{0.7}$Al$_{0.3}$O$_{1.15}$); Kyowa Chemical] (~5 g) to remove the metal-containing residues.

After the absorbent was separated by filtration (Whatman 113V), the filtrate was washed with water and evaporated to dryness to give the products, which were subsequently dried overnight.

Measurements

The MWD, M_n, and M_w/M_n ratios of the polymers were measured by size-exclusion chromatography (SEC) in chloroform at room temperature on three polystyrene gel columns (Shodex K-805L × 3) that were connected to a Jasco PU-980 precision pump, a Jasco 930-RI refractive index and 970-UV ultraviolet detectors. The columns were calibrated against 11 standard polystyrene samples (Polymer Laboratories; M_n = 580–1547000; M_w/M_n < 1.1) as well as the monomer. ^{1}H NMR spectra were recorded in CDCl$_3$ at 40 °C on a JEOL JNM-LA500 spectrometer, operating at 500.16 MHz. Polymers for ^{1}H NMR analysis were fractionated by preparative SEC (column: Shodex K-2002).

Acknowledgments

With appreciation M.S. and M.K. acknowledge the support from the New Energy and Industrial Technology Development Organization (NEDO) under the Ministry of International Trade and Industry (MITI), Japan, through the grant for "Precision Catalytic Polymerization" in the Project "Technology for Novel High-Functional Materials" (fiscal 1996–2000). Y.K. is grateful to the Japan Society for the Promotion of Sciences (JSPS) for the JSPS Research Fellowships for Young Scientists and also to the Ministry of Education, Science, Culture, and Sports, Japan for the partial support of this work by the Grant-in-Aid for Scientific Research (No. 3370).

References and Notes

1. For recent reviews on living radical polymerizations, see: (a) Georges, M. K.; Veregin, R. P. N.; Kazmaier, P. M.; Hamer, G. K. *Trends Polym. Sci.* **1994**, *2*, 66. (b) Davis, T. P.; Kukulj, D.; Haddleton, D. M.; Maloney, D. R. *Trends Polym. Sci.* **1995**, *3*, 365. (c) Malmstöm, E. E.; Hawker, C. J. *Macromol. Chem. Phys.* **1998**, *199*, 823. (d) Sawamoto, M.; Kamigaito, M. *Trends Polym. Sci.* **1996**, *4*, 371. (e) Colombani, D. *Prog. Polym. Sci.* **1997**, *22*, 1649. (f) *Controlled Radical Polymerization;* Matyjaszewski, K. Ed.; ACS Symposium Series 685; American Chemical Society: Washington DC, 1998. (g) Sawamoto, M.; Kamigaito, M. In *Synthesis of Polymers (Materials Science and Technology Series)*, Schlüter, A.-D., Ed.; Wiley-VCH, Weinheim, Germany, Chapter 6, 1999. (h) Sawamoto, M.; Kamigaito, M. *CHEMTECH* **1999**, *29* (6), 30. (i) Patten, T. E.; Matyjaszewski, K. *Adv. Mat.* **1998**, *10*, 901. (j) Matyjaszewski, K. *Chem. Eur. J.* **1999**, *5*, 3095. (k) Patten, T. E.; Matyjaszewski, K. *Acc. Chem. Res.* **1999**, *32*, 895.

2. Kato, M.; Kamigaito, M.; Sawamoto, M.; Higashimura, T. *Macromolecules* **1995**, *28*, 1721.
3. Ando, T.; Kato, M.; Kamigaito, M.; Sawamoto, M. *Macromolecules* **1996**, *29*, 1070.
4. Ando, T.; Kamigaito, M.; Sawamoto, M. *Tetrahedron* **1997**, *53*, 15445.
5. Nishikawa, T.; Ando, T.; Kamigaito, M.; Sawamoto, M. *Macromolecules* **1997**, *30*, 2244.
6. Simal, F.; Demonceau, A.; Noels, A. F. *Angew. Chem. Int. Ed.* **1999**, *38*, 538.
7. Wang, J.-S.; Matyjaszewski, K. *J. Am. Chem. Soc.* **1995**, *117*, 5614.
8. Wang, J.-S.; Matyjaszewski, K. *Macromolecules* **1995**, *28*, 7901.
9. Patten, T. E.; Xia, J.; Abernathy, T.; Matyjaszewski, K. *Science* **1996**, *272*, 866.
10. Percec, V.; Barboiu, B. *Macromolecules* **1995**, *28*, 7970.
11. Percec, V.; Barboiu, B.; Kim, H.-J. *J. Am. Chem. Soc.* **1998**, *120*, 305.
12. Haddleton, D. M.; Jasieczek, C. B.; Hannon, M. J.; Shooter, A. J. *Macromolecules* **1997**, *30*, 2190.
13. Ando, T.; Kamigaito, M.; Sawamoto, M. *Macromolecules* **1997**, *30*, 4507.
14. Matyjaszewski, K.; Wei, M.; Xia, J.; McDermott, N. E. *Macromolecules* **1997**, *30*, 8161.
15. Granel, C.; Dubois, Ph.; Jérôme, R.; Teyssié, Ph. *Macromolecules* **1996**, *29*, 8576.
16. Uegaki, H.; Kotani, Y.; Kamigaito, M.; Sawamoto, M. *Macromolecules* **1997**, *30*, 2249.
17. Uegaki, H.; Kotani, Y.; Kamigaito, M.; Sawamoto, M. *Macromolecules* **1998**, *31*, 6756.
18. Uegaki, H.; Kamigaito, M.; Sawamoto, M. *J. Polym. Sci. Part A Polym. Chem.* **1999**, *37*, 3003.
19. Ida, H.; Kamigaito, M.; Sawamoto, M. *Polym. Prep. Jpn.* **1998**, *47* (2), 149.
20. Percec, V.; Barboiu, B.; Neumann, A.; Ronda, J. C.; Zhao, M. *Macromolecules* **1996**, *29*, 3665.
21. Moineau, G.; Granel, C.; Dubois, Ph.; Jérôme, R.; Teyssié, Ph. *Macromolecules* **1998**, *31*, 542.
22. Lecomte, Ph.; Draiper, I.; Dubois, Ph.; Teyssié, Ph.; Jérôme, R. *Macromolecules* **1997**, *30*, 7631.
23. Kotani, Y.; Kamigaito, M.; Sawamoto, M. *Macromolecules* **1999**, *32*, 2420.
24. Kotani, Y.; Kamigaito, M.; Sawamoto, M. *Polym. Prepr. Jpn.* **1995**, *44* (7), 1274.
25. Kotani, Y.; Kamigaito, M.; Sawamoto, M. *Polym. Prepr. Jpn.* **1996**, *45* (2), 133.
26. Senoo, M.; Kotani, Y.; Kamigaito, M.; Sawamoto, M. *Macromolecules* **1999**, *32*, 8005.
27. Curran, D. P. In *Comprehensive Organic Synthesis*; Trost, B. M., Fleming, I., Eds.; Pergamon: Oxford, U.K., 1991; Vol. 4, p 715.
28. For example, see: *CRC Handbook of Chemistry and Physics*; Weast, R. C., Ed.; 68th edition; CRC Press: 1987; pp F169–F170.
29. Curran, D. P.; Bosch, E.; Kaplan, J.; Newcomb, M. *J. Org. Chem.* **1989**, *54*, 1826.
30. Su, M.-D.; Chu, S.-Y. *J. Am. Chem. Soc.* **1999**, *121*, 1045.
31. Oka, M.; Tatemoto, M. *Contemporary Topics in Polymer Science*, Plenum Press, Vol. 4, 1984, p 763.

32. Kato, M.; Kamigaito, M.; Sawamoto, M.; Higashimura, T. *Polym. Prepr. Jpn.* **1994**, *43* (2), 255.
33. Matyjaszewski, K.; Gaynor, S. G.; Wang, J. -S. *Macromolecules* **1995**, *28*, 2093.
34. Gaynor, S. G.; Wang, J. -S.; Matyjaszewski, K. *Macromolecules* **1995**, *28*, 8051.
35. Lansalot, M.; Farcet, C.; Charleux, B.; Vairon, J.-P.; Pirri, R. *Macromolecules* **1999**, *32*, 7354.
36. (a) Ueda, N.; Kamigaito, M.; Sawamoto, M. *Polym. Prepr. Jpn.* **1996**, *45*, 1267; **1997**, *46*, 149; **1998**, *47*, 134.   (b) Ueda, N.; Kamigaito, M.; Sawamoto, M. MACRO 98 *Preprints*, p 237.
37. Metzger, J. O.; Mahler, R. *Angew. Chem. Int. Ed. Engl.* **1995**, *34*, 902.
38. Devis, K.; O'Malley, J.; Paik, H.-J.; Matyjaszewski, K. *Polym. Prepr. (Am. Chem. Soc., Div. Polym. Chem.)* **1997**, *38* (1), 687.
39. Matyjaszewski, K.; Shipp, D. A.; Wang, J.-L.; Grimaud, T.; Patten, T. E. *Macromolecules* **1998**, *31*, 6836.
40. Haddleton, D. M.; Heming, A. M.; Kukulj, D.; Jackson, S. G. *J. Chem. Soc. Chem. Commun.* **1998**, 1719.
41. Hamasaki, S.; Kamigaito, M.; Sawamoto, M. *Polym. Prepr. Jpn.* **1998**, *47* (8), 1582.
42. The contribution of the iodine-transfer polymerization (*31*) may be clarified by the method of Fukuda (*43*). However, we do not estimate it significant, mainly due to the following reasons. In this Ru(II)-mediated system, the obtained M_n increased linearly with conversion from the initial stage, whereas the conversion-M_n profile usually shows a downward curvature in the iodine-transfer systems (*32–34*). Second, the MWDs (Figure 9) were narrower than those in the iodine-transfer polymerization. Third, the occurrence of halogen-exchange reactions (Figure 5) apparently shows the radical formation process assisted by Ru(II).
43. Ohno, K.; Goto, A.; Fukuda, T.; Xia, J.; Matyjaszewski, K. *Macromolecules* **1998**, *31*, 2699.

Chapter 13

Copper-Mediated Living Radical Polymerization Utilizing Biological and End Group Modified Poly(ethylene-*co*-butylene) Macroinitiators

David M. Haddleton, Adam P. Jarvis, Carl Waterson,
Stefan A. F. Bon, and Alex M. Heming

Department of Chemistry, University of Warwick,
Coventry, CV4 7AL, United Kingdom (D.M.Haddleton@warwick.ac.uk)

Summary : Copper mediated living radical polymerization can be used with a wide range of functional initiators to produce functional polymers. Block copolymers may be efficiently prepared using macroinitiators. This is demonstrated in this paper by synthesis and characterisation of a cholestrol based initiator and macroinitiators based on mono and difunctional polymers of ethylene and butylene. Polymerization of styrene and methacrylates using Schiff base ligands in conjunction with Cu(I)Br proceeds in a controlled manner yielding homopolymers, A-B diblock and A-B-A triblock (co)polymers of defined molecular weight and low polydispersity. Polymers based on methacrylic acid, 2-dimethylaminoethyl methacrylate and a random copolymer of methacrylic acid and methyl methacrylate have been synthesized by use of the cholestrol initiator to give resulting water soluble/dispersible polymers.

Introduction

Synthetic polymers containing biologically active moieties and amphiphilic segments are understandably receiving significant attention. Such polymers play important roles in many biological processes and their potential in applications for medicinal uses is now being realised (1-3). Living radical polymerization, and more specifically atom transfer polymerization, allows control of the synthesis of polymers whilst being inert to many types of useful functional groups. For example, Fukuda and co-workers (4) have synthesized glycopolymers by atom transfer polymerization, whilst Marsh *et al* have produced polymers incorporating a uridine monomer, 5'-methacryloyluridine (5).

Copper(I) mediated polymerization is proving a versatile technique for the synthesis of many different polymers (6-9) We have been ulitizing a range of Schiff base ligands used in conjunction with Cu(I)Br and an appropriate initiator has been established giving a versatile and extremely effective living polymerization system for acrylics and other vinyl monomers (10-13). Herein, we describe the synthesis and characterisation of a cholesterol derived initiator, for atom transfer polymerization which illustrates the versatility of this approach, giving a wide range of functionalised polymers. Block copolymer formation using macroinitiators synthesized by the esterification of Kraton L-1203™ and Kraton L-2203™, commercially available mono and dihydroxyl terminated copolymers of polyethylene and butylene, is also reported. These examples have been chosen so as to illustrate the potential diverse range of polymers which can be produced via the simple approach of transforming alcohols via esterification with 2-bromo-*iso*-butyrylbromide into initiators for living radical polymerization.

Experimental

General Information.

For general procedures and analysis techniques see previous publications (14). All reactions were carried out using standard Schlenk techniques under a nitrogen atmosphere. Methyl methacrylate and styrene were purified by passing down an activated basic alumina column so as to remove inhibitor, water and other protic impurities. Trimethylsilyl methacrylate (TMSMA) was purified by a trap-to-trap distillation over CaH_2. Kraton liquids L-1203 and L-2203 were obtained from Shell Chemicals, Belgium and used without further purification. All other reagents were used as received without further purification. Cu(I)Br (Aldrich, 98%) was purified according to the method of Keller and Wycoff.(15).

Synthesis of cholesteryl-2-bromoisobutyrate, 1

Cholesterol (2.0 g, 5.17 mmol) was dissolved in anhydrous pyridine (15 ml) with 4-dimethylaminopyridine (0.26 mmol, 0.031 g), a solution of 2-bromo-2-methyl propionyl bromide (7.75 mmol, 0.96 ml) was added dropwise with stirring, the reaction was left overnight at room temperature. The product was isolated by first removal of insolubles by filtration and a CH_2Cl_2 solution was washed with sodium bicarbonate and water. The organic layer was dried over magnesium sulfate, fliltered and the solvent removed to yield the product as a yellow solid which on washing with methanol resulted in a white solid. Yield = 2.20g (79.4%), pure by TLC. Calcd. for $C_{31}H_{51}BrO_2$: C, 69.51; H, 9.60. Found: C, 69.48; H 9.54.

Synthesis of Kraton L-1203 Macroinitiator, 2

Kraton L-1203 168.15 g (0.04 mol) was dissolved in anhydrous tetrahydrofuran 600 mL. Triethylamine 8.4 mL (0.06 mol) was added to the mixture followed by the addition with stirring of 2-bromo-2-methyl propionyl bromide 7.4 mL (0.06 mol). The reaction was allowed to stir overnight at room temperature. The product was isolated by filtration followed by rotary evaporation. The resulting viscous product was dissolved in $CHCl_3$ 500 mL and the solution was sequentially washed with saturated $NaHCO_3$ solution and water. The $CHCl_3$ layer was dried with $MgSO_4$ filtered and the solvent was removed to leaving a clear colourless viscous liquid. Yield = 165.9 g

Synthesis of the difunctional macroinitiator based on Kraton L-2203, 3

Kraton L-2203 192.14g (0.06 mol) was dissolved in anhydrous tetrahydrofuran 600 mL. Triethylamine 25.2 mL (0.18mol) was added to the mixture followed by the addition with stirring of 2-bromo-2-methyl propionyl bromide 23.1 mL (0.18 mol). The reaction was allowed to stir overnight at room temperature and the workup followed the same procedure as for Kraton L-1203 to give a pale yellow viscous liquid. Yield = 170.8 g.

Polymerization of Methyl Methacrylate with 1 as initiator; [MMA]/[I]/[Cu]/[L] = 100/1/1/2 in 66% toluene solution

Initiator **1**, (0.268 g, 0.5 mmol) Cu(I)Br (0.072 g, 0.5 mmol) and a magnetic follower were placed in a Schlenk tube. Deoxygenated toluene (10.6 mL) and *N-n*-propyl-2-pyridylmethanime (0.180 g, 1.15 mmol) were added. The solution was deoxygenated via three freeze pump thaw cycles and the solution heated to 90 °C. Deoxygenated inhibitor free MMA (10.6 mL) was then added (t = 0). Samples were removed periodically for analysis, via syringe. Polymerization of other monomers with **1** were carried out under similar conditions.

Polymerization of Methyl Methacrylate with 2 and 3 as initiator;
[MMA]/[I]/[Cu(I)]/[Ligand] = 90/1/1/2

Macroinitiator, **2** (4.35 g , 1 mmol) , Cu(I)Br (0.1433 g , 1 mmol) and a magnetic follower were placed in an oven dried Schlenk tube. Deoxygenated xylene 20.4 mL , N-*n*-octyl-2-pyridylmethanimine (0.4380 g , 2 mmol) and deoxygenated MMA 9.6 mL (90 mmol) were added to the Schlenk tube. The resulting solution was deoxygenated via four freeze pump thaw cycles and the reaction was placed in a thermostatically controlled oil bath at 90°C. Samples were removed periodically for conversion and molecular weight analysis. Conversions were carried out by gravimetry and molecular weight analysis was carried out by GPC analysis or ^{1}H NMR analysis. Polymerizations with other monomers utilised a similar regime as described above.

Quaternisation of Poly(2-(dimethylamino)ethyl methacrylate), P(DMAEMA)

To a solution of P(DMAEMA) in THF was added a 3-fold excess of methyl iodide at room temperature. The quaternised polymer precipitated from solution after stirring for 24 hours. The solid was filtered and purified by soxhlet extraction with THF and dried under vacuum for 24 hours.

Results and Discussion

Synthesis of initiators 1, 2 and 3

The cholesterol derived initiator, cholesteryl-2-bromoisobutyrate (**1**, scheme 1) was synthesized via reaction of 2-bromo-*iso*-butyrylbromide with cholesterol. Addition of the 2-bromoisobutyryl group was confirmed by ATR-FTIR and NMR. The Kraton derived macroinitiators (**2 & 3**, scheme 1) were synthesized via the esterification of Kraton L-1203 and Kraton L-2203. Introduction of the initiator species onto the hydroxyl termini of the Kraton L-2203 was clearly seen by both ATR-FTIR ($\upsilon_{C=O}$ 1738 cm^{-1} and loss of υ_{O-H} 3342 cm^{-1}), Figure 1. Esterification of the hydroxyl groups upon Kraton L-2203 was quantitative by ^{1}H NMR, Figure 2. ^{1}H NMR demonstrated two types of hydroxyl end termini. The triplet at δ 3.7 (A), HO-__CH$_2$__ (CH$_2$)$_3$ is due to an ethylene end terminus and the doublet at δ 3.3 (B), HO-__CH$_2$__CH (CH$_2$CH$_3$) corresponds to a butylene end terminus. Upon functionalisation both resonaces moved downfield to δ 4.22 and 3.91 respectively. Interestingly the monofunctional derivative only contained ethylene end termini resulting in a triplet resonance. It was also noted that the methyl signal of the ester groups was split into

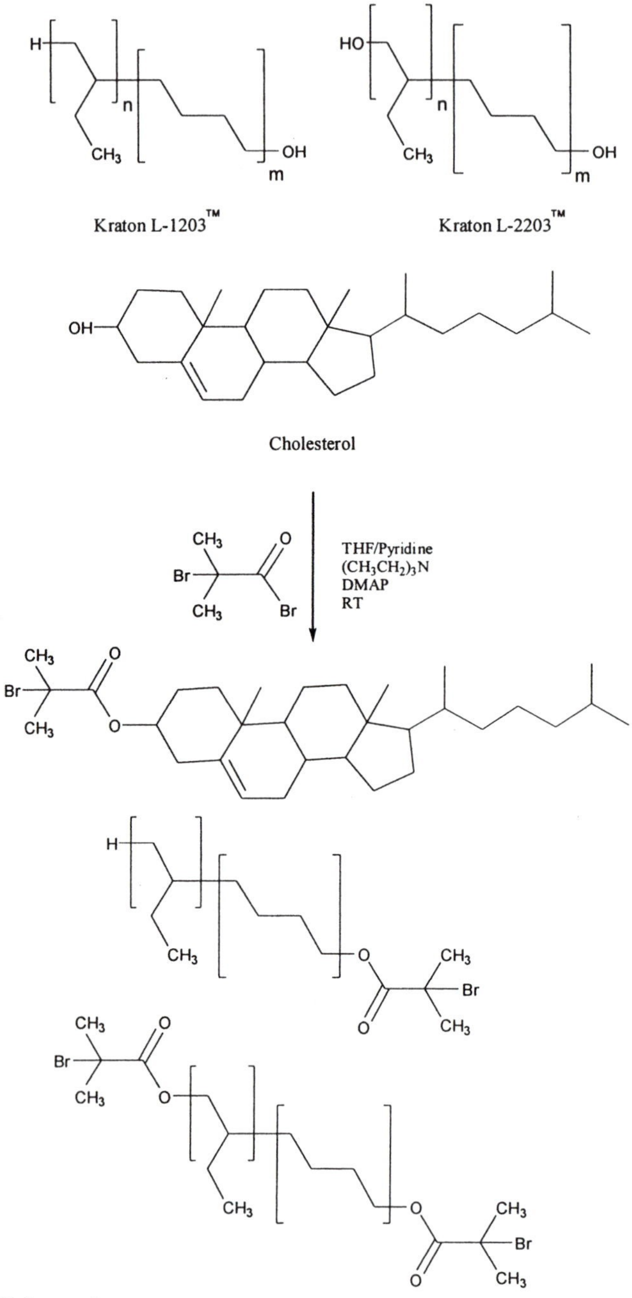

Kraton L-1203™   Kraton L-2203™

Cholesterol

THF/Pyridine
(CH₃CH₂)₃N
DMAP
RT

1

2

3

Scheme 1

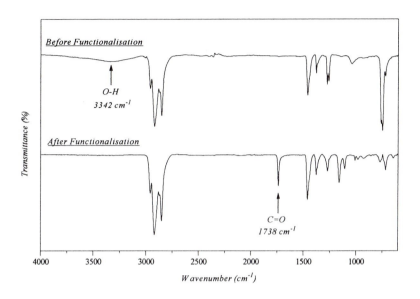

Figure 1. FT-IR Analysis of Kraton L-2203™ Before and After Functionalisation

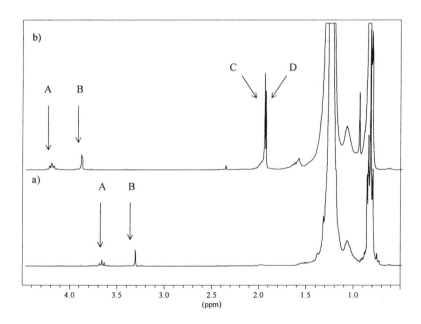

Figure 2. ^1H NMR Analysis of Kraton L-2203 Before (a) and After (b) Functionalization

two separate signals (C and D), again being attributed to two different end group environments.

Polymerization of methyl methacrylate initiated by 1

Polymerization of methyl methacrylate using a Schiff base ligand in conjunction with Cu(I)Br and **1** as an initiator proceeds in a controlled manner to give polymers with M_n close to that predicted with relatively narrow molecular weight distributions, Table 1. The incorporation of the cholesterol moiety within the polymer, including the retention of the vinyl group is confirmed by ^1H NMR, Figure 3.

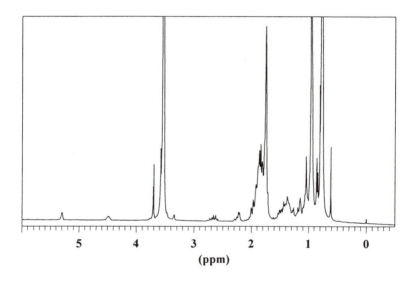

Figure 3. ^1H NMR spectrum of PMMA initiated by 1 showing the retention of the vinyl proton at δ 5.30.

The number average molar mass increases linearly with conversion following an apparent lack of control at the beginning of the reaction, Figure 4. This might be exlpained by insufficient Cu(II) at the beginning of the reaction or deviation away from the PMMA calibration in the SEC most evident at low mass. The ^{13}C NMR spectrum of the polymer shows the ω-terminal end group which, as expected after atom transfer polymerization, contains a tertiary bromide. Resonances at δ 58.93, 58.43 and 58.35 are characteristic of *meso* and *racemic* end group stereochemistry, corresponding to resonances obtained from a model compound, methyl-2-bromoisobutyrate, [(CH$_3$)$_2$C(CO$_2$Me)Br] δ 55.4 (16).

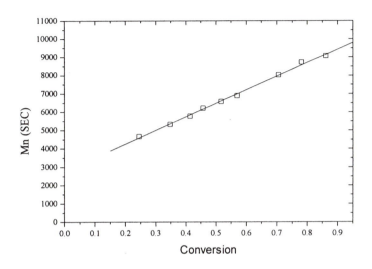

Figure 4. Evolution of molecular weight for the polymerization of MMA initiated by 1; [MMA]/[I]/[Cu]/[L] = 100/1/1/2

Table 1. Selected data for the atom transfer polymerization of vinyl monomers in toluene for initiator 1.

Monomer	Reaction time (hours)	Conversion	M_n^c (theory) (g/mol)	M_n (g/mol)	M_w/M_n
MMA[a]	6.85	93	9840	9510[d]	1.20
Styrene[b]	7.00	34	3940	4460[e]	1.23
DMAEMA[a]	2.00	100	3040	3050[f]	1.20
TMSMA[a]	2.73	84	2540	2910[f]	-
TMSMA/MMA[a]	4.83	98	2990	2450[f]	-

[a] *Polymerizations carried out at 90 °C.* [b] *Polymerizations carried out at 110 °C.* [c] M_n *(theory)* $= ([M_{monomer}]_0/[I]_0 \times MW_{monomer}) * conversion + MW_{initiator}$, *where* $MW_{monomer}$ *is the molecular weight of monomer, where* $MW_{initiator}$ *is the molecular weight of the cholesterol initiator and* $[M_{monomer}]_0/[I]_0$ *is the initial concentration ratio of monomer to initiator.* [d] M_n *obtained from GPC with molecular weights being calibrated using poly(methyl methacrylate) standards.* [e] M_n *obtained from GPC with molecular weights being calibrated using poly(styrene) standards.* [f] M_n *obtained from* ^1H *NMR spectroscopy.*

Polymerization of styrene initiated by 1

The Mn again increases linearly with conversion, and unlike the polymerization of methyl methacrylate, control is observed throughout the reaction, Figure 5. As expected, the rate of polymerization is slower than that observed for the polymerization of methyl methacrylate.

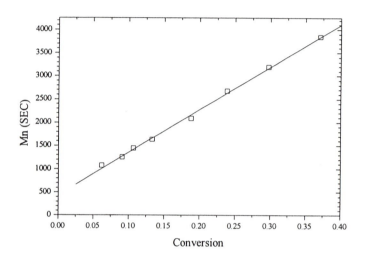

Figure 5. Evolution of M_n for the polymerization of styrene initiated by 1; [Styrene]/[I]/[Cu]/[L] = 100/1/1/2.

Synthesis of water soluble polymers initiated with 1

We are particuarly interested in the synthesis of amphiphilic block copolymers containing hydrophilic-hydrophobic moieties. These polymers form micelles in aqueous solution, with the hydrophobic blocks aggregating to form the core and the hydrophilic blocks in the solvated outer layer (17-19). Water soluble polymers were synthesized by two methods. The first was the polymerization of (2-(dimethylamino)ethyl methacrylate) (DMAEMA) using atom transfer polymerization. The P(DMAEMA) synthesized typically gave low polydispersity indexes (M_w/M_n = 1.20) with ^{1}H NMR spectroscopy giving excellent agreement for M_n with the theoretical values calculated, Table 1. An increase in the hydrophilicity of P(DMAEMA) is achieved by quaternisation with methyl iodide, Figure 6.

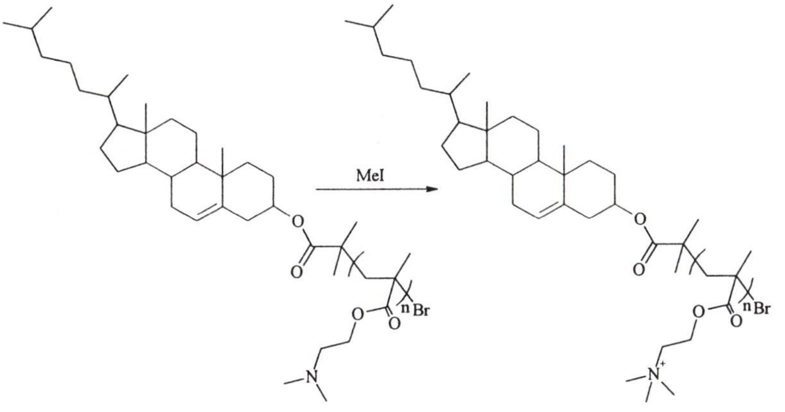

Figure 6. Quaternisation of P(DMAEMA) with methyl iodide.

^{1}H NMR spectroscopy confirmed quantitative quaternisation of the amine group in the polymer chain enhancing water solubility dramatically compared with the unquaternised polymers. The ^{1}H NMR spectrum of unquaternised P(DMAEMA) in CDCl$_3$ confirms the presence of both the cholesterol and DMAEMA units. On changing the solvent to D$_2$O, the cholesterol resonances are suppressed for both the quaternised and unquaternised polymers. This suggests the hydrophobic cholesterol moiety is poorly solvated; possibly indicating the existence of micelle formation.

The second method performed was polymerization of trimethylsilyl methacrylate (TMSMA) by atom transfer polymerization with **1**. Deprotection of the trimethylsilyl group to free acid was achieved via the hydrolysis of the polymer in methanol at 60 °C forming poly(methacrylic acid). Removal of the trimethylsilyl group was confirmed by ^{1}H NMR spectroscopy. The incorporation of methyl methacrylate into poly(methacrylic acid) to form a statistical copolymer was also studied to form polymers with varying degrees of hydrophilicity. The synthesis of a statistical copolymer containing 50 mol % TMSMA and 50 mol % MMA gave a composition and M_n (obtained from ^{1}H NMR spectroscopy) close to that expected (M_{ncalc} = 2520, M_{ntheo} = 2500). These poly(methacrylic acid) polymers form aggregates with hydrophilic-hydrophobic domains in basic aqueous solution.

Polymerization of MMA and BzMA initiated by 2

Polymerization of methyl methacrylate and benzyl methacrylate with **2** as macroinitiator proceeds in a controlled manner to give A-B block copolymers with M_n close to that predicted and narrow molecular weight distribution, Table 2. The kinetic plots for the polymerization of MMA and BzMA are shown in Figure 7. The rate is first order with respect to monomer concentration and is linear indicating a 'living'or 'pseudo-living' process. The M_n increases linearly with conversion and agree well with expected values. The initial points in the molecular weight plot are low in comparison to the theoretical M_n with this being due to Kraton L-1203 not being available with an hydroxyl functionality of exactly 1.0. As a consequence the samples taken from the reaction contain small amounts of the unfunctionalised Kraton. This observation has also been noted in the work of Batsberg *et al* for the A-B block copolymerization of styrene (9).

The incorporation of the PMMA block is demonstrated in Figure 8 where the GPC for the macroinitiator has shifted to higher mass upon block copolymerization with MMA. Kraton macroinitiators are viscous liquids and the Kraton-PMMA copolymer product was isolated as a white powder. The unfunctionalised Kraton was removed from the polymer by precipitation in 40 % ethanol /60 % hexane.

Table 2 Selected data for the ATP polymerization of vinyl monomers in xylene using the monofunctional macroinitiator Kraton L-1203 ™

Monomer	[M]:[2]	Time / mins	Conv / %	M_n (Theory)	M_n (Exp)	M_w/M_n
Kraton L-1203				4500	7170[b]	1.04
MMA	45 : 1	280	92.1	11310a	10650[b]	1.11
MMA[c]	45 : 1	280	92.1		11700[b]	1.11
MMA	90 : 1	280	93.5	15580a	14340[b]	1.15
MMA[c]	90 : 1	280	93.5		15520[b]	1.11
BzMA	50 : 1	150	96.6	15710a	13920[b]	1.15
BzMA[c]	50 : 1	150	96.6		14570[b]	1.16

[a] M_n (Theory) =([Monomer]$_0$/[Initiator]$_0$ x $M_{W\ monomer}$)x Conv + $M_{W\ kraton}$,$M_{W\ Kraton}$ is assumed to be 7170 as obtained against PMMA calibration. [b]M_n obtained from GPC against PMMA equivalent. [c]Following precipitation.

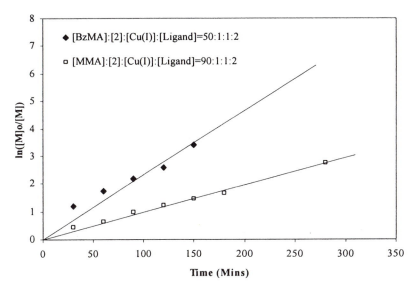

Figure 7. First order rate plots for the copper mediated atom transfer polymerization of methyl methacrylate (MMA) and benzyl methacrylate (BzMA) utilising [Monomer]:[2]:[Cu(I)]:[Ligand] = 90:1:1:2 and 50:1:1:2 respectively

Polymerization of MMA and DMAEMA initiated by 3

Polymerization of MMA and 2-di(methylethyl)amino methacrylate using **3** as macroinitiator also proceeded in a controlled manner to give A-B-A block copolymers with M_n close to that predicted and narrow molecular weight distribution, Table 3. The M_n increases linearly with conversion with values of M_n agreeing well with predicted values.

A value for k_p [pol *] of 2.17×10^{-4} s^{-1} for the 70 : 1 reaction agrees well with k_p [pol *] = 1.2×10^{-4} s^{-1} for the 140:1 reaction. The resulting A-B-A block copolymers have a relatively narrow molecular weight distribution with M_w/M_n typically less than 1.25. Block copolymerization with the water soluble monomer dimethylaminoethyl methacrylate (DMAEMA) was also investigated. The first order rate plots for the A-B-A block copolymer are linear indicating little termination occurs. The polymerization involving DMAEMA : **3** = 89:1 was carried out by halving the concentration of macroinitiator whilst keeping the concentration of monomer the same. The value of k_p [pol *] = 2.84×10^{-4} s^{-1} for the 44.5 : 1 reaction agrees well with k_p [pol *] = 1.46×10^{-4} s^{-1} for the 89:1 reaction which is approximately half. The polydispersity index of the resulting block copolymers is relatively narrow typically M_w/M_n =1.20

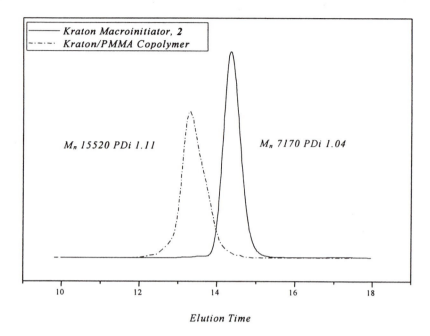

Figure 8. GPC overlay demonstrating A-B block copolymer formation

Table 3. Selected data for the ATP polymerization of vinyl monomers in xylene using the difunctional macroinitiator Kraton L-2203 ™

Monomer	[M]:[3]	Rxn Time (mins)	Conv (%)	M_n (Theory)	M_n (Exp)	M_w/M_n
Kraton L-2203				3500	7380[c]	1.11
MMA	140 : 1	240	78.5	18400[a]	17720[c]	1.20
After Ppt					18490[c]	1.16
MMA	70 : 1	240	95.5	14110[a]	14090[c]	1.23
After Ppt					14660[c]	1.16
DMAEMA	89 : 1	240	86.0	15530[b]	16410[d]	1.19
DMAEMA	44.5 : 1	240	98.2	10370[b]	11791[d]	1.20

[a] M_n (Theory) =([Monomer]$_0$/[Initiator]$_0$ x $M_{W\ monomer}$) x Conv + $M_{W\ kraton}$,$M_{W\ Kraton}$ is assumed to be 7380 as obtained against PMMA calibration. [b] $M_{W\ Kraton}$ is taken to be 3500. [c]M_n obtained from GPC against PMMA equivalent [d] M_n determined from ^{1}H NMR spectroscopy.

Conclusions

Esterification of alcohols containing a range of functionality e.g. biological molecules containing an OH functionality, produces initiators containing the 2-bromoisobutyrate fragment. This is a versatile and facile route for the synthesis of functionalised initiators for atom transfer polymerization. A range of functionalised polymers have been synthesized by initiation of cholesteryl-2-bromoisobutyrate **1**. Living polymerization of methyl methacrylate and styrene afford polymers of controlled molecular mass and narrow molecular mass distributions. Water soluble polymers containing the cholesterol moiety have been synthesized by the hydrolysis of poly(trimethylsilyl methacrylate) and quaternisation of poly(2-(dimethylamino)ethyl methacrylate).

Esterification of the hydroxyl end functionalities of Kraton L-1203 and Kraton L-2203 with 2-bromo-2-methylpropionyl bromide is a simple and effective route to generating mono and difunctional macroinitiators based on copolymers of poly(butylene-co-ethylene). The subsequent use of these macroinitiators in atom transfer polymerization of methacrylates mediated with a Schiff base ligand and Cu(I)Br has been shown to be an effective method for producing A-B and A-B-A block copolymers with controlled molecular weight and narrow polydispersity. Taken together the examples chosen show the potential of this chemistry to synthesise end functional polymers and block copolymers from macroinitiators.

Acknowledgement

We wish to thank BP (APJ), EPSRC (AMH, CW) for funding and Shell, Belgium for supplying the Kraton Liquid polymers

References

1. Spaltenstein, A.; Whiteside, G. M. *J. Am. Chem. Soc* **1991**, *113*, 686.
2. Wasserman, P. M. *Science* **1987**, *235*, 553.
3. Miyata, T.; Nakamae, K. *Trends in Polymer Science* **1997**, *5*, 198-206.
4. Ohno, K.; Tsujii, Y.; Fukuda, T. *J. Polym. Sci. Part a-Polym. Chem.* **1998**, *36*, 2473-2481.
5. Khan, A.; Haddleton, D. M.; Hannon, M. J.; Marsh, A. *In Preparation* **1999**.

6. (a) Shipp, D. A.; Wang, J.-L.; Matyjaszewski, K. *Macromolecules* **1998**, *31*. (b) Patten, T. E.; Matyjaszewski, K. *Accounts of Chemical Research* **1999**, *32*, 895. (c) Matyjaszewski, K. *Chemistry-a European Journal* **1999**, *5*, 3095.
7. Gaynor, S. G.; Matyjaszewski, K. *Macromolecules* **1997**, *30*, 4241.
8. Jankova, K.; Chen, X. Y.; Kops, J.; Batsberg, W. *Macromolecules* **1998**, *31*, 538-541.
9. Jankova, K.; Kops, J.; Xianyl, C.; Batsberg, W. *Macromol. Rapid Commun.* **1999**, *20*, 219.
10. Haddleton, D. M.; Jasieczek, C. B.; Hannon, M. J.; Shooter, A. J. *Macromolecules* **1997**, *30*, 2190-2193.
11. Haddleton, D. M.; Waterson, C.; Derrick, P. J.; Jasieczek, C.; Shooter, A. J. *Chem. Commun.* **1997**, 683.
12. Percec, V.; Kim, H. J.; Barboiu, B. *Macromolecules* **1997**, *30*, 8526-8528.
13. Haddleton, D. M.; Kukulj, D.; Duncalf, D. J.; Heming, A. H.; Shooter, A. J. *Macromolecules* **1998**, *31*, 5201.
14. Haddleton, D. M.; Crossman, M. C.; Hunt, K. H.; Topping, C.; Waterson, C.; Suddaby, K. G. *Macromolecules* **1997**, *30*, 3992.
15. Keller, R. N.; Wycoff, H. D. *Inorg. Synth.* **1947**, *2*, 1.
16. Haddleton, D. M.; Heming, A. H.; Kukulj, D.; Jackson, S. G. *J. Chem. Soc. Chem. Commun.* **1998**, 1719.
17. Butun, B. C.; Billingham, N. C.; Armes, S. P. *Chem. Commun.* **1997**, 671.
18. Butun, B. C.; Billingham, N. C.; Armes, S. P. *J. Am. Chem. Soc.* **1998**, *120*, 11818.
19. Baines, F. L.; Billingham, N. C.; Armes, S. P. *Macromolecules* **1996**, *29*, 3416.

Chapter 14

Atom Transfer Radical Copolymerization of Styrene and Butyl Acrylate

Grégory Chambard and Bert Klumperman[1]

Laboratory of Polymer Chemistry, Eindhoven University of Technology,
P.O. Box 513, 5600 MB Eindhoven, The Netherlands

Atom transfer radical polymerization of styrene and butyl acrylate has been investigated from a kinetic point of view. Attention is focused on the activation of the dormant species as well as on the termination that plays a role in these reactions. It has been shown that the activation of a styrene dormant species is much slower in methyl methacrylate compared to reported data obtained in styrene. Termination reactions seem to play a minor role in ATRP, although in the early stages of the reaction loss of functionality is observed. The reactivity ratios for the copolymerization of styrene and butyl acrylate are determined as well and are very similar to the ones for conventional free-radical polymerization. From simulations it is clear that the system is self-tuning in the sense that a difference in equilibrium constants between both dormant species does not lead to a change in radical ratios in the reaction mixture.

Introduction

Free-radical polymerization is one of the most widely used techniques to produce polymers. It is robust, since it is relatively inert towards impurities such as water, and can be applied with a wide variety of monomers. A great disadvantage, however, is the inability to control polymer properties, such as tacticity, and chain topology (*e.g.* in the case of block copolymers).

Since a decade, techniques have been developed that combine the robustness and flexibility of free-radical polymerization with the ability to keep control over polymer properties *(1,2,3,4)*. Atom transfer radical polymerization (ATRP) is one of the most promising *(3,4)*. Styrene and derivatives, as well as (meth)acrylates can be (co)polymerized *(5,6,7)*.

There are, however, still many features that remain underexposed up to now and need to be looked at in more detail. For instance, the amount of dead chains that are inevitably formed during the course of reaction is an important parameter when preparing *e.g.* block copolymers.

[1]Corresponding author.

Furthermore, copolymerization is only rarely reported in literature
(8,9,10,11,12) and needs to be closely examined.
This paper deals with the above-mentioned 'gaps' in literature. In
homopolymerizations of both styrene (S) and butyl acrylate (BA),
attention has been focused on kinetics and termination events.
Furthermore, the copolymerization of styrene and butyl acrylate has
been investigated from a kinetic point of view, taking knowledge on
both homopolymerizations into account.

Theory

The kinetics of ATRP rely on the reversible activation and deactivation
of a dormant species, an alkyl halide. The halide atom is transferred back
and forth from the alkyl halide to a transition metal complex, usually a
copper complex (scheme 1).

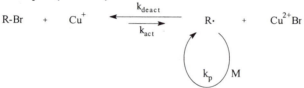

Scheme 1. Activation/deactivation process in ATRP.

When a dormant species is activated, it is able to propagate via normal
radical kinetics. To describe the monomer concentration, the following
expression can be used for most of the reaction time (13,14):

$$-\ln(1-X) = \frac{3}{2}k_p \cdot \left(\frac{K_a [R-Br]_0 [Cu^+]_0}{3k_t} \right)^{\frac{1}{3}} \cdot t^{\frac{2}{3}} = A \cdot t^{\frac{2}{3}} \quad (1)$$

In this equation, X stands for overall conversion and K_a for the
equilibrium constant for the reversible activation/deactivation process as
shown in scheme 1. For a given monomer, the evolution of ln(1-X) is
dependent on the concentrations of dormant species and catalyst, and on
the 'system constants' k_p, k_t and K_a.
It is very interesting to obtain information about the activation
parameter, k_{act}, in the kinetic scheme. Fukuda et al. introduced a very
simple and elegant method to determine this parameter (15). The method
is based on the monitoring of the concentration of a macroinitiator
species as a function of time with size exclusion chromatography (SEC).
A minor amount of radical initiator is added to ensure that the
macroinitiator, after it is activated, reacts irreversibly with the radicals
derived from this radical initiator. This causes a separation of the
original macroinitiator peak and the peak of the product of the trapped
macroinitiator radical in the SEC chromatograms.

$$\ln\left(\frac{S_0}{S_t} \right) = k_{act} \cdot [Cu^+] \cdot t \quad (2)$$

In this equation, S_0 stands for the area of under the macroinitiator peak in the SEC chromatogram at t=0, while S_t is this area at later times. When plotting the natural logarithm of the ratio of the two areas vs. time, this leads to a value of the activation rate parameter, k_{act}.

Experimental

General:
Styrene (S, Aldrich) and butyl acrylate (BA, Aldrich) were distilled prior to use. Xylene (Aldrich), CuBr (98%, Aldrich), tosylchloride (TsCl, 99%, Aldrich) were used as received. Di-4,4'-n-heptyl-2,2'-bipyridine (dHbpy) was synthesized according to a known synthesis route *(16)*.

Typical polymerization procedure:
Monomer is put into a 100 mL round-bottomed flask, together with xylene and dHbpy and purged for 1 hour with argon. The typical total reaction volume was always kept at around 20 mL. After this, an appropriate amount of CuBr was added, so that Cu:dHbpy was 1:2, and the solution was degassed for another half hour. Then, the mixture was heated up to reaction temperature (typically 110°C), after which the reaction was started by addition of previously degassed ethyl 2-bromoisobutyrate (EiB, for the homopolymerizations) or a solution of tosyl chloride (TsCl) in xylene (in the case of copolymerizations).
Samples were taken at different time intervals and monomer concentrations were determined by gas chromatography using a HP 5890 equipped with AT Wax column (Alltech, length 30m, film thickness 1.0m) with autosampler. In the case of copolymerizations, partial monomer conversions were calculated as well as the fraction of S (f_S) as a function of total conversion (X). Since these data are all calculated from the monomer concentrations measured, they all contain an error and therefore the X, f_S-data were analyzed using the nonlinear least squares method that takes errors in all variables into account *(17)*.
Molecular weights were determined by SEC at 40°C using tetrahydrofuran as solvent and polystyrene standards from Polymer Laboratories. Poly(butyl acrylate) molecular weights were corrected via the universal calibration principle using Mark-Houwink parameters *(18)*, while the molecular weights of the copolymers were calculated relative to polystyrene standards.

Results and Discussion

Homopolymerizations of styrene and butyl acrylate:
In figure 1 and 2, the kinetic plots are depicted for the homopolymerization of butyl acrylate and styrene, respectively.

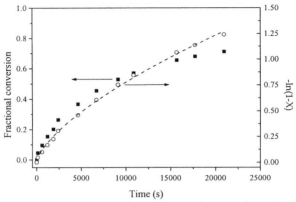

Figure 1. Kinetic plot for the polymerization of BA in xylene. [BA]=3.78 M, [Cu]=0.0168 M, [EiB]=0.0222M at 110°C. Dotted line is the best fit through the experimental data according to equation 1.

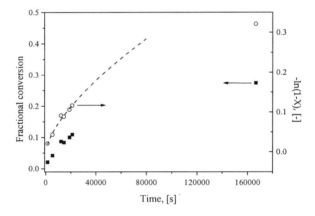

Figure 2. Kinetic plot for the polymerization of S in xylene. [S]=4.19 M, [Cu]=0.0153 M, [EiB]=0.0434 M at 110°C. Dotted line is the best fit through the experimental data according to equation 1.

When the $-\ln(1-X)$ data is fitted against $t^{2/3}$, we get a very good estimation of the set of constants (A) in equation 1. For BA, a value of $1.67 \cdot 10^{-3}$ $s^{-2/3}$ is obtained, while for S this value is $1.53 \cdot 10^{-4}$ $s^{-2/3}$. When we take the difference in k_p into account (about 1600 $L \cdot mol^{-1} \cdot s^{-1}$ for S and 76000 $L \cdot mol^{-1} \cdot s^{-1}$) together with the concentrations of copper complex and initiator and we assume that k_t is $5 \cdot 10^8$ $L \cdot mol^{-1} \cdot s^{-1}$ in both cases, we obtain a value of the equilibrium constant K_a of $5.85 \cdot 10^{-10}$ for S and $1.26 \cdot 10^{-11}$ for BA. For S, this value is an order of magnitude smaller than reported in literature (16). The difference can be found in the fact that we used another initiator and that the literature value was obtained for a bulk system. The first factor in particular is known to have a significant influence on the course of polymerization *(19,20)*. Furthermore, in their analysis Matyjaszewski *et al.* ignored the persistent radical effect *(14)* and calculated the activation/deactivation equilibrium constant assuming a constant deactivator concentration. The value for BA is almost 50 times lower than that of S, which is very large compared to the results of Arehart *(12)*. It must be noted, however, that their value for the equilibrium constant of BA was obtained at 100°C.

One of the drawbacks of the ATRP system is its oxygen sensitivity. Unlike conventional radical polymerization, where a small amount of oxygen in the reaction mixture is eliminated due to the constant production of radicals, a small amount of oxygen affects the whole course of reaction *(21)*. Oxygen can react with the Cu^+ complex in a simple redox reaction to yield a Cu^{2+} complex that is no longer able to activate the dormant species. Moreover, oxygen can also trap the radicals originating from the dormant species, although the concentration of the latter is much lower than the Cu^+ concentration. In both cases, the $[Cu^+]$ will decrease and, as a result, the reaction rate will drop. This is illustrated by figure 3, where a small amount of oxygen is introduced by a syringe during the course of reaction.

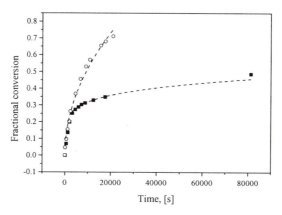

Figure 3. Kinetic plots for two polymerizations of BA in xylene. ■ *With oxygen introduced during the reaction ([BA]=3.78 M, [Cu]=0.0187 M, [EiB]=0.0186 M),* ○ *without oxygen ([BA]=3.78 M, [Cu]=0.0168 M, [EiB]=0.0222 M) at 110°C.*

It can clearly be seen that in the beginning both reactions proceed at the same rate. From the fifth data point on, however, the rate of the reaction where oxygen is introduced is drastically decreased. Even a minor amount of oxygen in the system will hamper any kinetic investigation on ATRP polymerizations. When the data from the fifth point on is fitted with equation 1, taking only into account the decreased $[Cu^+]$, this leads to an estimation of the set of constants equaling $1.94 \cdot 10^{-4}$ s$^{-2/3}$, while the reaction without oxygen yielded a constant of $1.69 \cdot 10^{-3}$ s$^{-2/3}$. This means that the concentration of Cu^+ has decreased with a factor of about 650. This illustrates that a small amount of oxygen, *e.g.* a contaminated syringe, will slow down the reaction rate to a very large extent. As well as in conventional free-radical polymerization, radicals are involved in ATRP and therefore also termination takes place. The amount of termination is dependent on the equilibrium constant in scheme 1. When K_a is large, this means that more radicals are present in the steady state situation and as a consequence, more termination will take place.

We have monitored the amount of termination that takes place in an ATRP of S by looking at the spectra in ^{1}H NMR. Attention is focused on the signals of the protons residing in the initiator fragment and the proton at the end of the dormant species chain, see figure 4.

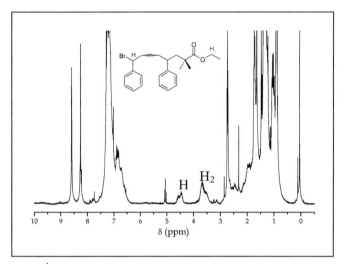

Figure 4. ^{1}H NMR spectrum of a PS dormant species. The peak at δ=4.5 ppm originates from proton at the end of the polymer chain, while the peak at δ=3.6 ppm comes from the two protons in the initiator fragment.

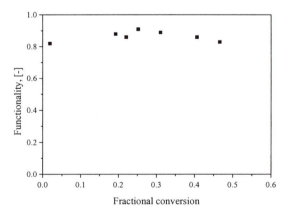

Figure 5. Fraction of chains functionalised with Br as a function of conversion for a polymerization of S at 110°C in xylene. [S]=4.33 M, [Cu]=0.0314 M, [EiB]=0.161 M.

The ratio of the initiator and the end-group peaks in the ^{1}H NMR spectrum are a measure for the fraction of chains that can still be activated since they still contain a bromine atom at the end. In figure 5, the fraction of chains that are still functionalised with Br is plotted for samples taken from a S ATRP at different conversions. In the beginning of the reaction, we see that the fraction of chains with a bromine end group has already decreased from 1 to about 0.85. This is logical, since in the beginning of the reaction there is not enough Cu^{2+} present yet and the bimolecular termination of the radicals is still competing with the deactivation reaction. When the persistent radical concentration has been built up, the fraction of terminated chains stays at a relatively constant level. However, the fraction of living chains is lower than reported by Matyjaszewski *et al. (22)*, who reported a steady-state concentration of Cu^{2+} of 4-6% relative to the initial Cu^+ concentration. The origin of this phenomenon is currently investigated in our laboratories. It has to be noted, however, that care has to be taken when interpreting the NMR data, since the error in the peak areas can be up to 5%.

Both polystyrene (PS, M_n=1585 g·mol^{-1}) and poly(butyl acrylate) (PBA, M_n=2200 g·mol^{-1}) were prepared and the PS was used to determine the activation rate parameter, k_{act}, with the use of the method described earlier employing equation 2. In figure 6, the SEC chromatograms are shown for the PS dormant species. PS was therefore dissolved in MMA, together with a known amount of copper complex and a small amount of cumyl hydroperoxide (CHP) to enhance the SEC resolution. Note that although the reaction was carried out at 110°C, the reaction mixture did not boil due to the high polymer concentration.

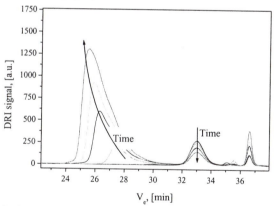

Figure 6. SEC chromatograms of PS in order to determine k_{act}. [PS]=0.0089 M, [Cu]=0.0053 M, [CHP]=0.0088 M in MMA at 110°C.

From the SEC chromatograms and the concentration of the copper complex, the activation rate parameter for S could be determined. Note that the chromatograms in figure 6 are obtained with a DRI-detector and that they have been corrected for differences in dn/dc, normalized and subsequently scaled with conversion. The decrease of the macroinitiator peak at V_e=33 min was monitored as a function of time. It is clear that the radicals originating from this macroinitiator are trapped by the radicals originating from CHP initiation and yield polymer that is well separated from the macroinitiator in SEC. For PS, k_{act} has a value of 0.03 ($\pm$ 0.003) L•mol^{-1}•s^{-1}. This is much lower than the 0.45 L•mol^{-1}•s^{-1} that has been found by Fukuda et al [15]., who determined this parameter in styrene. There is evidence that solvent can have a significant influence on the activation process, which could explain the discrepancy between our results and those reported in literature. We are currently verifying this by using a different and independent method [23].

The PS and PBA have also been used to make block copolymers. For chain extension of PS with BA the molecular weight as a function of conversion is plotted in figure 7, and a comparison of the calculated copolymer composition with the experimentally obtained is plotted in figure 8.

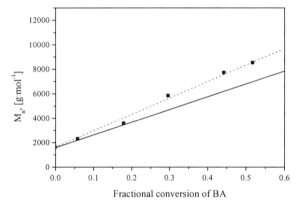

Figure 7. M_n as a function of conversion for a chain extension of PS with BA in xylene at 110°C. Solid line: theoretical prediction; dotted line: fit through data. [BA]=3.44 M, [Cu]=0.032 M, [PS]=0.042 M.

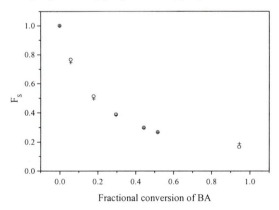

Figure 8. Fraction of S in the block copolymers as a function of conversion. O Expected from conversion data, + determined by NMR spectroscopy. Conditions as in figure 7.

From figure 7, it is clear that the molecular weight grows linearly with conversion. It is important to mention that these molecular weights are relative to polystyrene standards and that differences in detector responses have not been taken into account. This explains why the molecular weights of the block copolymers are systematically higher than theoretically expected. Figure 8 shows that the experimentally determined fraction of styrene in the block copolymer is in good agreement with the expected fraction of styrene in the block copolymer, calculated from conversion data.

Copolymerization of styrene and butyl acrylate:
The copolymerization kinetics for an ATRP system are completely analogous to the free-radical copolymerization kinetics *(24)*. For the determination of the reactivity ratios the differential copolymer composition equation can be derived:

$$F_1 = \frac{r_1 f_1^2 + f_1 f_2}{r_1 f_1^2 + 2 f_1 f_2 + r_2 f_2^2} \qquad (3)$$

By fitting of the copolymer composition at low conversion data as a function of initial monomer feed composition, the reactivity ratios can be derived from this equation. However, when performing ATRP experiments, we are not able to determine copolymer composition at low conversion, since the polymer chains are growing throughout the reaction. It is therefore necessary to make use of the integrated form of the copolymer equation *(25)*:

$$X = 1 - \left(\frac{f}{f_0}\right)^{\frac{r_2}{1-r_2}} \left(\frac{1-f}{1-f_0}\right)^{\frac{r_1}{1-r_1}} \left(\frac{f_0 - \delta}{f - \delta}\right)^{\frac{1-r_2}{2-r_1-r_2}} \qquad (4)$$

Where δ is a function of the reactivity ratios as well:

$$\delta = \frac{1 - r_1 r_2}{(1 - r_1)(1 - r_2)} \qquad (5)$$

The integrated copolymerization equation can also be used to estimate reactivity ratios *(26,27)* and has the advantage over the differential copolymerization equation that it can be applied up to high conversion as well. With gas chromatography or on-line techniques as Raman spectroscopy, the monomer feed ratio is determined at various conversions. Fitting equation 4 to these data with nonlinear least squares parameter estimation yields estimations of the reactivity ratios, r_1 and r_2.

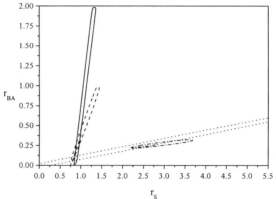

Figure 9. 95% Joint confidence intervals for four ATRP reactions of S and BA at 110°C. $-f_S=0.807$, $---f_S=0.691$, $\cdots f_S=0.397$, $\cdot-\cdot f_S=0.255$.

Four copolymerizations have been carried out at different initial monomer feed ratios. The initial monomer feed ratios were chosen where a large composition drift was expected, *i.e.* at $f_S=0.255$, 0.397, 0.691 and 0.807. As can be seen from figure 9, the shape of the joint confidence intervals is greatly dependent on the initial monomer feed composition. At high f_S, the r_S is well-determined, but the error in r_{BA} is large. When applying a low fraction of styrene in the monomer feed, the situation is exactly the opposite. To obtain a reliable joint confidence interval for both reactivity ratios, the sum of the squared residuals spaces of experiments at different initial monomer feed ratios must be combined. The combined sum of squared residuals space can be visualized, leading to the following estimation of reactivity ratios:

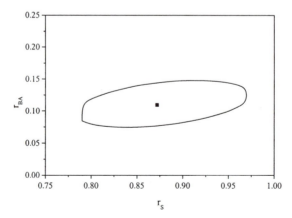

Figure 10. 95% Joint confidence interval resulting from the four joint confidence intervals presented in figure 8. $r_S=0.87$ and $r_{BA}=0.11$.

As can be seen, the reactivity ratios compare very well with the data for free-radical copolymerization data *(28)* and are also in good agreement with the values found by Matyjaszewski *et al.* *(10)*. However, a difference in the reactivity ratio for butyl acrylate, r_{BA}, is significant. An explanation for this observance still has to be found.

When the results for the copolymerization are evaluated using simulations, we can look at the influence of activation and deactivation differences on the copolymerization kinetics and therefore on the resulting r-values. In these simulations, only the activation/deactivation equilibrium reaction is considered, together with propagation and termination reactions. Transfer reactions are omitted in this approach and the termination rate constant, k_t, is assumed constant for reasons of simplicity. In the following figures, k_{act}^1, *i.e.* the activation rate parameter for PS, is varied while all other parameters are kept constant. The radical ratio is displayed in figure 11, while in figure 12 the ratio of the dormant species is shown.

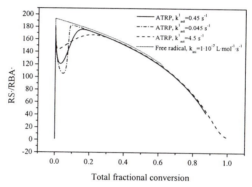

Figure 11. Radical ratios for simulations of ATRP copolymerizations with different activation rate parameters (k_{act}^1) for styrene, using $r_S=0.80$ and $r_{BA}=0.20$. For comparison, a conventional free-radical copolymerization with the same parameters has also been simulated.

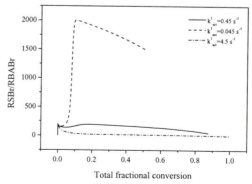

Figure 12. Dormant species ratios for simulations of ATRP copolymerizations for the same set of simulations. In all simulations, the set of parameters listed in table 1 is used.

Table 1. Parameters in ATRP simulations.

parameter	value
[S]	1.11 mol $\cdot$L^{-1}
[BA]	2.89 mol $\cdot$L^{-1}
[Cu$^+$]	0.05 mol $\cdot$L^{-1}
[I]	0.05 mol $\cdot$L^{-1}
k_p^{11}	1572 L $\cdot$mol^{-1} $\cdot$s^{-1}
k_p^{22}	76000 L $\cdot$mol^{-1} $\cdot$s^{-1}
r_S	0.20
r_{BA}	0.80
k_{act}^1	0.45, 4.5 or 0.045 L $\cdot$mol^{-1} $\cdot$s^{-1}
k_{deact}^1	1.1 $\cdot 10^7$ L $\cdot$mol^{-1} $\cdot$s^{-1}
k_{act}^2	0.45 L $\cdot$mol^{-1} $\cdot$s^{-1}
k_{deact}^2	1.1 $\cdot 10^7$ L $\cdot$mol^{-1} $\cdot$s^{-1}
k_t	1$\cdot 10^8$ L $\cdot$mol^{-1} $\cdot$s^{-1}

First of all, it should be stressed that the simulations are only intended to investigate any influence of the ATRP equilibrium for both dormant species on the resulting reactivity ratios.

From figure 11, it can be seen that no matter what the difference is between the activation of both dormant species, the resulting radical ratio will always be the same. At the same time, the dormant species ratio is different in all cases (figure 12). In other words, the ATRP system is tuning itself in this way, so that the resulting radical ratio in all systems will be the same. Since the copolymer composition is only governed by the reactivity ratios and the ratio of both radicals in the reaction mixture, it is clear that in all cases, the kinetics will be the same after a short period of time.

Although the ATRP system will self tune its radical ratios so that always the same ratio is attained, there is a short period in the beginning of the reaction where steady state has not yet been reached. This is due to the fact that the Cu^{2+} concentration has to be built up (note that the ATRP reaction is started without the presence of Cu^{2+} in the system) according to the persistent radical effect[13]. The reactivity ratios calculated from this data might be influenced by that short period where the ATRP system is not yet tuned. The data from the simulations have therefore been used to estimate the reactivity ratios by nonlinear least squares parameter estimation. The resulting joint confidence intervals are shown in figure 13.

From figure 13, it can be seen that a difference in activation rate parameters between the two dormant species could lead to a difference in estimation of the reactivity ratios in ATRP copolymerization. Although this has to be investigated in more detail, this could be an

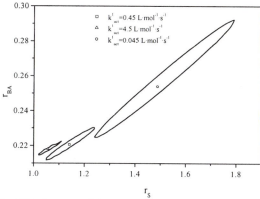

Figure 13. 95% Joint confidence intervals for ATRP simulations for different values of k^1_{act}.

explanation for the reactivity ratios as shown in figure 10, which are slightly different from the ones observed in conventional free-radical copolymerization.

Concluding remarks

From homopolymerizations of S and BA, the equilibrium constants have been determined from the $-\ln(1-X)$ vs. $t^{2/3}$ plot. The set of constants in equation 1 has been estimated for S to be $1.69 \cdot 10^{-3}$ $s^{-2/3}$, and for BA $1.53 \cdot 10^{-4}$ $s^{-2/3}$. From these data, together with the initial concentrations of initiator and Cu^+, the equilibrium constants could be calculated and they proved to be $5.85 \cdot 10^{-10}$ for S and $1.26 \cdot 10^{-11}$ for BA. These data should be supported by information on the activation rate parameters, k_{act}, for both dormant species. This work is currently carried out in our laboratory.

For the homopolymerization of styrene, it is shown that termination events predominantly take place in the beginning of the reaction. After the persistent radical concentration has been built up, termination becomes less pronounced. This is reflected in the fact that the degree of dormant species with a bromine end group stays at a constant level of about 85%.

For the copolymerization of S and BA in ATRP, reactivity ratios have been determined: $r_S=0.11$ and $r_{BA}=0.87$. These reactivity ratios correspond well to the ones found for conventional free-radical copolymerization. However, simulations show a dependence of reactivity ratios on a difference in activation rate parameters.

210

References

1. Moad, G; Rizzardo, E.; Solomon, D.H. *Macromolecules* **1982**, *15*, 909
2. Georges, M.K.; Veregin, R.P.N.; Kazmaier, P.M.; Hamer, G.K. *Macromolecules* **1993**, *26*, 2987
3. Kato, M.; Kamigaito, M.; Sawamoto, M.; Higashimura, T. *Macromolecules* **1995**, *28*, 1721
4. Wang, J.S.; Matyjaszewski, K. *Macromolecules* **1995**, *28*, 7901
5. Qiu, J.; Matyjaszewski, K. *Polym. Prepr.* **1997**, *38 (1)*, 711
6. Coca, S.; Matyjaszewski, K. *Polym. Prepr.* **1997**, *38 (1)*, 691
7. Haddleton, D.M.; Jasieczek, C.B.; Hannon, M.J.; Shooter, A.J. *Macromolecules* **1997**, *30*, 2190
8. Arehart, S.V.; Greszta, D.; Matyjaszewski, K. *Polym. Prepr.* **1997**, *38 (1)*, 705
9. Greszta, D.; Matyjaszewski, K. *Polym. Prepr.* **1997**, *38 (1)*, 709
10. Haddleton, D.M.; Crossman, M.C.; Hunt, K.H.; Topping, C.; Waterson, C.; Suddaby, K.G. *Macromolecules* **1997**, *30*, 3992
11. Kotani, Y.; Kamigaito, M.; Sawamoto, M. *Macromolecules* **1998**, *31*, 5582
12. Arehart, S.V.; Matyjaszewski, K. *Macromolecules* **1999**, *32*, 2221
13. Fischer, H. *J. Am. Chem. Soc.* **1986**, *108*, 3925
14. Fischer, H. *Macromolecules* **1997**, *30*, 5666
15. Ohno, K.; Goto, A.; Fukuda, T.; Xia, J.; Matyjaszewski, K. *Macromolecules* **1998**, *31*, 2699
16. Matyjaszewski, K.; Patten, T.E.; Xia, J. *J. Am. Chem. Soc.* **1997**, *119*, 674
17. Van den Brink, M.; Van Herk, A.M.; German, A.L. *J. Polym. Sci., Part A: Polym. Chem.* **1999**, *37*, 3793
18. Hutchinson, R.A.; Paquet, D.A.; McMinn, J.H.; Beuermann, S.; Fuller, R.E.; Jackson, C. *DECHEMA Monographs* **1995**, *131*, 467
19. Percec, V.; Kim, H.-J.; Barboiu, B.; Van der Sluis, M. *Polym. Prepr.* **1997**, *38 (2)*, 408
20. Wang, J.-L.; Grimaud, T.; Matyjaszewski, K. *Macromolecules* **1997**, *30*, 6507
21. Matyjaszewski, K.; Coca, S.; Gaynor, S.G.; Wei, M.; Woodworth, B.E. *Macromolecules* **1998**, *31*, 5967
22. Matyjaszewski, K.; Kajiwara, A. *Macromolecules* **1998**, *31*, 548
23. Chambard, G.; Klumperman, B.; German, A.L. *Macromolecules*, in preparation
24. Heuts, J.P.A.; Davis, T.P. *Macromol. Rapid Commun.* **1998**, *19*, 371
25. Meyer, V.E.; Lowry, G.G. *J. Polym. Sci., Part A* **1965**, *3*, 2843
26. Meyer, V.E. *J. Polym. Sci., Part A-1* **1966**, *4*, 2819
27. Plaumann, H.P.; Branston, R.E. *J. Polym. Sci., Part A: Polym. Chem.* **1989**, *27*, 2819
28. Chambard, G.; Klumperman, B.; German, A.L. *Polymer* **1999**, *40*, 4459

Chapter 15

The Copper Catalyst in Atom Transfer Radical Polymerizations: Structural Observations

Guido Kickelbick[1], Ulrich Reinöhl[2], Teja S. Ertel[2],
Helmut Bertagnolli[2], and Krzysztof Matyjaszewski[3]

[1]Institut für Anorganische Chemie, Technische Universität Wien,
Getreidemarkt 9/153, 1060 Wien, Austria
[2]Institut für Physikalische Chemie, Universität Stuttgart,
Pfaffenwaldring 55, 70569 Stuttgart, Germany
[3]Department of Chemistry, Carnegie Mellon University, Pittsburgh, PA 15213

X-ray structure determination and EXAFS analysis were used to determine the structure of copper complexes with 2,2'-bipyridine (bpy) and linear triamine N,N,N',N'',N''-pentamethyldiethylene-triamine (pmdeta) as ligands. In the case of the bpy system the $Cu^{(I)}$ complex was identified as a cationic tetrahedral species where the copper is surrounded by two bpy ligands. The structure of the anion depends on the particular catalyst and reaction medium but in some systems can be represented by a linear $[Br_2Cu]^-$. The $Cu^{(II)}$ species generated in a model reaction was a cationic trigonal bipyramidal complex where the metal is surrounded by two bpy ligands and one bromine. In the case of the linear triamine (pmdeta) the $Cu^{(I)}$ atom is surrounded by one pmdeta ligand and a bromine. From crystal structure analysis it was concluded that the $Cu^{(II)}$ complex shows a distorted square pyramidal structure with two very different Cu-Br distances.

Introduction

Controlled radical polymerization techniques are based on the quantitative initiation and regulation of the concentration of radicals during the polymerization reaction. Low concentration of radicals disfavors the termination reactions by combination or disproportionation. Two methods belong to the most promising in this field: (i) the nitroxide mediated systems rely on the cleavage of a C-O bond by the reversible formation a stable nitroxyl radical and a growing radical *(1)* and (ii) atom transfer radical polymerization (ATRP) where the radicals are formed by the reversible cleavage of a C-X bond under transition metal catalyzed conditions *(2,3)*.

Both methods reveal excellent control over molecular weights and polydispersities combined with the ability to prepare well-defined copolymers *(4)*. However, ATRP can be applied to a larger variety of monomers and use more diversified initiator molecules. Although ATRP and the atom transfer radical addition (ATRA) method have been known for a relatively long time, the mechanism of the process including structure of all intermediates is not yet fully understood.

The catalytic cycle in transition metal mediated ATRP involves the switching between two oxidation states of a transition metal compound. The lower oxidation state complex adds a halogen atom by homolytic cleavage of a C-X bond forming a radical and promoting the transition metal compound into a higher oxidation state (Scheme 1). The equilibrium is strongly shifted to the side of the so-called dormant species. The coordination compound affects the equilibrium constants and therefore the control in the reaction is strongly dependent on the metal and the ligands forming the catalytic active species. Metals that are most commonly known for usage in ATRP reactions are Cu *(2,3)*, Ru *(5)*, Fe *(6-8)*, Ni *(9, 10)*, Pd *(11)* and Rh *(12)* in combination with a variety of different ligands tuned for every metal and the polymerization process. Whether the catalyst is suitable for the polymerization reaction depends strongly on the right combination of metal and ligand. For the future development of new catalyst as well as the proper understanding of the mechanism it is important to know structural details of the different types of catalytically active species in the reaction. This paper concentrates on the copper/amine mediated reactions because those are among the most commonly used ATRP catalytic systems. Two ligands used in this study are 2,2'-bipyridine (bpy) derivatives and linear triamines. In the case of the bpy system the ratio between $Cu^{(I)}$ halide and ligand that have been used varies from 1:1 to 1:3. *(13)* However, there is no clear indication whether the different ratios have any influence on the mechanism of the reaction.

Scheme 1

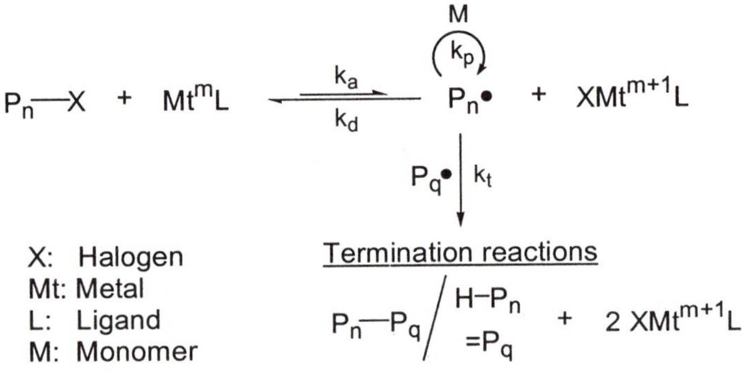

$$P_n\!-\!X \ + \ Mt^m L \ \underset{k_d}{\overset{k_a}{\rightleftharpoons}} \ \overset{\overset{M}{\left(k_p\right)}}{P_n{}^\bullet} \ + \ XMt^{m+1}L$$

$$P_q{}^\bullet \ \big| \ k_t$$

Termination reactions

X: Halogen
Mt: Metal
L: Ligand
M: Monomer

$$P_n\!-\!P_q \Big\langle \begin{matrix} H\!-\!P_n \\ =\!P_q \end{matrix} \ + \ 2\,XMt^{m+1}L$$

In this study X-ray crystallography and extended X-ray absorption fine structure (EXAFS) analysis were used to clarify the nature of the catalyst in the solid state as well as in solution. X-ray crystallography provided the first suggestions for structures of possible intermediates in ATRP. Also, we were able to crystallize compounds from reaction solutions and analyze their structures. To verify whether the results from the solid state can be transferred to solution reactions, EXAFS analysis was used.

Results and Discussion

Cu/bpy based systems

Possible structures involved in the mechanism of Cu/bpy mediated ATRP can be concluded from X-ray structures of related $Cu^{(I)}$ and $Cu^{(II)}$ complexes. $Cu^{(I)}$ surrounded by two bidentate ligands like bpy forms typically a $[(bpy)_2Cu]^+$ cation with a tetrahedral structure in a mixture of 1:2 between the copper halide and bpy. This structure type is found in literature accompanied by different counterions like $[ClO_4]^-$ *(14)* or $[Cl_2Cu]^-$ *(15)*. A bridged structure for the $Cu^{(I)}$ species is also possible where the halogen atoms are in bridging positions and only one bpy is bound to every copper center leading to a neutral dimer which is known to occur for bromine and iodine but not for chlorine atoms *(15)*. The $Cu^{(II)}$ species being part of the ATRP equilibrium should form a trigonal bipyramidal cation where the copper center is surrounded by two bpy ligands and one halogen atom. This structure type has many literature examples for chlorine, bromine and iodine as halogen atoms accompanied by a collection of different counterions. *(16)* Due to this large number of examples it appears to be a very stable class of compounds. As a typical representative of this class of complexes we selected $[(bpy)_2CuBr][BF_4]$ *(17)* which was used in our studies as a model compound. The proposed mechanism of a Cu/bpy mediated ATRP is shown in Scheme 2 based on the above-mentioned structures.

Scheme 2

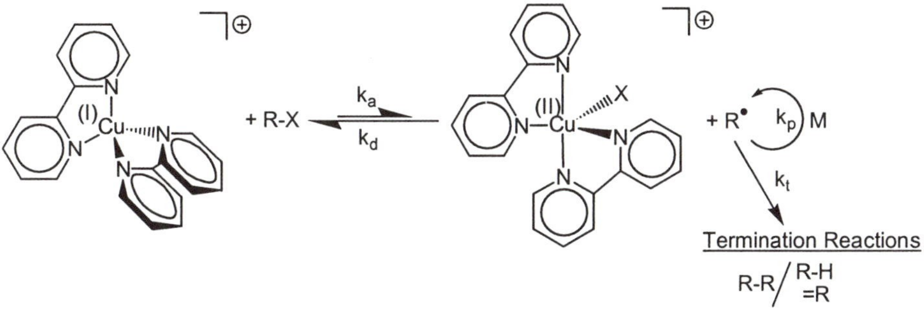

Some information on the structure of amine coordinated copper complexes in ATRP was available from UV model studies *(18)*. However, the structure of counterions and their role in the equilibrium remained obscure. Originally it was assumed that the counterion is a pure halide anion, which is released from the copper halide used in the reaction. This explanation is also supported by the stoichiometry of the reaction where in the typical ATRP case one eq. of CuX was mixed with two eq. of bpy. However, no example for [(bpy)$_2$Cu]$^+$ and only a few examples for [(bpy)$_2$CuX]$^+$ cations in combination with a simple halide as counterion are found in the literature (19-21). A first hint of the kind of anions involved in ATRP was a crystal structure from crystals precipitated after an ATRP reaction from the solution which showed a trigonal bipyramidal [dNbpy$_2$CuBr]$^+$ cation with a linear [Br$_2$Cu]$^-$ anion (Figure 1).

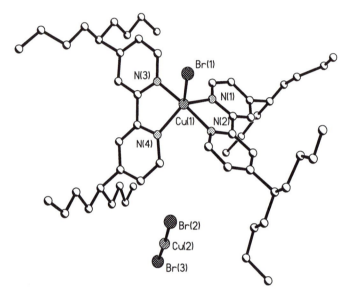

Figure 1. Crystal structure of the complex [dNbpy$_2$CuBr]$^+$[Br$_2$Cu]$^-$ prepared from crystals which precipitated from an ATRP reaction solution of methyl acrylate.

However, all the data above is based on crystal structure observations and therefore only typical for the solid state. To extend these observations to solution an additional technique was needed which is capable to give structural information for the catalytic active systems in solution without interacting with the observed species. For this purpose EXAFS analysis was used. This technique provides information about the distances and the number of the surrounding atoms through the analysis of the fine structure at the absorption edge of an atom.(22) Its big advantage is the applicability in solid state as well as in solution. Especially strong X-ray absorbers like copper and bromine, both typical atoms for the catalytic active ATRP species, can be examined with a high reliability. Therefore, this technique seemed to be ideal for the investigation of ATRP systems.

The first step in the analysis was to compare experimental EXAFS data for bond lengths with data from single crystal X-ray diffraction. Therefore model compounds were crystallized and their crystal data proved by single crystal X-ray structure determination. As model compounds [(bpy)$_2$Cu][ClO$_4$] (14) and [(bpy)$_2$CuBr][BF$_4$]

(17) were used. The experimental Cu-N and Cu-Br distances show a good agreement with literature values (Table I).

Table I. Comparison of experimental bonding distances [pm] from EXAFS measurements and single crystal X-ray structures for [(bpy)$_2$Cu]$^+$ and [(bpy)$_2$CuBr]$^+$.

	Cu-N EXAFS	Cu-N crystal *(14)*	Cu-Br EXAFS	Cu-Br crystal *(17)*
[(bpy)$_2$Cu][ClO$_4$]	198.0	202.0, 202.1		
[(bpy)$_2$CuBr][BF$_4$]	201.0	199.5, 199.6, 206.8, 211.4	241.0	241.9

The experiments in solution were carried out following standard ATRP procedures. Styrene and methyl acrylate were used as solvents to verify possible interactions between the coordination compound and the monomer, which are able to influence the structure of the catalyst. Substituted 4,4'-di(5-nonyl)-2,2'-bipyridine (dNbpy) was used as ligand for a better solubility of the complex in the nonpolar monomers. The results of the measurements are shown in Table II.

Table II. Results from EXAFS measurements from different ligand to copper halide mixtures, number of backscattering atoms and their distances [pm].

Mixture	Monomer	Cu-N		Cu-Br	
		Nr. of N atoms	Distance	Nr. of Br atoms	Distance
CuBr + dNbpy	St	4	201.0	2	225.0
CuBr + 2 dNbpy	St	4	203.0	2	227.0
CuBr + dNbpy	MA	4	200.0	2	226.0
CuBr + 2 dNbpy	MA	4	203.0	2	229.0
CuBr$_2$ + 2 dNbpy	MA	4	208.0	3 or 4	234.0
				1	243.0
CuBr + 2 dNbpy polymerization @ 95°C	St	4	201.0	2	224.0
CuBr + 2 dNbpy polymerization @ 80°C	MA	4	202.0	2	225.0

The first question was whether different ratios between CuBr and dNbpy affect the structure of the complex in solution. For both ratios 1:1 and 1:2 between CuBr and dNbpy two different types of Cu species were observed. One Cu atom surrounded by 4 nitrogen atoms in a Cu-N distance from 200.0 pm to 203.0 pm and one Cu atom surrounded by two bromine atoms in a distance from 225.0 pm to 229.0 pm. This result fits well with the data from a $[(dNbpy)_2Cu]^+$ cation and a $[Br_2Cu]^-$ anion. The most important conclusion from this data is that the predominant structure of complexes is the same whether one equivalent of CuBr is mixed with one or two equivalents of bpy. Therefore in the case of a 1:2 mixture usually used in ATRP it is possible that free bpy is present in solution. Figure 2 shows a comparison between Fourier-filtered experimental and calculated $k^3\chi(k)$ function and the corresponding Fourier functions for a mixture of one eq. CuBr and two eq. dNbpy in styrene at room temperature. The curve for the simulated data for a mixture of the model compounds $[(dNbpy)_2Cu]^+$ and $[Br_2Cu]^-$ fits very well with the experimental plot which demonstrates the reliability of the model.

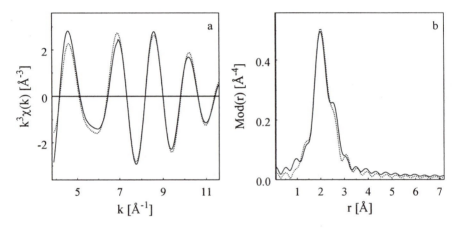

Figure 2. Fourier-filtered experimental (dotted line) and calculated (solid line) $k^3\chi(k)$ function and the corresponding Fourier transformations Mod(r) for a mixture of one eq. CuBr, two eq. dNbpy in styrene at RT, measured at the Cu K-edge; k = 3.8-11.8 Å^{-1}.

In Figure 3 the simulated line is divided in the functions of the anionic species and the cationic species alone. The superposition of these functions results in the simulated curve of the mixture. The Fourier-transformed functions (b parts of the figures) display directly the distances of all backscattering atoms, which is in this case a mixture of different species. The comparison of EXAFS data in styrene and methyl acrylate, which has a potentially larger interaction with the transition metal (oxygen lone pairs), reveals that there is no influence of the monomer on the complex.

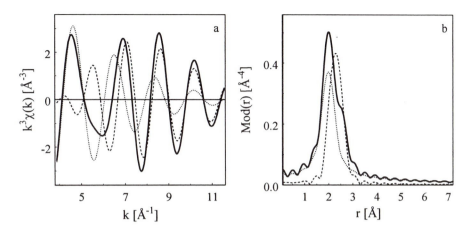

Figure 3. Simulation of the $k^3\chi(k)$ function and the corresponding Fourier transformations Mod(r) [(dNbpy)$_2$Cu]$^+$ cation (doted line) [Br$_2$Cu]$^-$ anion (dashed line) and the superposition of both (solid line).

The analysis of the mixture of CuBr$_2$ with 2 equivalents of dNbpy shows 4 Cu-N distances of 208.0 pm and 3 or 4 Cu-Br distances of 234.0 pm and one Cu-Br distance of 243.0 pm. In this case it was not possible to determine the exact coordination number of the Cu atom. The data fits, as expected, with the trigonal bipyramidal [dNbpy$_2$CuBr]$^+$ cation. The remaining Cu-Br distances belong to the corresponding anions which matches well with the literature known distorted tetrahedral [Br$_4$Cu]$^{2-}$ or planar [Br$_3$Cu]$^-$ anions (23). Unfortunately, EXAFS can not provide a quantitative information on the proportion of Br$^-$ and [Br$_2$Cu]$^-$, [Br$_3$Cu]$^-$, [Br$_4$Cu]$^{2-}$ anions. Figure 4 shows the good fit of Fourier-filtered experimental (dotted line) and calculated (solid line) $k^3\chi(k)$ function as well as the corresponding Fourier transformations Mod(r) for a mixture of one eq. Cu$_2$Br and two eq. dNbpy in methyl acrylate at room temperature. Additionally the corresponding composition of the simulated curve of the mixture from the cationic [(dNbpy)$_2$CuBr]$^+$ and anionic fraction [Br$_4$Cu]$^{2-}$ is shown in Figure 5.

Polymerization reactions of styrene and methyl acrylate also were investigated. No differences between the 1:1 and 1:2 mixtures of CuBr and dNbpy were observed by EXAFS.

There is a clear indication that the ATRP equilibrium is strongly shifted to the side of the Cu$^{(I)}$ species. Although a Cu$^{(II)}$ species is definitely formed during the process, as directly observed by EPR (24), we were not able to detect this species by EXAFS analysis due to its low concentration (<10%). No differences in the polymerization reactions based on effects of the different monomers were observed which was similar to the model studies above, hence an interaction of the monomer with the catalyst is not likely. Differences in kinetics between polymerizations of different monomers therefore should have other sources.

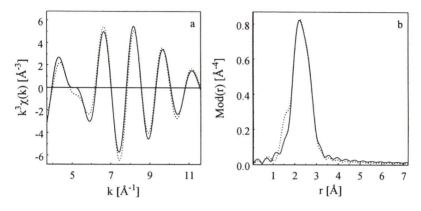

Figure 4. Fourier-filtered experimental (dotted line) and calculated (solid line) $k^3\chi(k)$ function and the corresponding Fourier transformations Mod(r) for a mixture of one eq. $CuBr_2$, two eq. dNbpy in methyl acrylate at RT, measured at the Cu K-edge; k = 3.8-11.8 Å^{-1}.

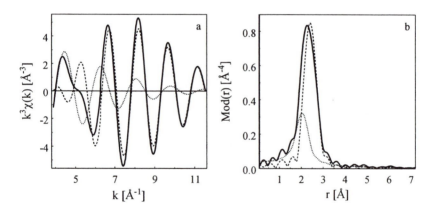

Figure 5. Simulation of the $k^3\chi(k)$ function and the corresponding Fourier transformations Mod(r) [(dNbpy)$_2$CuBr]$^+$ cation (doted line) [Br$_{3(4)}$Cu]$^{(2)-}$ anion (dashed line) and the superposition of both (solid line).

Cu/pmdeta based systems

Beside the bpy type ligands the linear triamine N,N,N',N'',N''-pentamethyldiethylenetriamine (pmdeta) is an efficient ligand for the copper mediated ATRP. Therefore this system was also examined by EXAFS methods. In the case of the linear triamine usually a 1:1 ratio between ligand and copper halide is applied in ATRP reactions *(25)*. In general the polymerizations with this type of catalyst are faster than those with the bpy based system and the catalyst is much more sensitive to oxidation. Assuming the Cu$^{(I)}$ complex has again a tetra-coordinated copper center the structural consideration is that the compound is a neutral (pmdeta)CuBr species. No structure of a discrete complex between Cu$^{(I)}$ and pmdeta is known in literature. Although we tried to crystallize such a compound we did not yet succeed. The Cu$^{(I)}$ compound seems to be too labile probably due to an open coordination site. However, a mixed Cu$^{(I)}$/Cu$^{(II)}$ system was characterized structurally where the Cu$^{(I)}$ complex is stabilized by CuCl$_2$ *(26)*. Compared to Cu$^{(I)}$ the Cu$^{(II)}$Br$_2$/pmdeta complex is stable and a crystal structure analysis was prepared (Figure 6). The Cu$^{(II)}$ in the structure is surrounded in a distorted square pyramidal way by three nitrogen atoms with Cu-N distances of 208.6 pm, 210.3 pm and 210.4 pm and two Br atoms with distances of 244.6 pm and 264.4 pm. The difference in the bond lengths of the two Br atoms (19.8 pm) can be one origin of the higher reactivity of the complex in ATRP because the switching of one Br atom is improved due to the weaker Cu-Br bond. Contrary to the bpy based system no rehybridization is needed to go from the Cu$^{(I)}$ species to the Cu$^{(II)}$ species and vice versa which is another probable reason for the enhanced reactivity of this complex.

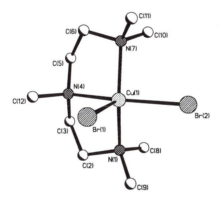

Figure 6. Crystal structure of a (pmdeta)CuBr$_2$ complex precipitated from an ATRP reaction solution for an methyl acrylate polymerization.

We were able to reproduce the crystallization of the compound from a solution of CuBr$_2$ and pmdeta in methyl acrylate and used the crystals to calibrate the EXAFS measurements. The results of these solid-state measurements are shown in Table III. Again the EXAFS data shows a very good agreement with X-ray structure data.

Table III. Comparison of atom distances [pm] from EXAFS measurements and X-ray data of (pmdeta)CuBr$_2$.

	Cu-N EXAFS	Cu-N crystal	Cu-Br EXAFS	Cu-Br crystal
(pmdeta)CuBr$_2$	209.0	208.6, 210.3, 210.4	244.0, 266.0	244.6, 264.4

Scheme 3 displays the proposed mechanism for the pmdeta system.

1:1 mixtures of CuBr and pmdeta in styrene and methyl acrylate were examined in EXAFS measurements. Both systems show three Cu-N distances of 214.0 pm and one Cu-Br distance of 233.0 pm (Table IV).

Scheme 3

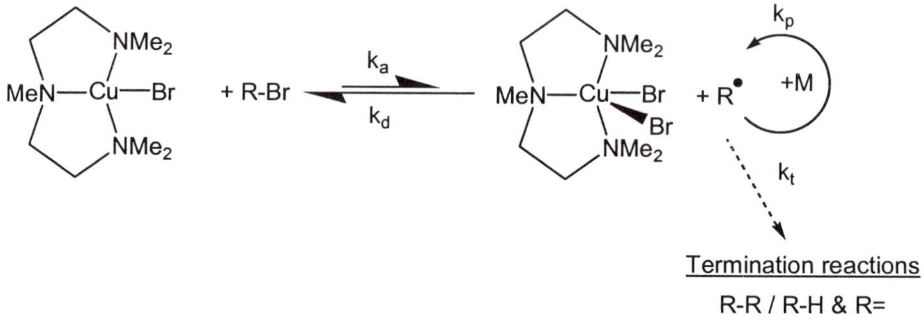

Termination reactions

R-R / R-H & R=

Table IV. Comparison of atom distances [pm] from EXAFS measurements of different CuBr/pmdeta systems.

Mixture	Monomer	Cu-N		Cu-Br	
		Nr. of N atoms	Distance	Nr. of Br atoms	Distance
CuBr + pmdeta	St	3	214.0	1	233.0
CuBr + pmdeta	MA	3	214.0	1	233.0
CuBr + pmdeta polymerization @ 60°C	MA	3	213.0	1	232.0

Again as in the case of bpy based system no difference between styrene and methyl acrylate as solvent was observed meaning that there is no direct interaction between the solvent and the copper species in solution. Also a bridging situation can be excluded from the analysis of the bromine K-edge which should have two copper atoms in its surrounding but in fact shows only one copper in the experiment. As already mentioned above, there is no data from X-ray structure for a (pmdeta)Cu[(I)]Br complex available. Comparisons with bridged Cu[(I)]Br structures of the H-substituted amine ligands (27) from crystal structures are possible but not very meaningful because the bond distances for the Cu-N bonds are different due to the lower basicity of RNH_2 group compared to a $RNMe_2$ group and smaller steric hindrances resulting in shorter bonds. Also Cu-Br distances are different due to the bridging structure.

The polymerization reaction shows no significant difference to the model $Cu^{(I)}$ systems therefore once more the $Cu^{(I)}$ species should be the predominant one and the $Cu^{(II)}$ species in the equilibrium could not be detected by EXAFS due to its low concentration, although nearly 5% of $Cu^{(II)}$ species has been directly measured by EPR *(24)*.

Conclusions

X-ray structure determination and EXAFS analysis showed the differences in Cu mediated ATRP with different ligands coordinating the copper. The classical bpy system revealed that the $Cu^{(I)}$ species in the ATRP equilibrium is a tetrahedral complex where the copper is surrounded by two dNbpy ligands as cation accompanied by a $[Br_2Cu]^-$ anion. The most likely $Cu^{(II)}$ complex in the catalytic equilibrium is a trigonal bipyramidal compound where the $Cu^{(II)}$ is surrounded by two dNbpy ligands and a bromine atom. From X-ray (Fig. 1) and UV studies it appears that in the presence of a large excess of Cu(I) species the anions have the $[Br_2Cu]^-$ structure *(18)*.

The linear triamine ligand pmdeta behaves differently due to the fact that it is a tridentate system and forms a neutral (pmdeta)CuBr compound. The corresponding $Cu^{(II)}$ complex is a distorted square pyramidal system with two very different Cu-Br bond length. This may be an origin of the higher reactivity of this catalyst due to a faster exchange of Br atoms between growing chains and a catalyst.

Acknowledgments. We thank Tomislav Pintauer for fruitful discussions. We wish to thank the Fonds zur Förderung der wissenschaftlichen Forschung (FWF), Austria, the Fonds der Chemischen Industrie and the National Science Foundation for financial support, as well as the HASYLAB at DESY, Hamburg (Germany) for providing synchrotron radiation. Work was done (partially) at SSRL which is operated by the Department of Energy, Office of Basic Energy Sciences (USA).

Literature Cited

1. Hawker, C. J. *Acc. Chem. Res.* **1997**, *30*, 373.
2. Matyjaszewski, K.; Patten, T. E.; Xia, J.; Abernathy, T. *Science* **1996**, *272*, 866.
3. Matyjaszewski, K.; Patten, T. E.; Xia, J. *J. Amer. Chem. Soc* **1997**, *119*, 674.
4. a) Patten, T. E.; Matyjaszewski, K. *Adv. Mater.* **1998**, *10*, 901; b) Patten, T. E.; Matyjaszewski, K. *Acc. Chem. Res.* **1999**, *32*, 895; c) Matyjaszewski, K. *Chem. Eur. J.* **1999**, *5*, 3095.
5. Kato, M.; Kamigaito, M.; Sawamoto, M.; Higashimura, T. *Macromolecules* **1995**, *28*, 1721.
6. Ando, T.; Kamigaito, M.; Sawamoto, M. *Macromolecules* **1997**, *30*, 4507.
7. Matyjaszewski, K.; Wei, M.; Xia, J.; McDermott, N. E. *Macromolecules* **1997**, *30*, 8161.

8. Moineau, G.; Dubois, P.; Jerome, R.; Senninger, T.; Teyssie, P. *Macromolecules* **1998**, *31*, 545.

9. Granel, C.; Dubois, P.; Jerome, R.; Teyssie, P. *Macromolecules* **1996**, *29*, 8576.

10. Uegaki, H.; Kotani, Y.; Kamigaito, M.; Sawamoto, M. *Macromolecules* **1997**, *30*, 2249.

11. Lecomte, P.; Drapier, I.; Dubois, P.; Teyssie, P.; Jerome, R. *Macromolecules* **1997**, *30*, 7631.

12. Moineau, G.; Granel, C.; Dubois, P.; Jerome, R.; Teyssie, P. *Macromolecules* **1998**, *31*, 542.

13. Examples for 1:1: Ref. 3; Davis, K.; Paik, H.; Matyjaszewski, K. *Macromolecules* **1999**, *32*, 1767; Wang, J. L.; Grimaud, T.; Matyjaszewski, K. *Macromolecules* **1997**, *30*, 6507; 1:2 ratio: Ref. 1-4; 1:3 ratio: DiRenzo, G. M.; Messerschmidt, M.; Mühlhaupt, R.; *Macromol. Rapid Commun.* **1998**, *19*, 381.

14. Munakata, M.; Kitagawa, S.; Asahara, A.; Masuda, H. *Bull. Chem. Soc. Jpn.* **1987**, *60*, 1927.

15. Skelton, B. W.; Waters, A. F.; White, A. H. *Aust. J. Chem.* **1991**, *44*, 1207.

16. Typical examples are: Jensen, W. P.; Jacobson, R. A. *Inorg. Chim. Acta* **1981**, *50*, 189; Harrison, W. D.; Hathaway, B. J.; Kennedy, D. *Acta Crystallogr., Sect. B* **1979**, *35*, 2301; Harrison, W. D.; Kennedy, D. M.; Power, M.; Sheahan, R.; Hathaway, B. J. *J. Chem. Soc., Dalton Trans.* **1981**, 1556; Freckmann, B.; Tebbe, K.-F. *Acta Crystallogr., Sect. A* **1981**, *37*, C228; Barclay, G. A.; Hoskins, B. F.; Kennard, C. H. L. *J. Chem. Soc.* **1963**, 5691.

15. Hathaway, B. J.; Murphy, A. *Acta Crystallogr., Sect.B* **1980**, *36*, 295.

16. Qiu, J.; Pintauer, T.; Gaynor, S.; Matyjaszewski, K. *Polym. Prepr. (Am. Chem. Soc., Div. Polym. Chem.)* **1999**, *40(2)*, 420.

17. Khan, M. A.; Tuck, D. G. *Acta Crystallogr., Sect. B* **1981**, *37*, 1409.

18. Stephens, F. S.; Tucker, P. A. *J. Chem. Soc., Dalton Trans.* **1973**, 2293.

19. Barclay, G. A.; Hoskins, B. F.; Kennard, C. H. L. *J. Chem. Soc.* **1963**, 5691.

20. Bertagnolli, H.; Ertel, T. S. *Angew. Chem. Int. Ed. Engl.* **1994**, *33*, 45.

21. a) Trouelan, P.; Levebvre, J.; Derollez, P. *Acta Crystallogr., Sect. C (Cr. Str. Comm.)* **1984**, *40*, 386; b) M. Sano, T. Maruo, H. Yamatera, *J. Chem. Phys.* **1986**, *84*, 66.

22. a) Kajiwara, A.; Matyjaszewski K. *Macromolecules* **1998**, *31*, 548; b) Kajiwara, A.; Matyjaszewski K. *Polymer J.* **1999**, *31*, 70.

23. Xia, J.; Matyjaszewski, K. *Macromolecules* **1997**, *30*, 7697.

24. Breeze, S. R.; Wang, S. *Inorg. Chem.* **1996**, *35*, 3404.

25. Boeyens, J. C. A.; Dobson, S. M.; Mboweni, R. C. M. *Acta Crystallogr., Sect. C (Cr. Str. Comm.)* **1991**, *47*, 186.

Chapter 16

Atom Transfer Radical Polymerization Mediated by Ruthenium(II)–Arene Complexes

François Simal, Dominique Jan, Albert Demonceau[1], and Alfred F. Noels

Laboratory of Macromolecular Chemistry and Organic Catalysis, C.E.R.M.,
University of Liège, Sart-Tilman (B.6a), B–4000 Liège, Belgium

Several ruthenium-based catalysts of the formula [RuX$_2$(arene)(L)] have been synthesized, and relative catalyst activities were determined by monitoring the atom transfer radical polymerization (ATRP) of methyl methacrylate. The following order of increasing activity was determined: X = I < Br < Cl and L = PPh$_3$ << PCy$_2$Ph < PiPr$_3$ < PCy$_3$ ≈ P(cC$_5$H$_9$)$_3$. Additional studies indicated that the release of the arene ligand is a prerequisite for ATRP.

Transition metal catalyzed atom transfer radical polymerization (ATRP) is one of the most efficient methods to control radical polymerization (1, 2). ATRP is based on the reversible formation of radicals from alkyl halides in the presence of transition metal complexes, and is a direct extension to polymers of the Kharasch reaction (ATRA, atom transfer radical addition) (Scheme 1) (3). Sawamoto *et al*. (4) and Matyjaszewski *et al*. (5, 6) were the first to report on ATRP, by using catalyst systems based on ruthenium and copper, respectively. The copper-based systems usually contain nitrogen ligands (bipyridine, multidentate amines, and Schiff bases), whereas the first ruthenium-based system, RuCl$_2$(PPh$_3$)$_3$, requires the addition of a Lewis acid, *i.e.*, MeAl(o,o'-di-*tert*-butylphenoxide)$_2$ or Al(OiPr)$_3$ for being active. Very recently, however, Ru(Ind)Cl(PPh$_3$)$_2$ (Ind = indenyl) has been reported to smoothly polymerize methyl methacrylate in the absence of added Lewis acid (7).

Sawamoto's ruthenium complexes for controlled radical polymerization

[1]Corresponding author.

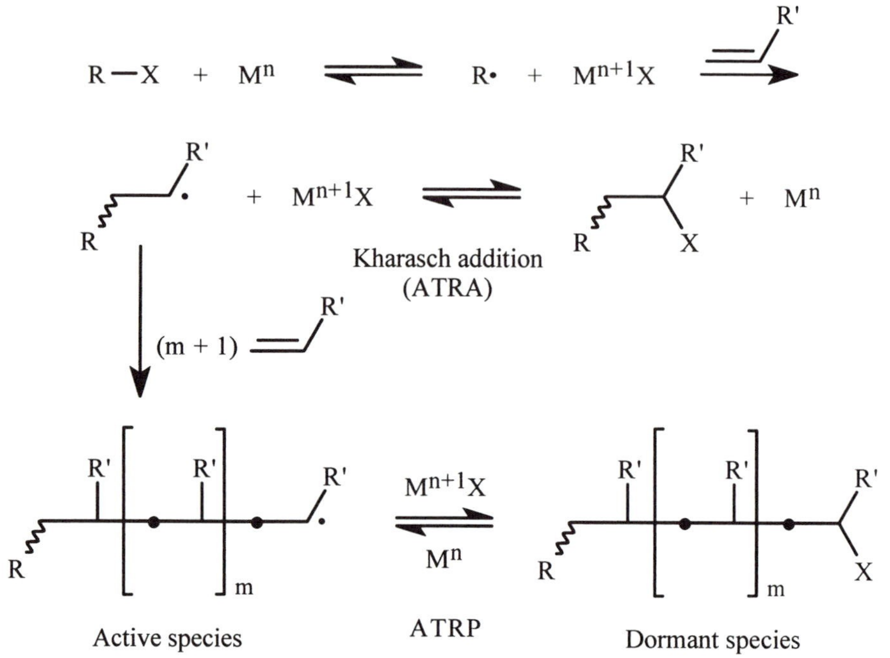

$$R{-}X \ + \ M^n \ \rightleftharpoons \ R\bullet \ + \ M^{n+1}X \ \xrightarrow{\quad R'\quad}$$

Kharasch addition
(ATRA)

$(m + 1)$

M^{n+1}X

M^n

Active species   ATRP   Dormant species

Scheme 1

In addition to copper(I) and ruthenium(II), nickel(II), iron(II), and related transition metal complexes are known nowadays to catalyze controlled radical polymerizations of (meth)acrylic, styrenic, and other monomers and to allow precision polymer synthesis. Despite its numerous advantages, ATRP still suffers from severe limitations. For instance, polymerizations carried out at temperatures below 50 °C and enhanced polymerization rates for a lower catalyst loading are needed and, last but not least, ATRP of some industrially important monomers (vinyl acetate, vinyl chloride, ...) and control of the polymer tacticity are still out of reach. We believe that a key to this challenges lies in the design of the components of the process (metal center, ligands, and initiator leaving group), so that we can tune the catalyst/initiator binary initiating systems in accordance with the structure and reactivity of monomers, the inherent stability of the growing radicals they produce, and other factors.

We reported recently on the exceptional efficacy of new catalytic systems based on $[RuCl_2(p\text{-cymene})(PR_3)]$ complexes (p-cymene = 4-isopropyltoluene) for promoting the controlled free-radical polymerization of vinyl monomers *without* cocatalyst activation (8, 9). Several advantages of these single-component catalyst systems are noteworthy:

1. $[RuCl_2(\eta^6\text{-}p\text{-cymene})(PR_3)]$ complexes are air-stable, saturated 18-electron species, which are readily obtained from commercially available $[RuCl_2(\eta^6\text{-}p\text{-cymene})]_2$ upon addition of a phosphine (Scheme 2).

2. $[RuCl_2(p\text{-cymene})(PR_3)]$ complexes are highly soluble in common organic solvents including heptane and methanol (the solvents used for precipitation of

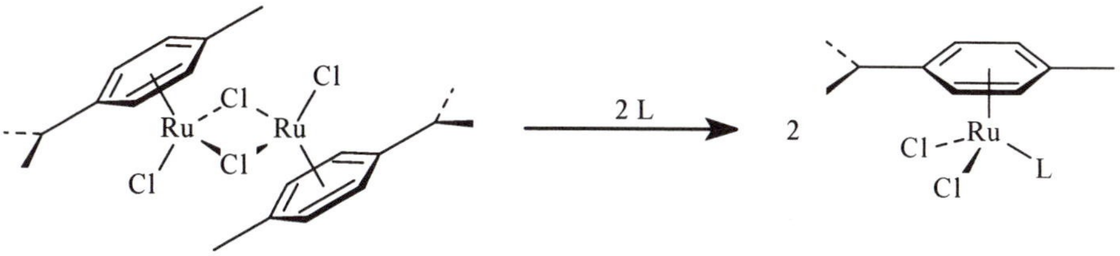

Scheme 2

polymethacrylates and polystyrenes, respectively) and neat monomers ((meth)acrylates, styrenes, ...). This yields white polymers as opposed to slightly colored polymers precipitated from reaction mixtures of nickel-, iron-, and the first generation copper-mediated polymerization reactions.

3. These complexes allow substantial possible variations of their basic structural motive (ligand L, halogen, and arene), and hence a fine tuning of the stereoelectronic parameters of those ligands, a prerequisite for investigating the mechanism as well as to further explore the preparative scope of these highly promising versatile polymerization catalysts.

The present contribution is aiming at illustrating how variation of ligands L, halogens, and arenes influences ATRP of methyl methacrylate considered as model of a vinyl monomer.

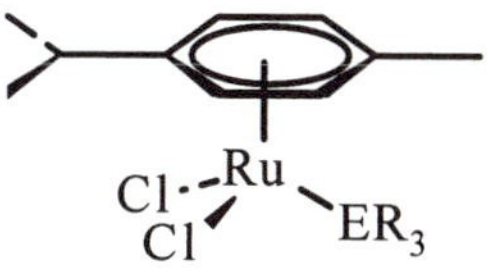

ER_3 :

1.a	PPh_3
1.b	PPh_2Cy
1.c	$PPhCy_2$
1.d	PCy_3
1.e	PMe_3
1.f	$PnBu_3$
1.g	$PiPr_3$
1.h	$P(cC_5H_9)_3$
1.i	$P(CH_2C_6H_5)_3$

1.j $AsCy_3$ **1.k** $SbCy_3$

Chemical Engineering of [RuX₂(arene)(L)] Complexes

Influence of the ligand L

Assessment of the performance of these complexes in ATRP of methyl methacrylate revealed a strong correlation with the nature of the chosen phosphine (Table I). It appeared that only phosphines which are both strongly basic and which possess a well-defined steric bulk ($160° < \theta < 170°$ where θ is the cone angle of the phosphine ligand, Scheme 3) give rise to ruthenium-phosphine catalyst systems presenting both high activity and high control of the polymerization of methyl methacrylate (high initiation efficiency f, and low molecular weight distribution $M_w/M_n = 1.1$-1.2 (Figure 1)). It is typically the case of tricyclohexylphosphine (**1.d**), triisopropylphosphine (**1.g**), tricyclopentylphosphine (**1.h**) and, to a lesser extent, dicyclohexylphenylphosphine (**1.c**).

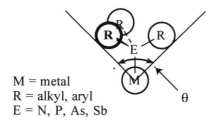

M = metal
R = alkyl, aryl
E = N, P, As, Sb

Scheme 3

In line with these observations, [RuCl₂(*p*-cymene)(ECy₃)] (E = P (**1.d**), As (**1.j**), and Sb (**1.k**)) complexes were then used to mediate ATRP of MMA (Table I). We observed that AsCy₃ exhibited considerable similarities to PCy₃, although the latter provided somewhat narrower polymer polydispersities than the former. On the other hand, tricyclohexylstibine exhibited significant differences since the polydispersities were then rather broad ($M_w/M_n = 1.9$-2.0).

To investigate the effect of nitrogen- and carbon-based ligand co-ordination, analogous ruthenium-*p*-cymene complexes with piperidine (**2.a**), 4-cyanopyridine (**2.b**), and isocyanides (**2.c-e**)

Cl--Ru
Cl   ᐟL   L : H—N⟨⟩   **2.a**   C=N-*n*Bu   **2.c**

                                          C=N-*t*Bu   **2.d**

                  N⟨⟩—CN **2.b**   C=N-Cy   **2.e**

were then examined as potential catalysts for ATRP. Our efforts were met with minimal success, regardless of the ligand used (Table I).

Table I. ATRP of Methyl Methacrylate Catalyzed by
[RuCl$_2$(p-cymene)(L)] Complexes^a

Ligand	pK_a	$\theta, °$	$T_D, °C^b$	yield, %	M_n^c	M_w/M_n	f^d
1.a	2.73	145	213	20	25 000	1.6	0.3
1.b	5.05	153	211	58	41 000	1.25	0.55
1.c	7.38	161-162	189	90	60 500	1.10	0.6
1.d	9.7	170	163	100	41 500	1.12	0.95
1.e	8.65	118	216	26	157 000	1.75	0.07
1.f	8.43	132	198	44	36 000	1.6	0.5
1.g	9.0	160	172	80	40 500	1.10	0.8
1.h	-	-	165	99	66 000	1.12	0.6
1.i	6.0	165	223	30	21 000	1.6	0.55
1.j	-	166	212	91	39 000	1.25	0.93
1.k	-	161	219	90	105 000	1.9	0.35
2.a	11.28	121	-	51	122 000	1.65	0.17
2.b	1.90	91.9	-	0	-	-	-
2.c	-	-	-	1	-	-	-
2.d	-	68-70	-	3	-	-	-
2.e	-	-	-	17	120 000	1.9	0.06
[RuCl$_2$(p-cymene)]$_2$				20	395 000	1.7	0.02

^{a} [MMA]$_0$:[initiator]$_0$:[Ru]$_0$ = 800:2:1 (initiator, ethyl 2-bromo-2-methyl-propionate; temperature, 85 °C; reaction time, 16 h; for experimental details, see Reference 8). ^{b} Temperature at which the p-cymene ligand is liberated as determined by TGA. ^{c} Determined by size-exclusion chromatography (SEC) with PMMA calibration. ^{d} f (initiation efficiency) = $M_{n,theor.}/M_{n,exp.}$ with $M_{n,theor.}$ = ([MMA]$_0$/[initiator]$_0$) x MW$_{MMA}$ x conversion.

Attempting to polymerize MMA in the presence of ruthenium-piperidine (**2.a**) and ruthenium-cyclohexylisocyanide (**2.e**) resulted in the sluggish formation of uncontrolled PMMA, while the other three complexes (**2.b-d**) were inactive (9).

Influence of the Halogen

An interesting trend was observed when the halogens were varied (Table II). Comparing [RuX$_2$(p-cymene)(PCy$_3$)] complexes, it is easily seen that going down the series from Cl to Br to I corresponds to a decrease in catalyst activity. It should be noted that in going from Cl to Br, catalyst activity is depressed only slightly, while changing to I has a precipitous effect. These observations are puzzling, as they suggest that the more electron withdrawing and smaller halogens generate more active catalysts.

**Table II. ATRP of Methyl Methacrylate Catalyzed by
[RuX₂(*p*-cymene)(PCy₃)] Complexes^a**

Halide, X	T_D, °C^b	yield, %	M_n^c	M_w/M_n	f^d
Cl	163	100	41 500	1.12	0.95
Br	171	83	52 000	1.25	0.65
I	182	73	45 000	1.65	0.65

$^{a,\ b,\ c,\ d}$ See Table I.

While consistent trends are observed throughout this series of catalysts, it appears that the steric and electronic effects of the phosphines upon catalyst activity are opposite to those observed for the halogens. Phosphines, which are larger and more electron donating, and likewise halogens, which are smaller and more electron withdrawing, lead to more active catalysts.

Influence of the Arene Ligand

The superior activity of PCy₃-containing complexes was confirmed with three different arene-ruthenium chlorides where the arene ligands were, respectively, *p*-cymene, benzene, and tetraline (Table III).

**Table III. ATRP of Methyl Methacrylate Catalyzed by
[RuCl₂(arene)(PCy₃)] Complexes^a**

Arene	T_D, °C^b	yield, %	M_n^c	M_w/M_n	f^d
p-Cymene	163	100	41 500	1.12	0.95
Benzene	174	95	38 000	1.13	0.99
Tetraline	209	85	40 000	1.15	0.85

$^{a,\ b,\ c,\ d}$ See Table I.

They all polymerized methyl methacrylate in a controlled fashion (M_w/M_n = 1.12-1.15). However, the *p*-cymene complex was found to be more active than the respective catalysts with benzene and tetraline ligands.

Polymerizability of Various Vinyl Monomers

The high efficiency of [RuCl₂(*p*-cymene)(PCy₃)] for the ATRP of MMA prompted us to investigate the performance of this catalyst in the polymerization of various vinyl monomers (Table IV).

Table IV. ATRP of Various Monomers Catalyzed by
[RuCl$_2$(*p*-cymene)(PCy$_3$)]^a

Monomer	initiator^b	yield, %	M_n^c	M_w/M_n	f^d
Methyl methacrylate	A	100	41 500	1.12	0.95
t-Butyl methacrylate	A	80	33 500	1.2	0.95
Isobornyl methacrylate	A	70	25 000	1.2	1.1
2-(Dimethylamino)ethyl methacrylate	A	90	30 500	1.85	1.2
Methyl acrylate	B	99	39 000	1.8	0.95
n-Butyl acrylate	B	80	37 500	1.95	0.85
Styrene	C	64	28 500	1.3	0.9
Vinyl acetate	A	0	-	-	-

$^{a,\ c,\ d}$ See Table I. ^{b} **A**, ethyl 2-bromo-2-methylpropionate, 85 °C; **B**, ethyl 2-bromopropionate, 85 °C; **C**, (1-bromoethyl)benzene, 110 °C.

Other methacrylates and styrene have also been successfully polymerized, although with a somewhat lesser control. 2-(Dimethylamino)ethyl methacrylate, methyl and *n*-butyl acrylate, however, were polymerized in an uncontrolled fashion, while vinyl acetate, a substrate known to be reluctant to undergo ATRP, was not polymerized under the same reaction conditions.

Furthermore, the best catalyst systems for ATRP of MMA have been shown to be also the best for ATRP of styrene (Figure 2). In both reactions, the same stereoelectronic requirements for the phosphine ligand of the ruthenium complex (sterically demanding basic phosphines) have been demonstrated.

Mechanistic Aspects

Controlled Radical Polymerization

With the most efficient catalyst systems, all the criteria of living polymerization are fulfilled. Indeed, the plots of ln([M]$_0$/[M]) versus time (Figure 3) and of M_n versus monomer conversion (Figure 4) are linear. Furthermore, control of MMA polymerization was confirmed by the addition of a second equivalent of MMA feed to the completely polymerized system. This second polymerization is also quantitative, and only a slight increase in polydispersity is observed (8).

Polymerizations apparently proceed by a radical mechanism, as indicated by the following results. For example, the polymerizations were perfectly inhibited by the addition of 1,1-diphenyl-2-picrylhydrazyl (DPPH) or galvinoxyl, which are well-known inhibitors for radical reactions. Furthermore, ^1H NMR analysis showed that the PMMA thus obtained were predominantly syndiotactic, very similar in steric structure to PMMA radically prepared with AIBN in toluene at 85 °C (8).

Ruthenium-Arene Catalysts

18-electron complexes of the general formula $[RuCl_2(\eta^6\text{-arene})(PR_3)]$ are also very active catalysts in olefin metathesis, including ring-opening metathesis polymerization (ROMP) (10), photoinduced ring-opening metathesis polymerization (PROMP) (11, 12), and ring-closing metathesis (RCM) (13). Loss of the arene ligand has been suggested to be a key step in these processes. Furthermore, the high efficiency of cationic ruthenium allenylidene complexes $[Ru=C=C=CPh_2(Cl)(PR_3)(p\text{-}cymene)]PF_6$ (R = iPr and Cy) for promoting non-conjugated diene and terminal yne-ene ring-closing metathesis (Scheme 4) has also been suggested to result from the displacement of the arene ligand (14).

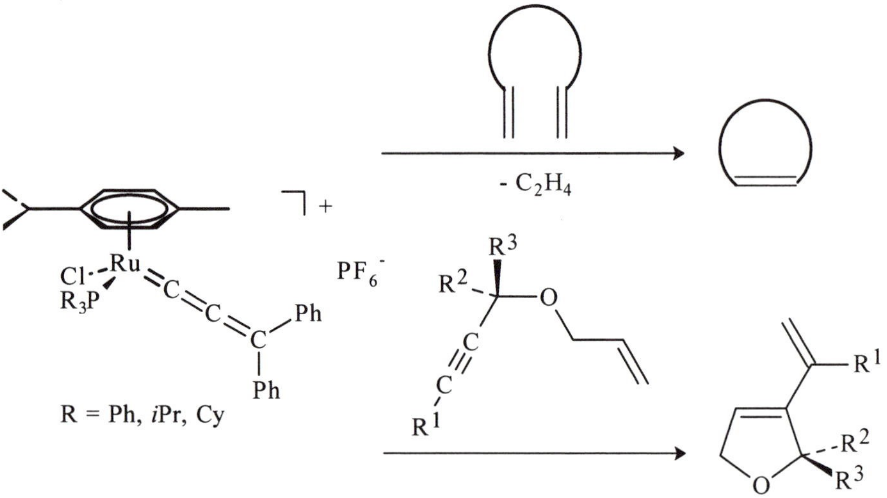

Scheme 4

The following results indicate that in ATRP the 18-electron catalyst precursor also loses the arene ligand to generate a highly coordinatively unsaturated ruthenium species.

1. ^{1}H NMR spectroscopy demonstrated the very fast liberation of the *p*-cymene ligand from [RuCl$_2$(*p*-cymene)(PCy$_3$)] at 85 °C in toluene-*d$_8$* and chlorobenzene-*d$_5$*.

2. The catalytic activity of [RuX$_2$(arene)(ER$_3$)] complexes parallels fairly well the lability of the arene ligand, as quantified by standardized thermogravimetric measurements (Tables I, II, and III, and Figure 5).

3. In line with previous observations (Table III), the catalytic activity of three [RuCl$_2$(arene)(PCy$_3$)] complexes increases in the order [RuCl$_2$(tetraline)(PCy$_3$)] (k_{app} = 3.328 x 10^{-5} s^{-1}) < [RuCl$_2$(benzene)(PCy$_3$)] (k_{app} = 5.203 x 10^{-5} s^{-1}) < [RuCl$_2$(*p*-cymene)(PCy$_3$)] (k_{app} = 10.725 x 10^{-5} s^{-1}) (Figure 3), paralleling the ease to displace the *p*-cymene group.

4. In order to support these observations, we synthesized novel ruthenium-arene complexes containing phosphines with pendant aryl groups [RuCl$_2$(*p*-cymene)-(PCy$_2$(CH$_2$)$_3$Ar)], **3.a** and **4.a**. The three methylene spacer allows an easy access to the corresponding chelated complexes, **3.b** and **4.b**, by simple heating in an inert solvent such as chlorobenzene.

3.a (R = H)
4.a (R = CH$_3$)

3.b (R = H)
4.b (R = CH$_3$)

Table V. ATRP of Methyl Methacrylate Catalyzed by Complexes 3 and 4^a

Complex	T_D, °C^b	yield, %	M_n^c	M_w/M_n	f^d
3.a	183	95	39 500	1.15	0.96
3.b	296	10	310 000	1.9	0.01
4.a	185	96	39 000	1.15	0.98
4.b	305	0	-	-	-

$^{a, b, c, d}$ See Table I.

With non chelated complexes **3.a** and **4.a** as catalyst precursors, ATRP proceeded efficiently, yielding 95% of PMMA with a number average molecular mass, M_n, of about 40 000, and a polydispersity index, M_w/M_n, of approximately 1.15. In sharp contrast to complexes **3.a** and **4.a**, the chelated analogues **3.b** and **4.b** were inefficient for promoting MMA polymerization under the same reaction conditions. For instance, complex **3.b** led to 10% MMA conversion and the polymerization was no longer controlled (Table V). This striking difference in activity may be related to the lability of the *p*-cymene ligand.

Thermogravimetric analyses of complexes **3.a** and **4.a** indicate that *p*-cymene release occurred at 183 and 185 °C, respectively. The high lability of the arene ligand in these complexes was also confirmed by ^{1}H NMR spectroscopy at 85 °C. On the other hand, chelated complexes **3.b** and **4.b** displayed much greater stability. They have been shown by ^{1}H, ^{13}C, and ^{31}P NMR spectrocopy to be stable for at least 8 h at 85 °C in deuterated aromatic solvents, and no labilization of the arene fragment was observed.

We may therefore suggest that *p*-cymene displacement is a key requirement in the activation process of these catalytic systems, giving rise formally to a 12-electron species. At this point, the stabilization of such highly unsaturated species appears to be of utmost importance. The observation that basic and bulky phosphines afford the most efficient catalyst systems for ATRP of methyl methacrylate, accounts for such stabilization effect. The question also arises about the possible co-ordination of the monomer during ATRP. ^{31}P NMR of a solution of complex **4.a** in a MMA:MMA-d_8 mixture (4 : 1) at 85 °C revealed that the intensity of the peak at δ 24.88 characteristic of complex **4.a** rapidly decreased to the benefit of a new signal at δ 40.98 culminating at 40% after 2-3 h, and then to a second signal at δ 28.36 (Figure 6). The latter absorption was assigned unambiguously to chelated complex **4.b**, whereas the former at δ 40.98 was tentatively assigned to a ruthenium-methacrylate complex, as suggested by the appearance of a new signal at δ 31.79 when [RuCl$_2$(*p*-cymene)(PCy$_3$)] (δ 24.7 (10)) was treated under the same conditions.

Conclusions

In exploring the reactivity of many analogous [RuX$_2$(arene)(PR$_3$)] catalysts for ATRP of methyl methacrylate, it was found that the ligand effects upon catalyst activity were exactly opposite for the phosphines and the halogens. Larger and more electron donating phosphines produced more active catalysts, while smaller and more electron withdrawing halogens likewise produced more active catalysts. Less bulky ligands such as piperidine, 4-cyanopyridine, and isonitriles led to inefficient catalysts. As indicated by several experimental facts, the mechanism would involve the dissociation of the arene ligand, giving rise formally to a 12-electron species at which activation of the alkyl halide (or polymer halide) bond could take place.

References

1. Sawamoto, M.; Kamigaito, M. Living Radical Polymerization, in Materials Science and Technology. A Comprehensive Treatment; Cahn, R. W.; Haasen, P.; Kramer, E. J. Editors; Synthesis of Polymers; Schlüter, A.-D., Volume Editor; Wiley-VCH, 1999, p 163.

2. Sawamoto, M.; Kamigaito, M. *Chemtech*, June 1999, 30.
3. Curran, D. P. *Comprehensive Organic Synthesis*, Trost, B. M.; Fleming, I. Eds.; Pergamon: Oxford, UK, 1991; Vol. 4, p 175.
4. Kato, M.; Kamigaito, M.; Sawamoto, M.; Higashimura, T. *Macromolecules* **1995**, *28*, 1721.
5. Wang, J.-S.; Matyjaszewski, K. *J. Am. Chem. Soc.* **1995**, *117*, 5614.
6. Wang, J.-S.; Matyjaszewski, K. *Macromolecules* **1995**, *28*, 7901.
7. Takahashi, H.; Ando, T.; Kamigaito, M.; Sawamoto, M. *Macromolecules* **1999**, *32*, 3820.
8. Simal, F.; Demonceau, A.; Noels, A. F. *Angew. Chem.* **1999**, *111*, 559; *Angew. Chem. Int. Ed.* **1999**, *38*, 538.
9. Simal, F.; Hallet, L.; Sebille, S.; Demonceau, A.; Noels, A. F. *Macromol. Symp.* **2000**, under press.
10. Demonceau, A.; Stumpf, A. W.; Saive, E.; Noels, A. F. *Macromolecules* **1997**, *30*, 3127.
11. Karlen, T.; Ludi, A.; Mühlebach, A.; Bernhard, P.; Pharisa, C. *J. Polym. Sci. Polym. Chem. Ed.* **1995**, *33*, 1665.
12. Hafner, A.; Mühlebach, A.; van der Schaaf, P. A. *Angew. Chem.* **1997**, *109*, 2213; *Angew. Chem. Int. Ed. Engl.* **1997**, *36*, 2121.
13. Fürstner, A.; Ackermann, L. *J. Chem. Soc., Chem. Commun.* **1999**, 95.
14. Fürstner, A.; Picquet, M.; Bruneau, C.; Dixneuf, P. H. *J. Chem. Soc., Chem. Commun.* **1998**, 1315.
15. Picquet, M.; Bruneau, C.; Dixneuf, P. H. *J. Chem. Soc., Chem. Commun.* **1998**, 2249.
16. Picquet, M.; Touchard, D.; Bruneau, C.; Dixneuf, P. H. *New J. Chem.* **1999**, 141.

Chapter 17

Polychloroalkanes as ATRP Initiators: Fundamentals and Application to the Synthesis of Block Copolymers from the Combination of Conventional Radical Polymerization and ATRP

M. Destarac[1-3], B. Boutevin[2], and Krzysztof Matyjaszewski[1]

[1]Department of Chemistry, Carnegie Mellon University,
4400 Fifth Avenue, Pittsburgh, PA 15213
[2]Laboratoire de Chimie Macromoléculaire,
Ecole Nationale Supérieure de Chimie de Montpellier,
8 rue de l'Ecole Normale, 34296 Montpellier Cedex 5, France

Polychloroalkanes CCl_nR_{4-n} (n = 2, 3 or 4) are tested as initiators in Cu-catalyzed atom transfer radical polymerization (ATRP) of methyl methacrylate (MMA) and methyl acrylate (MA), using the CuCl / 2,2'-bipyridine catalyst. 1,1,1-Trichloroalkanes, $RCCl_3$, are good initiators. For all the R groups tested, M_n increases with conversion and polydispersities are low ($1.1 < M_w/M_n < 1.3$). CCl_4 induces a multifunctional initiation, with lower than predicted M_n values. This deviation, previously observed with Ru and Ni-based catalysts, has been explained by the doubling of chains resulting from the activation of a central $-CCl_2-$ moiety, followed by a β-scission. Trichloromethyl-terminated vinyl acetate and vinylidene fluoride telomers are synthesized and used to initiate ATRP of various monomers, leading to PVOAc and PVDF-based block copolymers.

Polychloroalkanes CCl_nR_{4-n} (n=2, 3 or 4) have been used for a long time in redox-catalyzed radical addition and telomerization (1). Their reactivity increases with increasing number of chlorine atoms per molecule. Because of their moderate reactivity, very few examples of initiation by mono- and dichlorocompounds can be found in the literature. On the other hand, trichloroalkanes and especially carbon tetrachloride combined with various transition metals were proved to be efficient initiating systems for the synthesis of low molecular weight telomers with high yields, most of time 1:1 adducts (1). In the case of 1,1,1-trichloroalkanes, $RCCl_3$, it is known

---

[3]Current address: Rhodia, Centre de Recherches d'Aubervilliers, 52 rue de la Haie Coq, 93308 Aubervilliers Cedex, France.

that if the presence of a trichloromethyl group is necessary to ensure a reasonable reactivity, the nature of the neighboring substituent R has an important impact on the overall reactivity of the system (i.e yield of 1:1 adduct). When applied to Cu-catalyzed ATRP (2,3) of styrene, the presence of the 2,2'-bipyridine ligand (bpy) adjusts the activity of the catalyst in such a way that 1,1,1-trichloroalkanes are fast and nearly quantitative initiators, even those weakly reactive in radical addition (3a,c,d). Moreover, a multifunctional initiation has been evidenced for some initiators containing an electron-withdrawing substituent in the α-position to the trichloromethyl group, e.g. Cl or CO_2CH_3. The first section of this chapter deals with the case of MMA and acrylates.

Trichloromethyl-terminated polymers were obtained by two different ways: first through the use of a CCl_3-terminated azo initiator (3c), second with the telomerization of some monomers (vinyl acetate -VOAc- (4) and vinylidene fluoride -VDF- (5)) with chloroform, leading to well-defined $CCl_3(M)_nH$ telomers. A CCl_3-terminated poly(n-butyl acrylate) was synthesized using a CCl_3-terminated azo initiator (3c), and successfully used as a macroinitiator for ATRP of styrene to synthesize block copolymers. In the second part of this chapter, we describe the use of PVOAc and PVDF-based block copolymers by ATRP initiated by VOAc and VDF telomers, respectively.

Initiation with Polychloroalkanes. Fundamentals.

MMA.

A series of CCl_nR_{4-n} -type initiators (Table I) was chosen to initiate ATRP of MMA in 50 vol.-% 1,2-dimethoxybenzene at 90°C. Figure 1 represents the evolution of M_n and M_w/M_n with conversion. It shows that all the initiators, in conjunction with the CuCl / bpy complex, promote a controlled polymerization. In all cases, the molecular weight increases during polymerization, and polydispersities are low ($M_w/M_n < 1.3$ at high conversion). Polymerizations are fast (60-80% conversion after two hours according to initiator). CCl_3CH_2OH (1), $CCl_3C_8H_{17}$ (2) and $CCl_3CH_2CF_2Cl$ (3) initiate polymerization with an efficiency factor f (defined as $M_{n,th}$ / $M_{n,GPC}$) equal to 0.70, 0.73 and 0.87, respectively. In order to explain this partial initiation, the initiator consumption was followed by GC. 56% of 2 is consumed after 12% monomer conversion in half an hour (M_n=4100 g / mol). After one hour, the analysis of a second sample shows 36% monomer conversion and no residual peak corresponding to 2 (M_n=7340 g / mol). This means that slow initiation is not responsible for the observed incomplete generation of chains from the initiator. A significant amount of radicals is presumably deactivated irreversibly in the early stages of polymerization via radical-radical coupling. Initiations by CCl_4 (4) and $CCl_3CO_2CH_3$ (5) show an evolution of M_n with values matching theoretical predictions ($M_{n,th}=([M]_0/[RCCl_3]_0)*$(monomer conversion)+ $(MW)_{RCCl3}$) during the first half of the polymerization; then, the control over M_n is gradually lost with final values that are lower than expected. These results are consistent with works of Sawamoto (for 4 (6) and 5 (6c)) and Teyssie (4) (7) for

236

other transition metal catalysts. As opposed to initiators **1**, **2** and **3**, a nearly quantitative initiation can be assumed considering the very good correlation between experimental and theoretical M_n below 4000 g / mol (Figure 1). In order to shed light on the peculiar effect of **4** and **5** on M_n, a deeper investigation of the initiation by CCl_4 has been conducted.

Table I. Initiator description

Code	Structure
1	CCl_3CH_2OH
2	$CCl_3C_8H_{17}$
3	$CCl_3CH_2CF_2Cl$
4	CCl_4
5	$CCl_3CO_2CH_3$
6	$CH_3CCl_2CH_3$
7	$HCCl_2CH_2OH$
8	$CCl_3CH_2C(CH_3)ClCO_2CH_3$

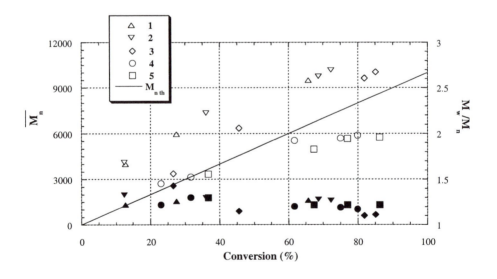

Figure 1. *Evolution of number average molecular weight, M_n, and polydispersity, M_w/M_n, with conversion in ATRP of MMA initiated by various $RCCl_3$ initiators. $RCCl_3$ / CuCl / bpy / MMA = 1 / 0.5 / 1 / 100 (50 vol.- % 1,2-dimethoxybenzene). $T=90^0C$. Opened symbols relate to molecular weights and the filled ones to polydispersities.*

Low molecular weight PMMA samples ($[M]_0$ / $[\text{Initiator}]_0 = 10$) were synthesized with CCl_4 and a model compound, 2,4,4,4-tetrachloro-2-methyl butyrate $CCl_3CH_2C(CH_3)ClCO_2CH_3$ (**8**) as initiators. These products were purified and analyzed by ^{13}C NMR spectroscopy.

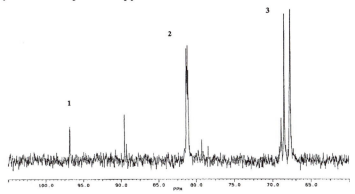

Figure 2. *^{13}C NMR analysis of an oligoMMA initiated with CCl_4 (NMR solvent: acetone d-$_6$). CCl_4 / $CuCl$ / bpy / MMA = 1 / 1 / 3 / 10, in bulk. $T = 130^0C$. $M_n=750$, M_w / $M_n = 1.38$.*

A ^{13}C NMR spectrum of the oligoMMA initiated by CCl_4 is shown in Figure 2. In agreement with the spectrum of **8**, the peak **1** at $\delta = 97$ ppm has been attributed to a trichloromethylated chain end ($\underline{C}Cl_3$-MMA-) and the series of peaks **3** between 67 and 69 ppm to a monochlorinated chain end (–MMA-Cl). The broad signal **2** centered at 81.5 ppm represents a central dichlorinated carbon atom –MMA-$\underline{C}Cl_2$-MMA- (a neighboring structure, -MMA-$\underline{C}Cl_2$-CH_2CCl_2-, has been characterized in a previous publication (8). The chemical shift corresponding to the central –CCl_2- moiety was reported to be located at 80.1 ppm). This result indicates that once oligomeric halides $CCl_3(MMA)_nCl$ are generated from CCl_4, the trichloromethyl terminal groups is rapidly activated by the metal catalyst. At this stage, polymerization proceeds with two propagating sites per chain.

Results described in Figure 1 show that M_n increases linearly with conversion and roughly matches theoretical values up to about half of the polymerization. Beyond this point, another reaction contributes to generate new polymer chains. In order to investigate the possible initiation pathways once a difunctional initiation is established, two dichloroalkanes were tested as initiators: 2,2-dichloropropane $CH_3CCl_2CH_3$ (**6**) and 2,2-dichloroethanol $HCCl_2CH_2OH$ (**7**). Indeed, these compounds do initiate polymerization, as shown in Table II. However, initiation is very slow, leading to much higher M_n than expected. Thus, once difunctional initiation is established and although slow compared to the activation of chain ends, an additional radical generation is likely to occur on the central –CCl_2- moiety, with a rate close to that of **6** or **7**. If this type of central radical is able to initiate polymerization, a three-directional propagation is established, i.e a branching point is formed. The ratio of the intrinsic viscosity *[η]* of a branched molecule to that of a linear molecule with the same molecular weight for a uniform three-arm PMMA was reported as equal to 0.90 by Hatada et al. (9). Therefore, it seems unlikely that a change in hydrodynamic volumes between linear and branched PMMA could explain the M_n profile. A possibility is for the newly generated radical **9** (cf. Scheme 1), which has a structure analogous to those formed in copolymerization of methacrylate macromonomers (10), to generate a new PMMA growing chain and an α-chloro macromonomer **10** by β-scission (Scheme 1). The proposed mechanism is supported by the ^{1}H NMR analysis of an oligoMMA initiated by **8** (Figure 3). The series of peaks between 5 and 5.3 ppm has been assigned to the vinyl protons of **10** ($C\underline{H}_2=CClCH_2-$), which is in excellent correlation with works of Rizzardo et al. on vinyl acetate analogs (signals at 5.15-5.3 ppm were attributed to vinyl protons $C\underline{H}_2=CClCH_2-PVOAc$) (11).

Table II. ATRP of MMA initiated by 6 and 7. Initiator / CuCl / bpy / MMA = 1 / 0.5 / 1 / 30 (50 – vol. % 1,2-dimethoxybenzene). T=90°C.

Initiator	t (hrs)	Initiator Conv. (%)	MMA Conv. (%)	$M_{n\,th}$ [a]	M_n	M_w/M_n
6	1	8.2	2.4	185	34320	1.51
6	9	18.1	75.6	2380	100100	1.71
7	1	5	2.2	180	38600	1.30
7	9	14.4	81.7	2570	120300	2.06

[a] $M_{n\,th}=([M]_0/[Initiator]_0)*$ (monomer conversion)$+ (MW)_{Initiator}$

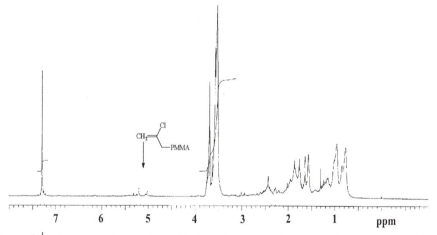

Figure 3. ^{1}H NMR spectrum of an oligoMMA initiated by $CCl_3CH_2C(CH_3)ClCO_2CH_3$ **8**. **8** / CuCl / bpy / MMA =1 / 1 / 3 / 10, in bulk. T = 130°C. M_n=780, M_w/M_n=1.44. NMR solvent: $CDCl_3$.

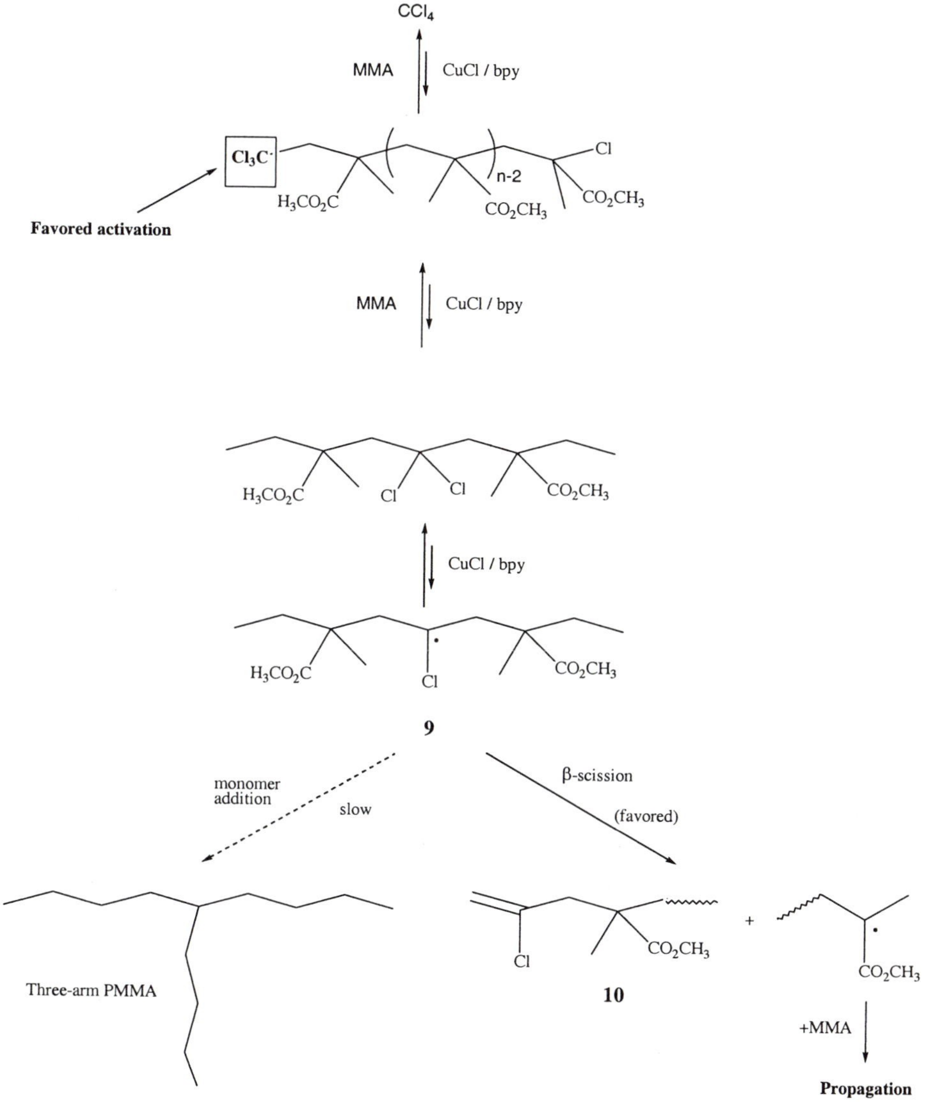
CCl4
MMA
CuCl / bpy
Cl3C·
n-2
Cl
H3CO2C
CO2CH3
CO2CH3
Favored activation
MMA
CuCl / bpy
H3CO2C
Cl
Cl
CO2CH3
CuCl / bpy
H3CO2C
Cl
CO2CH3
9
monomer addition
slow
β-scission
(favored)
Three-arm PMMA
Cl
CO2CH3
10
+
CO2CH3
+MMA
Propagation

Scheme 1.

Acrylates.

The use of RCCl₃ initiators in polymerization of MA has been investigated. Figure 4 shows the evolution of M_n and M_w/M_n with conversion for an initiation by **2** and **3**. The very good correlation between experimental and theoretical M_n throughout polymerization shows that **2** and **3** are fast and nearly quantitative initiators for the polymerization of MA. Polydispersities are low at high conversions (M_w/M_n <1.3). It is noticeable that polymerizations are quite slow (several days), when compared to polymerizations initiated by alkyl halides (*12*). Slow polymerization can be explained by a Cu (II) concentration higher than when initiated by alkyl halides, caused by a faster initiation (*3d*). ESR studies showed that less Cu (II) is present during polymerization of acrylates than for that of MMA (*13*) and styrene (*14*). Acrylate based-systems are more sensitive to Cu (II), which explains the dramatic decrease of rates observed for RCCl₃ initiators. To prevent this drawback, Cu (0) metal turnings were added to remove the excess of Cu (II) (*15*). This resulted in a considerable increase of the polymerization rate (more than three times faster) while maintaining an excellent control over molecular weights and polydispersities (Figure 4).

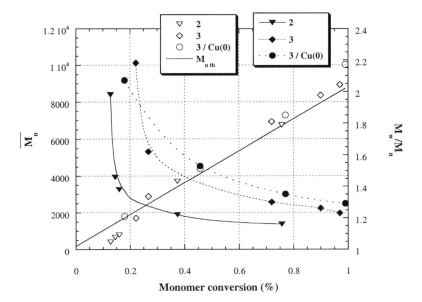

Figure 4 . *Dependence of M_n and M_w/M_n on conversion for the polymerization of MA by ATRP initiated by **2** and **3** and catalyzed by CuCl / bpy. RCCl₃ / CuCl / bpy / MA = 1 / 0.5 / 1 /100. T=90⁰C. 50 Vol. -% in 1,2-dimethoxybenzene for an initiation by **2**, 10 w. -% in the same solvent with **3** as an initiator. 6 weight-equivalents of Cu (0) turnings with respect to CuCl were used. Open symbols relate to molecular weights and the filled ones to polydispersities.*

Synthesis of Trichloromethyl-Terminated Polymers by Radical Polymerization and their Use as ATRP Macroinitiators for the Synthesis of Block Copolymers.

Far before the emergence of controlled radical polymerization, a first step towards control of radical polymerization was achieved by many research teams over the past twenty years by incorporating well-defined terminal functionalities. By using adequate multifunctional radical initiators or functional transfer agents, a wide range of functional polymers were used as starting materials to synthesize block copolymers (*16*). Among these methods, telomerization has been widely investigated in our group (*1*). We recently reported that radical telomerization of VOAc (*4*) and VDF (*5*) with chloroform leads to well-defined CCl_3-functionalized telomers. In the following section, we report the synthesis of block copolymers from α-trichloromethylated PVOAc and PVDF ATRP macroinitiators (Scheme 2).

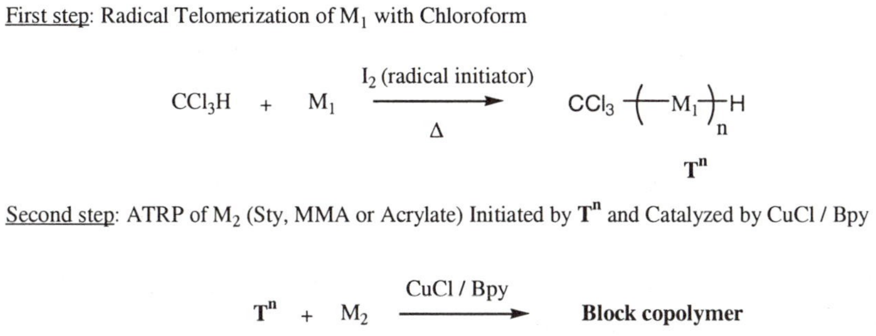

First step: Radical Telomerization of M_1 with Chloroform

$$CCl_3H \quad + \quad M_1 \quad \xrightarrow[\Delta]{I_2 \text{ (radical initiator)}} \quad CCl_3 \left(\!-M_1\!\right)_n H$$

$$T^n$$

Second step: ATRP of M_2 (Sty, MMA or Acrylate) Initiated by T^n and Catalyzed by CuCl / Bpy

$$T^n \quad + \quad M_2 \quad \xrightarrow{CuCl / Bpy} \quad \textbf{Block copolymer}$$

Scheme 2.

Vinyl Acetate
A detailed kinetic study of the radical telomerization of VOAc with chloroform and structural analysis of the related $CCl_3(CH_2CHOAc)_nH$ telomers showed that well-defined structures were obtained up to relatively high molecular weights (*4*). $CCl_3CH_2CH_2OAc$ (T^1. In T^n, n represents the number average degree of polymerization calculated from ^{1}H NMR) is an appropriate model compound to study the ability of higher molecular weight telomers to initiate ATRP of styrene efficiently. Indeed, reactions initiated by T^1 show a linear increase of M_n with conversion with a small downward deviation from theory at high conversions, presumably due a little amount of side reactions (Figure 5). Moreover, experimental molecular weights match theoretical values at low monomer conversion, characteristic of a fast and nearly quantitative initiation. Polydispersities are consistently low (M_w/M_n around 1.3- Figure 5). The apparent polymerization rate is first order with respect to monomer (*4b*), which means that the concentration of growing radicals is constant throughout polymerization. For an initiation with T^9, the polydispersity drops from a relatively high initial value - M_w / M_n = 1.90 - to 1.37 after only 14% conversion, characteristic of a fast initiation. When ATRP of St is initiated with T^{62}, although molecular weights increase from that of macroinitiator, M_n values are lower than expected (Figure 5).

242

However, after precipitation in hexane (*18*), a combination of UV and RI detections for GPC analysis (Figure 6) indicate a high purity of the copolymer. Also, polydispersities decrease significantly with monomer conversion (Figure 5).

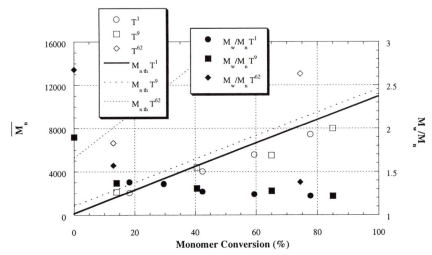

Figure 5. *Evolution of M_n and M_w/M_n with monomer conversion in ATRP of styrene initiated by T^n and catalyzed by CuCl / Bpy. T^1 and T^9) $[M]_0 = 8.7$ M, $[T]_0 = [CuCl]_0 = 1/3 [Bpy]_0 = 0.87$ M. T = 130°C. T^{62}) $[M]_0 = 4.35$ M, $[T]_0 = [CuCl]_0 = 0.5 [Bpy]_0 = 2.18.10^{-2}$ M. 50 vol.-% in diphenyl ether. T = 130°C. Opened symbols relate to molecular weights and the filled ones to polydispersities.*

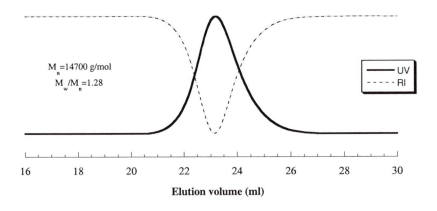

Figure 6. *UV and refractive index traces of a precipitated PVOAc-b-PS copolymer (initiated with T^{62}). Reactions conditions of Figure 6 (18)*

Vinylidene Fluoride

Poly(vinylidene fluoride) (PVDF) is an important engineering polymer whose remarkable piezoelectrical properties and weatherability gave rise to a broad range of applications. Because it is an expensive material, attention has been paid in the past to its ability to be blended with other polymers. However, the literature is very limited regarding PVDF blends (*19*) because of the lack of appropriate compatibilizers- i.e PVDF-based block or graft copolymers. Very recently, the combination of telomerization and ATRP was used for the synthesis of PSt-b-PVDF-b-PSt block copolymers starting from a BrCF$_2$-terminated VDF difunctional telomer (*20*). The results described in the latter report indicate very long reaction times (several days), uncontrolled molecular weights and large polydispersities. In this communication, we report the preparation of CCl$_3$(VDF)$_n$H telomers and their use as new ATRP initiators, giving access to PVDF-based block copolymers according to the two-step procedure described in Scheme 2.

The telomerization of VDF with chloroform has been studied in detail (*21*). The structural elucidation of the telomers by NMR confirms that the telomer structures are mainly controlled by a proton transfer from chloroform to the vinylidene fluoride radical (*21b*). As for VOAc telomers, the 1:1 adduct (1,1,1-trichloro-2,2-difluoropropane CCl$_3$CH$_2$CF$_2$H) is an appropriate ATRP model initiator. Its isolation from a low molecular weight telomer distribution was attempted. The purest distillation fraction was contaminated by 17% of diadduct. We chose to consider the CCl$_4$/VDF 1:1 adduct (*22*) **3** as a model initiator. In Cu-catalyzed redox telomerization, radical generation from the trichloromethyl end of **3** is highly favored compared with that of the CF$_2$Cl end (*22*), which means that little or none radical generation is expected via the latter end when T^1 is used to initiate ATRP. Figure 7 shows the evolution of molecular weights and polydispersities with monomer conversion for ATRP of styrene, MMA and MA initiated by **3** and catalyzed by CuCl/bpy. M$_n$ increases linearly with conversion for all monomers, and values are very close to those predicted, assuming a quantitative initiation and no transfer (M$_{n,th}$ = ([M]$_0$/[**3**]$_0$)*(monomer conversion)+(M.W)$_3$). Polydispersities are low at high monomer conversion (1.1<M$_w$/M$_n$<1.3 for all monomers).

ATRPs of St, MMA, MA and n-butyl acrylate (BA) have been initiated with VDF telomers with DP$_n$ values equal to 5, 11 and 16. Results depicted in Table III show that for all the T^n / monomer pairs, M$_n$ increases from that of the VDF telomer during polymerization and values are quite close to the theoretical ones. Starting from fractionated VDF telomers (by distillation or precipitation) with narrow molecular weight distributions (1.03< M$_w$/M$_n$ <1.11- measured by GPC in DMF calibrated with PMMA standards), polydispersities remain very low during polymerization (typically M$_w$/M$_n$<1.2 at high monomer conversion- Table III). Moreover, no residual trace of the starting PVDF could be detected by GPC (DMF as an eluent) when superimposing the initial VDF telomer (T^{11}) and the derived block copolymer traces (Figure 8). This

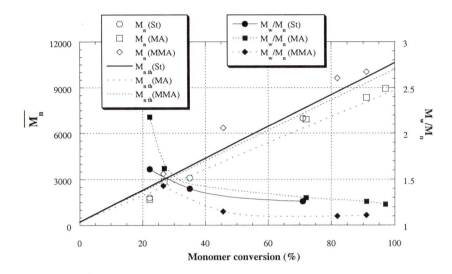

Figure 7. *Evolution of M_n and M_w/M_n with conversion in ATRP of St, MMA and MA initiated by T^I and catalyzed by CuCl / Bpy. T^I / CuCl / Bpy / Monomer = 1 / 0.5 / 1 /100 for MMA (50 vol. -% in 1,2-dimethoxybenzene, T=90⁰C), MA (10 wt. -% in 1,2-dimethoxybenzene, T=90⁰C), and St (10 wt. -% in 1,2-dimethoxybenzene, T=130⁰C). Open symbols relate to molecular weights and the filled ones to polydispersities.*

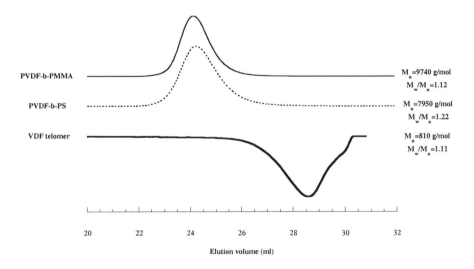

Figure 8. *GPC traces of T^{II} and corresponding block copolymers with St (Table III, entry 5) and MMA (Table III, entry 9).*

Table III. ATRP initiated by VDF Telomers. M_n and M_w/M_n vs. Conversion Relationship. T" / CuCl / Bpy / Monomer = 1 / 0.5 / 1 / 100 for all monomers.

Momomer (concn (M) in solvent, T°C)	VDF telomer (10^2x concn (M))	Time (h)	$M_{n\ th}$	M_n^a	M_w/M_n	Conversion (%)
MMA (4.7 M DMB[b], 90)	T[1] (4.70)	2	8520	9620	1.10	85.1
MMA (4.7 M acetone, 90)	T[5] (4.70)	1	2430	3065	1.18	19.6
MMA (4.7 M acetone, 90)	T[5] (4.70)	16.5	9470	10210	1.13	90.0
MMA (4.7 M acetone, 90)	T[11] (4.70)	1.75	4420	6200	1.13	37.0
MMA (4.7 M acetone, 90)	T[11] (4.70)	17	9520	9740	1.12	89.1
MMA (4.7 M acetone, 90)	T[16] (4.70)	3.33	7540	8290	1.15	64.0
Sty (7.98 M DMB, 130)	T[1] (7.98)	9.66	7610	7000	1.27	71.0
Sty (4.35 M DPE[c], 130)	T[11] (4.35)	0.5	2470	2110	1.27	16.0
Sty (4.35 M DPE, 130)	T[11] (4.35)	15.75	8705	7950	1.22	75.8
BA (3.49 M acetone, 90)	T[5] (3.49)	1	1910	1140	1.21	11.3
BA (3.49 M acetone, 90)	T[5] (3.49)	17	2110	1900	1.25	13.9
BA (3.49 M acetone, 90)	T[5] (3.49)	140	3700	4440	1.19	25.7
BA (3.49 M acetone, 90)[d]	T[5] (3.49)	140	9590	10600	1.30	73.1
MA(10.2 M DMB, 90)	T[1](10.2)	42.25	8050	8360	1.26	91.0

[a] measured by GPC: THF eluent for samples initiated with T[1], DMF eluent otherwise (PMMA standards for MMA, PS standards for St , MA and BA) [b] 1,2-dimethoxybenzene [c] diphenyl ether [d] addition of 6 weight-equivalents of Cu (0) turnings with respect to CuCl

result strongly supports the formation of pure block structures. In the case of acrylates (MA and BA), polymerizations are very slow (ATRP of BA reaches about 25% conversion after 140h). Slow polymerization can be explained by a Cu (II) concentration higher than when initiated by alkyl halides, generated by a faster initiation as mentioned in the first part of this chapter. Here again, Cu (0) metal turnings were added (in the case of BA-Table III, Entry 13) to remove the excess of Cu (II) (15). This resulted in a considerable increase of the polymerization rate while maintaining quite good control over molecular weights and polydispersities.

Conclusions

1,1,1-Trichloroalkanes and CCl_4 are good initiators for ATRP of St (3d), MMA and MA catalyzed by the CuCl / bpy catalyst. CCl_4 and other α-activated $RCCl_3$-type initiators are multifunctional initiators. In the case of MMA, the previously observed deviation of M_n with Ru and Ni-based catalysts has been explained by the doubling of chains resulting from the activation of a central $-CCl_2-$ moiety, followed by a β-scission. ATRP of MA is slow but can be strongly accelerated by adding some Cu (0) metal, without affecting the control of the reaction.

246

A dual method comprising telemerization of a first monomer M_1 (vinyl acetate or vinylidene fluoride) with chloroform, followed by ATRP of a second monomer M_2 initiated by the resulting $CCl_3(M_1)_nH$ telomers has been developed. By varying $[CCl_3H]_0$ / $[M_1]_0$ and $[M_2]_0$ / $[T^n]_0$ ratios in the telemerization and ATRP steps, respectively, chain length of both blocks and copolymer composition can be easily predetermined. Because of the excellent ability of the trichloromethyl group to initiate ATRP and the versatility of telemerization, it appears that this dual process shows great promise for the preparation of new block copolymers comprising one block from a monomer whose polymerization is controlled by none of the known CRP methods.

Acknowledgements

The authors are grateful to Elf Atochem NA for financial support of this project.

References

1. Boutevin, B.; Pietrasanta, Y. "Telomerization" in *Comprehensive Polymer Science*; Allen, G.; Bevington, J. C.; Eastmond, A. L. (Eds.), Pergamon Press, Oxford, Vol. 3, 185, **1989**.

2. Matyjaszewski, K., Ed. *Controlled Radical Polymerization*; ACS Symposium Series 685; American Chemical Society: Washington, DC, **1998**.

3. (a) Destarac, M.; Bessière, J-M.; Boutevin, B. *Polym.Prepr. (Am. Chem. Soc., Div. Polym. Chem.)* **1997**, *38 (1)*, 677. (b) Destarac, M.; Bessière, J-M.; Boutevin, B. *Macromol. Chem., Rapid. Commun.* **1997**, *18*, 967. (c) Destarac, M.; Boutevin, B. *Polym.Prepr. (Am. Chem. Soc., Div. Polym. Chem.)* 1998, *39 (2)*, 568. (d) Destarac, M.; Bessière, J-M.; Boutevin, B. *J. Polym. Sci A: Polym. Chem.* **1998**, *36*, 2933. (e) Destarac, M.; Alric, J.; Boutevin, B. *Polym. Prepr. (Am. Chem. Soc., Div. Polym. Chem.)* **1998**, *39 (1)*, 308.

4. (a) Destarac, M.; Pees, B.; Bessière, J-M.; Boutevin, B. *Polym.Prepr. (Am. Chem. Soc., Div. Polym. Chem.)* **1998**, *39 (2)*, 566. (b) Destarac, M.; Pees, B.; Boutevin, B. Submitted to *Macromol. Chem. Phys.*

5. (a) Ameduri, B.; Duc, M.; Kharroubi, M.; Boutevin, B. *Polym.Prepr., Am. Chem. Soc., Div. Polym. Chem.* **1998**, *39 (2)*, 845. (b) Duc, M.; Ameduri, B.; Boutevin, B. Submitted to J. Polym. Sci A: Polym. Chem.

6. (a) Kato, M.; Kamigaito, M.; Sawamoto, M.; Higashimura, T. *Macromolecules* **1995**, *28*, 1721. (b) Kotani, Y.; Kato, M.; Kamigaito, M.; Sawamoto, M. *Macromolecules* **1996**, *29*, 6979. (c) Ando, T.; Kamigaito, M.; Sawamoto, M. *Tetrahedron* **1997**, *53, 45*, 15445.

7. Granel, C.; Dubois, Ph.; Jérôme, R.; Teyssié, Ph. *Macromolecules* **1996**, *29*, 8576.

8. Alfiguigui, C.; Ameduri, B.; Bessière, J-M.; Boutevin, B. *Eur. Polym. J.* **1996**, *32, 2*, 135.

9. Hatada, K.; Nishiura, T.; Kitayama, T.; Tsubota, M. *Polym. Bull.* **1996**, *36*, 399.

10. Krstina, J.; Moad, G.; Rizzardo, E.; Winzor, C. L.; Berge, C. T.; M. Fryd. *Macromolecules* **1995**, *28*, 5381.

11. Rizzardo, E.; Chong, Y.K.; Evans, R.A.; Moad,G.; Thang, S.H. *Macromol. Symp.* **1996**, *111*, 1.

12. Wang, J-S.; Matyjaszewski. K. *Macromolecules* **1995**, *28*, 7901.

13. Kajiwara, A.; Matyjaszewski, K. *Macromol.Rapid.Commun.* **1998**, *19*, 319.

14. Matyjaszewski, K.; Kajiwara, A. *Macromolecules* **1998**, *31*, 548

15. Matyjaszewski, K.; Coca, S.; Gaynor, S. G.; Wei, M.; Woodworth, B. E *Macromolecules* **1997**, *30*, 7348.

16. Ameduri, B.; Gramain, P.; Boutevin, B. *Adv. Polym. Sci.* **1997**,*127*, 87.

17. In the 5-6 K range, M_n values of PVOAc measured by GPC calibrated with PSt standards are nearly 10% underestimated. M. Destarac, unpublished results.

18. It is worth mentioning that the precipitation hardly affects M_n and M_w/M_n values (M_n=13050 and M_w/M_n=1.38 before precipitation; M_n=14700 and M_w/M_n=1.28 after precipitation).

19. Ouhadi. T.; Fayt, R.; Jérôme, R.; Teyssié, Ph. *J. Appl. Polym. Sci.* **1986**, *32*, 5647.

20. Ying, S.; Zhang, Z.; Shi, Z. *Polymer* **1999**, *40*, 1341.

21. (a) Ameduri, B.; Duc, M.; Kharroubi, M.; Boutevin, B. *Polym.Prepr., Am. Chem. Soc., Div. Polym. Chem.* **1998**, *39 (2)*, 845. (b) Duc, M.; Ameduri, B.; Boutevin, B. *J. Polym. Sci A: Polym. Chem.* In press.

22. Boutevin, B.; Furet, Y.; Lemanach, F.; Vial-Revillon, F. *J. Fluor. Chem.* **1990**, *47*, 95.

Chapter 18

Controlled Radical Polymerization Utilizing Preformed Metal Complexes Based on Substituted Bipyridines and Terpyridines

U. S. Schubert, G. Hochwimmer, C. E. Spindler, and O. Nuyken

Lehrstuhl für Makromolekulare Stoffe, Technische Universität München, Lichtenbergstrasse 4, D–85747 Garching, Germany

Metal complexes based on different substituted dimethyl-2,2'-bipyridines and 5,5"-dimethyl-2,2':6',2"-terpyridine were used in combination with aluminium isopropoxide for the controlled radical polymerization of styrene, acrylate and methacrylate derivatives. Depending on the substitution pattern and the central metal ion the effectivity of the systems revealed distinct differences. The control over molecular weight, molecular weight distribution and reaction rate depends strongly on the choice of central metal ion and the ligand system. Best results concerning control of molecular weight and molecular weight distribution for styrene and acrylates were obtained using a combination of $[Cu(II)(4,4'-dimethyl-2,2'-bipyridine)_3](PF_6)_2$, aluminium isopropoxide, (1-bromoethyl)benzene as initiator and acetonitrile as solvent at 75°C. Furthermore, the polymers obtained were colorless with a very low content of residual metal ions. Using this polymerization system, defined block copolymers and other special architectures are accessible.

Radical polymerization processes are responsible for more than half of the overall production of polymers worldwide and have therefore an enormous impact for the industrial economies. However, in conventional radical polymerization processes control of molecular weight and especially molecular weight distribution is rarely possible (1). Therefore tailor-made polymeric materials and special architectures like block copolymers are usually not accessible. This is mainly due to chain transfer and termination processes. In order to overcome the drawbacks of these common procedures and to open new avenues for the design of polymers and block copolymers via radical methods, controlled ("living") free radical polymerization techniques have attracted much interest in the last years. Besides nitroxide-mediated polymerization (2 - 9) and triazolinyl-based polymerization (10, 11), metal-mediated polymerization utilizing e.g. copper(I), ruthenium(II), nickel(II), iron(II) or rhenium(V) (12 - 37) shows a great potential for industrial application. Especially the so called **atom transfer radical polymerization** (ATRP) system allows the controlled polymerization of styrene and acrylate type monomers, the synthesis of block copolymer or other polymer architectures (38 - 40). Based on a system containing alkyl halides and *in situ* generated complexes of bidentate, tridentate or polydentate ligands, e. g. bipyridine or multifunctional amino compounds with copper(I) ions, polymers with narrow molecular weight distribution, predetermined molecular weight or block copolymers become accessible.

However, besides this important advantages of the ATRP system there are also some serious drawbacks: (1) Polymers contain relatively high amount of metal ions (e. g. copper); (2) Polymers are often colored, however, colorless polymers can be obtained by reducing the amount of copper and ligands (41); (3) The fine-tuning of the "catalytic system" is rather difficult, e. g. in regard to reaction rates or polydispersities; (4) Addition of solid materials is necessary; (5) For homogeneous polymerization's bipyridines or terpyridines with long alkyl chains have to be synthesized or addition of co-solvents is necessary (42). Furthermore, the use of different counterions has a distinct influence on the polymerization results (see, e.g. (43 - 45)). In order to overcome some of these disadvantages we utilized a different approach: Well-defined preformed complexes (instead of *in situ* generated species) of dimethyl substituted 2,2'-bipyridines and 5,5"-dimethyl-2,2':6':2"-terpyridine with hexafluorophosphate counter ions showing good solubility characteristics in moderate polar solvents were used in combination with aluminium isopropoxide and alkylhalides like (1-bromoethyl)benzene for controlled polymerization of styrene, methacrylates and acrylates at 75°C.

Experimental Section

Polymerization

Experimental details for the general polymerization procedures can be found elsewhere (46, 47). A typical procedure was as follows: $[(CH_3)_2CHO]_3Al$ (0.292 mmol) and monomer (51 mmol) were added to a solution of the copper bipyridine complex (0.073 mmol) in CH_3CN (1 mL) under argon. After addition of (1-bromoethyl)benzene (0.073 mmol) the reaction mixture was degassed by three freeze/pump/thaw cycles. Then the mixture was heated to 75°C and kept at this temperature during polymerization. After distinct time intervals samples (0.1 mL) were taken under argon. The weight of the samples were measured. Afterwards the samples were quenched and dissolved in a CH_2Cl_2 solution (0.5 mL) of benzochinone (3% in CH_2Cl_2). The solution was dropped into MeOH (10 mL) resulting in a precipitation of polystyrene. After centrifugation of the probes the polystyrene was dried *in vacuo* at 40°C. By comparison of the probe weight with the weight of the dried polystyrene the conversion was calculated.

Metal Complexes

The metal complexes were prepared by complexation of the corresponding free ligands 4,4'-dimethyl-2,2'-bipyridine, 5,5'-dimethyl-2,2'-bipyridine, 6,6'-dimethyl-2,2'-bipyridine and 5,5"-dimethyl-2,2':6',2"-terpyridine with suitable metal salts (such as Cu(II) acetate or $[Cu(I)(CH_3CN)_4](PF_6)$) as described elsewhere. $[Cu(II)(4,4'-dimethyl-2,2'-bipyridine)_3](PF_6)_2$ and $[Cu(II)(5,5'-dimethyl-2,2'-bipyridine)_3](PF_6)_2$ see (46), $[Cu(I)(6,6'-dimethyl-2,2'-bipyridine)_2](PF_6)$ see (48), $[Cu(II)(5,5"-dimethyl-2,2':6',2"-terpyridine)_2](PF_6)_2$ see (49).

Characterization

Gel permeation chromatography (GPC) analysis was performed on a Waters Liquid Chromatograph system using Shodex GPC K-802S columns, Waters Differential Refractometer 410 and Waters UV Absorption Detector 486 with chloroform as eluent. Calibration was conducted with polystyrene standards. UV/VIS measurements were recorded using a Varian Cary 3. Atomic absorption spectroscopy was performed on a Varian SpectrAA 400 using GTA-96.

Results and Discussion

General Remarks

Our main focus in the field of controlled radical polymerization concerns a partial improvement of the established ATRP systems. In detail, we were interested to achieve the following goals: (A) Preparation of colorless poly-styrenes, poly(methacrylate)s and poly(acrylate)s with low contents of residual copper ions after a simple precipitation procedure; (B) First steps towards a potential fine-tuning of the catalytic system; (C) Utilization of soluble and homogeneous metal complexes without synthesis of heterocyclic ligands with long alkyl chains.

Common ATRP works with a ratio of 2:1:1:100 (bipyridine:copper(I) salt:alkyl halide:monomer) (see e. g. (24)), which are added as solids or liquids into the polymerization vessel. It is assumed that the necessary bipyridine metal complexes are formed more or less quantitatively in this mixture. In contrast to this approach we chose to skip the assembly step in the vessel and utilize already preformed N-heterocyclic metal complexes. Besides the advantage to use clearly defined metallo-organic systems it also allows the tuning of the solubility characteristics of the complexes by exchanging the counter ions. In our experiments we used hexafluorophosphate counter ions due to their solubility in moderate polar solvents like acetonitrile or even styrene. Furthermore, the use of preformed metal complexes will be a key factor for the simple and quantitative removal of the metal ions from the final polymer due to the good solubility characteristics of the corresponding metal complexes (in contrast to the common procedures, no solid compounds are dispersed in the polymerization mixture). In order to test the possibilities of a potential fine-tuning of the catalytic systems we introduced methyl substituents in different positions. In first experiments utilizing copper complexes of 4,4'-dimethyl-2,2'-bipyridine, 5,5'-dimethyl-2,2'-bipyridine, 6,6'-dimethyl-2,2'-bipyridine and 5,5"-dimethyl-2,2':6',2"-terpyridine ligands (Figure 1). Normally concentrated solutions in acetonitrile were titrated into the reaction.

In order to initiate the radical polymerization we utilized alkylhalides like (1-bromoethyl)benzene as initiator in combination with aluminium isopropoxide for the controlled polymerization of styrene, methacrylates and acrylates at 75°C (Scheme 1).

252

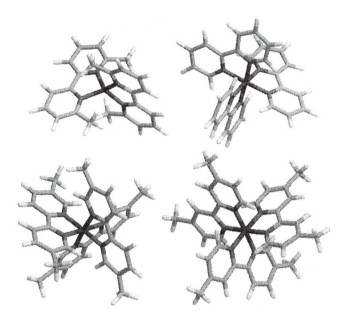

Figure 1. Schematic representation of the utilized preformed metal complexes in the solid state (HyperChem 5.0, MM+): Top left: [Cu(I)(6,6'-dimethyl-2,2'-bipyridine)₂](PF₆), top right: [Cu(II)(5,5''-dimethyl-2,2':6',2''-terpyridine)₂] (PF₆)₂; bottom left: [Cu(II)(5,5'-dimethyl-2,2'-bipyridine)₃](PF₆)₂; bottom right: [Cu(II)(4,4'-dimethyl-2,2'-bipyridine)₃](PF₆)₂.

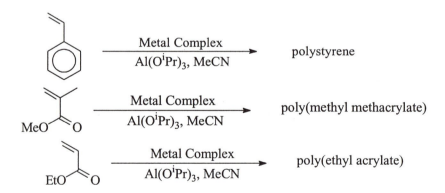

Scheme 1. Polymerization of styrene, methyl methacrylate and ethyl acrylate utilizing the preformed metal complexes.

The aluminium isopropoxide was first only added in order to allow polymerizations to take place at low temperatures (see also similar results for the polymerization of MMA with $RuCl_2(PPh_3)_3$ systems (12)). However, recent results indicate that the aluminium isopropoxide also interacts with the used copper(II) complexes in order to reduce copper(II) to copper(I) (50). This could be demonstrated e. g. by UV/VIS spectroscopy measurements as well as by visual observations: The original clear blue color of the [Cu(II)(4,4'-dimethyl-2,2'-bipyridine)$_3$](PF$_6$)$_2$ complex in acetonitrile changes to a green color after addition of the initiator (1-bromoethyl)benzene. After addition of the aluminium isopropoxide a brown color can be observed most probably indicating a reduction process. Very recently, we were also able to isolate the intermediate catalytic species from the polymerization mixture in form of single crystals. First results reveal a copper complex species with only two bipyridine ligands (50). Furthermore, other reduction agents could be used instead. Further detailed investigation in this direction are currently in progress.

Controlled Polymerization of Styrene with Preformed Metal Complexes

In first experiments we could demonstrate that a polymerization system consisting of a preformed [Cu(II)(4,4'-dimethyl-2,2'-bipyridine)$_3$](PF$_6$)$_2$ complex combined with aluminium isopropoxide, (1-bromoethyl)benzene and a small amount acetonitrile as solvent for the complex was able to polymerize styrene in a way typical for controlled radical polymerizations. The polymers obtained during polymerization, showed above 10% conversion, monomodal and rather narrow molecular weight distributions with polydispersities between 1.1 and 1.4 typical for living systems. The results of the GPC investigations for samples taken at different times are shown in Figure 2.

The dependency of the number average molecular weights of the obtained polystyrene samples as function of conversion show a linear relationship between the number average molecular weight of the polymers (M_n) and monomer conversion (Figure 3) (46). Very promising result were observed concerning the copper content and the color of the resulting polymers after a simple precipitation step. The polystyrene samples revealed to be completely colorless. These findings were supported by the relatively low content of residual metal ions. Both UV/VIS spectroscopy titration experiments and atomic absorption spectroscopy (AAS) measurements gave values of less than 200 ppm copper in the samples (control measurements using the same method and instrumentation with polstyrene from autopolymerization revealed 150 ppm copper content). Furthermore, the aluminum content is lower than 2.5 ppm (flame AAS). Controlled polymerization of styrene using conventional ATRP methods utilizing bipyridine/metal salt mixtures revealed a copper content of 2000 to 2500 ppm.

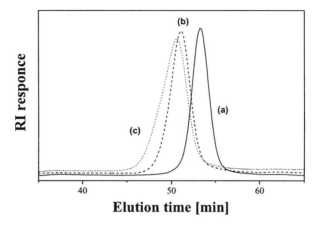

*Figure 2. GPC curves (chloroform as eluent) for samples obtained during poly-
merization of styrene in the presence of [Cu(II)(4,4'-dimethyl-2,2'-
bipyridine)₃](PF₆)₂ ([(CH₃)₂CHO]₃Al: 0.24 mmol; monomer: 42 mmol; ratio
[Cu(II)(4,4'-dimethyl-2,2'-bipyridine)₃](PF₆)₂:(1-bromoethyl)benzene = 1:2;
ratio [Cu(II)(4,4'-dimethyl-2,2'-bipyridine)₃](PF₆)₂:monomer = 1:700).
Conversion of 19.4% (a), 42.2% (b) and 49.1% (c).*

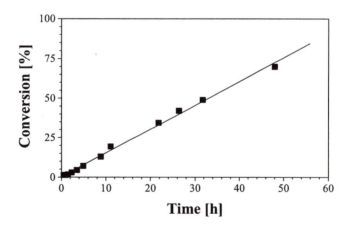

*Figure 3. Dependence of monomer conversion as function of time in the poly-
merization of styrene at 75°C using [Cu(II)(4,4'-dimethyl-2,2'-
bipyridine)₃](PF₆)₂ ([(CH₃)₂CHO]₃Al:0.292 mmol; styrene: 51 mmol; ratio
[Cu(II)(4,4'-dimethyl-2,2'-bipyridine)₃](PF₆)₂:(1-bromoethyl)benzene = 1:1;
ratio [Cu(II)(4,4'-dimethyl-2,2'-bipyridine)₃](PF₆)₂:monomer = 1:700).*

In order to develop methods for a potential fine-tuning of these controlled polymerization processes we started to vary the substitution patterns of the utilized ligands. Therefore we investigated further the polymerization behavior of [Cu(II)(5,5'-dimethyl-2,2'-bipyridine)$_3$](PF$_6$)$_2$ and [Cu(I)(6,6'-dimethyl-2,2'-bipyridine)$_2$](PF$_6$) complexes which differ in the position of the methyl substituents on the ligands. Therefore the complexation constants and the stability of the complexes show distinct differences (it should be pointed out that the 6,6'-dimethyl-2,2'-bipyridine ligand was used as copper(I) complex whereas the other ligands where used as copper(II) complexes). Furthermore, we also investigated a corresponding terpyridine copper(II) complex ([Cu(II)(5,5"-dimethyl-2,2':6',2"-terpyridine)$_2$](PF$_6$)$_2$) (see Figure 1 for a structural overview of the different metal complexes used).

Figure 4 demonstrates the different reaction rates of the investigated complexes. Comparing these results it could be preliminarily concluded that there are significant differences in the polymerization rate based on the methyl substitution position. Up to now the [Cu(II)(4,4'-dimethyl-2,2'-bipyridine)$_3$] (PF$_6$)$_2$ complex works best concerning reaction rate, control over molecular weight and molecular weight distribution. Similar linear relationships between molecular weights of the polystyrenes on conversion and conversion on time could be observed when the [Cu(II)(4,4'-dimethyl-2,2'-bipyridine)$_3$](PF$_6$)$_2$ complex was replaced by [Cu(II)(5,5'-dimethyl-2,2'-bipyridine)$_3$](PF$_6$)$_2$ in the reaction mixture. However, in this case the polymerization rate drops to the half. The observed polydispersities vary between 1.2 and 1.4. On the other side, narrow molar mass distributions could be found in this case already at low conversions. The behavior of the [Cu(II)(5,5"-dimethyl-2,2':6',2"-terpyridine)$_2$] (PF$_6$)$_2$ complex revealed within the range of the previous two concerning the relationships between molecular weights of the polystyrenes on conversion and conversion on time (Figure 4). In preliminary results a very different behavior was observed in the case of the [Cu(I)(6,6'-dimethyl-2,2'-bipyridine)$_2$](PF$_6$) complex. Under identical experimental conditions neither a control of molecular weight nor a control of the molecular weight distribution could be observed.

Furthermore, metal complexes with different metal ions like [Ru(II)(2,2'-bipyridine)$_2$(4,4'-dimethyl-2,2'-bipyridine)](PF$_6$)$_2$ or [Ni(II)(2,2'-bipyridine)$_3$] (PF$_6$)$_2$ were applied for the polymerization of styrene in combination with [(CH$_3$)$_2$CHO]$_3$Al and (1-bromoethyl)benzene. However, these systems did not show any control over M_n or M_w/M_n.

Controlled Polymerization of Acrylate and Methacrylate Type Monomers

After the successful experiments in the controlled radical polymerization of styrene using the system [Cu(II)(4,4'-dimethyl-2,2'-bipyridine)$_3$](PF$_6$)$_2$, aluminium isopropoxide, (1-bromoethyl)benzene and traces of acetonitrile we

256

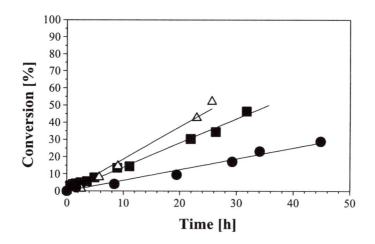

Figure 4. Dependence of monomer conversion as function of time in the polymerization of styrene utilizing different preformed metal complexes: [Cu(II)(4,4'-dimethyl-2,2'-bipyridine)₃](PF₆)₂ (Δ), [Cu(II)(5,5'-dimethyl-2,2'-bipyridine)₃](PF₆)₂ (●); [Cu(II)(5,5''-dimethyl-2,2':6',2''-terpyridine)₂](PF₆)₂ (■) ([(CH₃)₂CHO]₃Al: 0.24 mmol; monomer: 42 mmol; ratio metal complex:(1-bromoethyl)benzene = 1:2; ratio metal complex:monomer = 1:700).

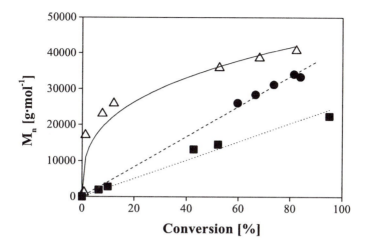

Figure 5. Dependence of the molecular weights as function of monomer conversion in the polymerization of methyl methacrylate (Δ), ethyl acrylate (●) and styrene (■) at 75°C using [Cu(II)(4,4'-dimethyl-2,2'-bipyridine)₃](PF₆)₂ ([(CH₃)₂CHO]₃Al: 0.24 mmol; monomer: 42 mmol; ratio metal complex:(1-bromoethyl)benzene = 1:2; ratio [Cu(II)(4,4'-dimethyl-2,2'-bipyridine)₃](PF₆)₂:monomer = 1:700).

checked the transfer of this method to other industrially interesting monomers. Therefore we extended our experiments towards methyl methacrylate and ethyl acrylate (47) (Scheme 1). Both for methyl methacrylate and ethyl acrylate a higher polymerization rate could be observed than for styrene. For example after 20 h 58% of the methyl methacrylate and 85% of the ethyl acrylate monomers were polymerized compared with only 38% in the case of styrene (Figure 5).

The molecular weight of the obtained polymers increased linearly with conversion for styrene and ethyl acrylate, whereas in the case of methyl methacrylate different behavior could be observed. In the last case the molecular weight on conversion relationship increased sharply up to 10% conversion and then flattered off increasingly.

The obtained polydispersity values (M_w/M_n) were lower than 1.4 in all cases. The best controlled reaction was observed in the polymerization of styrene with values around 1.1, followed by the polydispersities of the poly(methyl methacrylate) between 1.1 and 1.25. The polydispersities of the resulting polyethyl acrylate were larger with values from 1.2 to 1.4. In Figure 6 the GPC traces of poly(ethyl acrylate) samples taken at different conversion values are shown.

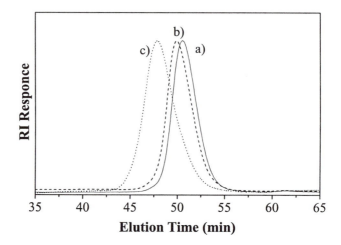

*Figure 6. GPC curves (chloroform as eluent) for samples obtained during poly-
merization of methyl methacrylate in the presence of [Cu(II)(4,4'-dimethyl-2,2'-
bipyridine)₃](PF₆)₂ ([(CH₃)₂CHO]₃Al: 0.24 mmol; monomer: 42 mmol; ratio
[Cu(II)(4,4'-dimethyl-2,2'-bipyridine)₃](PF₆)₂:(1-bromoethyl)benzene = 1:2;
ratio [Cu(II)(4,4'-dimethyl-2,2'-bipyridine)₃](PF₆)₂:monomer = 1:700):
(a) 7.8%, (b) 12.2%, (c) 82.2%.*

Preparation of Block Copolymers

One main focus in the present research of controlled radical polymerization
procedures is the production of well-defined block copolymers. It was shown
above with the system [Cu(II)(4,4'-dimethyl-2,2'-bipyridine)₃](PF₆)₂, aluminium
isopropoxide, (1-bromoethyl)benzene and acetonitrile as solvent, that
homopolymers of styrene and acrylate type monomers with a predetermined
molecular weight and a narrow molecular weight distribution are accessible.
Therefore it was of interest to extend this system to block copolymers. Generally
we synthesized a polystyrene block first using a monomer conversion of around
80% (removal of the residual monomer *in vacuo*). After addition of the acrylate
type monomers the polymerization was continued. Figure 7 shows the GPC
curves of polystyrene and the corresponding poly(styrene-*b*-ethyl acrylate) which
was received after 95% conversion of the added ethyl acrylate. Compared to the
GPC-chromatogram of the polystyrene homopolymer a clear shift towards higher
molecular weights can be observed. The shoulder of the GPC curve is a clear

indication of residual homo polystyrene, formed by unavoidable side reactions such as termination via radical combination or disproportionation and transfer. It is also possible to synthesize a poly(styrene-*b*-methyl methacrylate). However, in this case a high content of polystyrene homopolymer and poly(styrene-*b*-methyl methacrylate) with only short MMA segments were obtained. A similar result was found in the synthesis of poly(styrene-*b*-methyl methacrylate) with [Cu(II)(5,5"-dimethyl-2,2':6',2"-terpyridine)$_2$](PF$_6$)$_2$ as catalyst.

Conclusion

In the present paper we summarize our current research approach in controlled radical polymerization. We have found that a mixture of preformed [Cu(II)(4,4'-dimethyl-2,2'-bipyridine)$_3$](PF$_6$)$_2$ complex in combination with aluminium isopropoxide, an alkyl halide and some solvent for the metal complex is quite suitable for the controlled homogeneous radical polymerization of

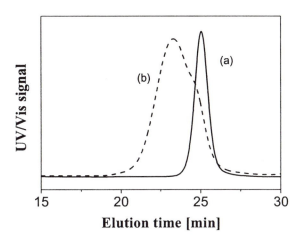

Figure 7. GPC curves (chloroform as eluent) for samples obtained during block copolymerization of ethyl acrylate with polystyrene in the presence of [Cu(II)(4,4'-dimethyl-2,2'-bipyridine)$_3$](PF$_6$)$_2$ ([(CH$_3$)$_2$CHO]$_3$Al: 0.29 mmol; styrene: 51 mmol; ratio [Cu(II)(4,4'-dimethyl-2,2'-bipyridine)$_3$](PF$_6$)$_2$:(1-bromoethyl)benzene = 1:2; ratio [Cu(II)(4,4'-dimethyl-2,2'-bipyridine)$_3$] (PF$_6$)$_2$:monomer = 1:700): (a) GPC trace after consumption of styrene, (b) GPC trace after addition of ethyl acrylate.

styrene, methacrylate and acrylate type monomers. Utilizing this system some drawbacks of the conventional ATRP method could be diminished. Especially the production of non-colored polymers with low remaining metal content, the application of soluble metal complexes, homogeneous polymerization conditions without application of long chain substituted bipyridines and the potential fine-tuning properties of the metal complexes should be pointed out. For all three classes of monomers control over molecular weight and low molecular weight distribution could be observed. Finally, some block copolymers were prepared. Further research will focus on a better understanding of the mechanistic details in the applied system and in the preparation of novel functional polymer architectures.

Acknowledgement

The research was supported partly by the *Bayerisches Staatsministerium für Unterricht, Kultus, Wissenschaft und Kunst* and the *Fonds der Chemischen Industrie*. We thank *Reilly Tar & Chem. Corp.* for contributing some of the ligands and C. Eschbaumer for help in the synthesis of substituted bipyridines.

References

1. The Chemistry of Free-Radical Polymerization; Moad, G.; Solomon, D. H.; Pergamon: Oxford, **1995**.
2. Georges, M. K.; Veregin, R. P. N.; Kazmaier, P. N.; Hamer, G. K. *Polym. Mater. Sci. Eng.* **1993**, *68*, 6.
3. Georges, M. K.; Veregin, R. P. N.; Kazmaier, P. N.; Hamer, G. K.; Saban, M. *Macromolecules* **1994**, 27, 7228.
4. Hawker, C. J. *J. Am. Chem. Soc.* **1994**, *116*, 11185.
5. Benoit, D.; Grimaldi, S.; Finet, J. P.; Tordo, P.; Fontanille, M.; Gnanou, Y. *Polym. Prep. (Am. Chem. Soc., Div. Polym. Chem.)* **1997**, *38*, 729.
6. Fukuda, T.; Terauchi, T.; Goto, A.; Tsujii, Y.; Miyamoto, T.; Shimizu, Y. *Macromolecules* **1996**, *29*, 3050.
7. Gabaston, L. I.; Furlong, S. A.; Jackson, R. A.; Armes, S. P. *Polymer* **1999**, *40*, 4505.
8. Hawker, C. J.; Hedrick, J. L. *Macromolecules* **1995**, *28*, 2993.
9. Leduc, M. R.; Hawker, C. J.; Dao, J.; Frechet, J. M. J. *J. Am. Chem. Soc.* **1996**, *118*, 11111.
10. Colombani, D.; Steenbock, M.; Klapper, M.; Müllen, K. *Macromol. Rapid Commun.* **1997**, *18*, 243.
11. Steenbock, M.; Klapper, M.; Müllen, K.; Bauer, C.; Hubrich, M. *Macromolecules* **1998**, *31*, 5223.

12. Uegaki, H.; Kotani, Y.; Kamigaito, M.; Sawamoto, M. *Macromolecules* **1997**, *30*, 2249.

13. Ueda, J.; Matsuyama, M.; Kamigaito, M.; Sawamoto, M. *Macromolecules* **1998**, *31*, 557.

14. Ando, T.; Kato, M.; Kamigaito, M.; Sawamoto, M. *Macromolecules* **1996**, *29*, 1070.

15. Nishikawa, T.; Kamigaito, M.; Sawamoto, M. *Macromolecules* **1999**, *32*, 2204.

16. Granel, C.; Dubois, P.; Jerome, R.; Teyssie, P. *Macromolecules* **1996**, *29*, 8576.

17. Husseman, M.; Malmstrom, E. E.; McNamara, M.; Mate, M.; Mecerreyes, D.; Benoit, D. G.; Hedrick, J. L.; Mansky, P.; Huang, E.; Russell, T. P.; Hawker, C. J. *Macromolecules* **1999**, *32*, 1424.

18. Arvanitopoulos, L. D.; King, B. M.; Huang, C.-Y.; Harwood, H. J. *Polym. Prep. (Am. Chem. Soc., Dov. Polym. Chem.)* **1997**, *38*, 752.

19. Wayland, B. B.; Basickes, L.; Mukerjee, S.; Wei, M. L.; Fryd, M. *Macromolecules* **1997**, *30*, 8109.

20. Kotani, Y.; Kamigaito, M.; Sawamoto, M. *Macromolecules* **1999**, *32*, 2420.

21. Arvanitopoulos, L. D.; Greuel, M.; Harwood, H. J. *Proc. Int. Symp. Macromol.* **1994**, 35.

22. Wang, J.-S.; Matyjaszewski, K. *Macromolecules* **1995**, *28*, 7901.

23. Patten, T.; Xia, J.; Abernathy, T.; Matyjaszewski, K. *Science* **1996**, *272*, 866.

24. Matyjaszeski, K.; Patten, T.; Xia, J. *J. Am. Chem. Soc.* **1997**, *119*, 674.

25. Matyjaszewski, K.; Shipp, D.A.; Wang, J.-S.; Grimaud, T.; Patten, T. E. *Macromolecules* **1998**, *31*, 6836.

26. Gao, B.; Chen, X.; Iván, B.; Kops, J.; Batsberg, W. *Polym. Bull.* **1997**, *39*, 559.

27. Percec, V.; Barboiu, K.; Neumann, A.; Ronda, J. C.; Zhao, M. *Macromolecules* **1996**, *29*, 3665.

28. Grimaud, T.; Matyjaszewski, K. *Macromolecules* **1997**, *30*, 2216.

29. Kickelbick, G.; Matyjaszewski, K. *Macromol. Rapid Commun.* **1999**, *20*, 341.

30. Kickelbick, G.; Paik, H. J.; Matyjaszewski, K. *Macromolecules* **1999**, *32*, 2941.

31. Qiu, J.; Gaynor, S. G.; Matyjaszewski, K. *Macromolecules* **1999**, *32*, 2872

32. Xia, J. H.; Matyjaszewski, K. *Macromolecules* **1999**, *32*, 2434.

33. Haddleton, D. M.; Jasieczek, C. B.; Hannon, M. J.; Shooter, A. J. *Macromolecules* **1997**, *30*, 2190.

34. Haddleton, D. M.; Duncalf, D. J.; Kukulj, D.; Crossman, M. C.; Jackson, S. G.; Bon, S. A. F.; Clark, A. J.; Shooter, A. J. *Eur. J. Inorg. Chem.* **1998**, 1799.

35. Haddleton, D. M.; Crossman, M. C.; Dana, B. H.; Duncalf, D. J.; Heming, A. M.; Kukulj, D.; Shooter, A. J. *Macromolecules* **1999**, *32*, 2110.
36. DiRenzo, G. M.; Messerschmidt, M.; Mülhaupt, R. *Macromol. Rapid Commun.* **1998**, *19*, 381.
37. Destarac, M.; Bessière, J. M.; Boutevin, B. *Macromol. Rapid Commun.* **1997**, *18*, 967.
38. Patten, T. E.; Matyjaszewski, K. *Adv. Mater.* **1998**, *10*, 901.
39. Matyjaszewski, K. *Chem. Eur. J.* **1999**, *5*, 3095.
40. Patten, T. E.; Matyjaszewski, K. *Acc. Chem. Res.* **1999**, *32*, 895.
41. Xia, J.; Gaynor, S. G.; Matyjaszewski, K. *Macromolecules* **1998**, *31*, 5958
42. Matyjaszewski, K.; Nakagawa, Y.; Jasieczek, C. B. *Macromolecules* **1998**, *31*, 1535.
43. Coca, S.; Davis, K.; Miller, P.; Matyjaszewski, K. *Polymer Preprints* (1) **1997**, *38(1)*, 689.
44. Matyjaszewski, K.; Wei, M.; Xia, J.; Gaynor, S. G. *Macromol. Chem. Phys.* **1998**, *199*, 2289.
45. Woodworth, B. E.; Metzner, Z.; Matyjaszewski, K. *Macromolecules* **1998**, *31*, 7999.
46. Schubert, U. S.; Hochwimmer, G.; Spindler, C. E.; Nuyken, O. *Macromol. Rapid Commun.* **1999**, *20*, 351.
47. Schubert, U. S.; Hochwimmer, G.; Spindler, C. E.; Nuyken, O. *Polym. Bull.*, **1999**, *43*, 319.
48. Hochwimmer, G.; Nuyken, O.; Schubert, U. S. *Macromol. Rapid Commun.* **1998**, *19*, 309.
49. Schubert, U. S.; Spindler, C. E.; Eschbaumer, C.; Nuyken, O. *Polym. Preprint* **1999**, *40(2)*, 416.
50. Schubert, U. S.; Hochwimmer, G.; Kickelbick, G.; Filep, A.; Spindler, C. E.; Nuyken, O. in preparation.

Chapter 19

Reverse Atom Transfer Radical Polymerization Using AIBN or BPO as Initiator

Wenxin Wang and Deyue Yan[1]

College of Chemistry and Chemical Technology, Shanghai Jiao Tong University, 800 Dongchuan Road, Shanghai 200240, Peoples Republic of China

The reverse atom transfer radical polymerization (ATRP) of methyl acrylate can be realized with the initiating system, AIBN/CuCl$_2$/bpy, in bulk by means of a new polymerization procedure. The reverse ATRP of styrene is also accessible using BPO/CuCl$_2$/bpy as the initiating system. The initiation mechanism in reverse ATRP with BPO is different from that with AIBN due to the redox reaction between BPO and Cu^I generated from the reaction of radicals with Cu^{II}. The "living"/controlled radical polymerization of methyl methacrylate and styrene can be implemented using BPO/Cu^{I}Cl/bpy as the initiating system in solvents at ambient temperature (40 ℃). The mechanism suggested the initiation is a redox reaction, and the propagation is identical with that of ATRP.

In recent years, the "living"/controlled radical polymerization has attracted much attention. Several approaches have been reported[1-6], among which atom transfer radical polymerization (ATRP) was considered as the most effective way to carry out controlled radical polymerization. The transition-metal-catalyzed atom transfer radical addition, ATRA, gives a unique and efficient method for carbon-carbon bond formation in organic synthesis[7]. Matyjaszewski[6,8], Teyssie[9], Sawamoto[10], and Percec[11] have successfully introduced this approach into polymerization chemistry, and developed the novel type of 'living'/controlled radical polymerization, i.e., atom transfer radical polymerization (ATRP), in which an alkyl halide, R-X, was used as the initiator and a transition-metal species complexed with suitable ligand(s), Mt^n/Lx, as the catalyst. In addition, Matyjaszewski and coworker[12] further reported that a 'living'/controlled polymerization was also accessible using a conventional radical initiator (azodiisobutyronitrile, AIBN) and a transition-metal compound at higher oxidation state (e.g.,Cu^{II}Cl$_2$) complexed with suitable ligands (e.g., 2.2'-bipyridine, bpy) as a catalyst. The latter was called reverse ATRP. As a "classical" radical initiator (AIBN) instead of an alkyl halide was used, many polymer chemists have showed great interest in reverse ATRP. Matyjaszewski and coworker[13] pointed out

[1]Corresponding author.

that reverse ATRP of styrene and methyl acrylate can be carried out under homogeneous conditions by using alkyl substituted bipyridine ligands, such as 4,4'-nonyl-2,2'-bipyridine (dNbipy). Teyssie and coworkers[14] implemented the reverse ATRP of MMA using FeCl$_3$ and AIBN as the initiating system in the presence of tripenylphosphine. Most recently, Yan et al.[15] developed a new procedure for reverse ATRP of methyl acrylate (MA) in bulk; Yan[16, 17] and Matyjaszewski[18] further investigated the reverse ATRP using BPO/Cu^{II}X$_2$/Lx or BPO/Cu^{I}X/Lx initiating system. It is interesting to narrate some results on these reverse ATRP topics.

Reverse ATRP of MA Initiated by AIBN/Cu^{II}Cl$_2$/bpy

The initiating system AIBN/Cu^{II}Cl$_2$/bpy developed by Matyjaszewski[12] can initiate the "living"/controlled radical polymerization of styrene in bulk, however, it does so only under the high mole ratio of AIBN to CuCl$_2$. In addition, the "living"/controlled polymerization of methyl acrylate can't be implemented by the similar procedure. Fortunately, the "living"/controlled polymerization of methyl acrylate can be realized with AIBN/Cu^{II}Cl$_2$/bpy initiating system by the aid of a new polymerization approach. Table I shows the characterization data of poly(methyl acrylate) generated from reverse ATRP catalyzed by AIBN/Cu^{II}Cl$_2$/bpy in bulk. The mole ratio of Cu^{II}Cl$_2$ to AIBN is not a critical parameter for the 'living' /controlled radical polymerization using the new polymerization procedure. Little influence of the [Cu^{II}Cl$_2$]/[AIBN] ratio on the controlled nature of the polymerization has been observed. Different ratios of Cu^{II}Cl$_2$ to AIBN used in this work always result in well-controlled radical polymerizations and provide high initiation efficiency and polymers with low polydispersity (for instance, M_w/M_n = 1.36).

Figure 1 shows that the measured molecular weight linearly increases with the monomer conversion, and matches the theoretical value calculated from eq. 1 for bulk polymerization of MA initiated by AIBN in the presence of 2 molar equiv. Cu^{II}Cl$_2$ and 6 molar equiv. bpy at 100 ℃.

$$M_{n,th} = ([M]_0/[BPO]) \times (MW) \times conversion \tag{1}$$

where (MW) denotes the molecular weight of the monomer.

Table I Characterization Data of Poly(methyl acrylate)*

[AIBN] (mol/L)	[CuCl$_2$] (mol/L)	[bpy] (mol/L)	Time (h)	Conv (%)	$M_{n,th}$	$M_{n,sec}$	M_w/M_n
0.0155	0.031	0.093	12	93	28630	29900	1.46
0.0155	0.062	0.186	18	89	27400	28600	1.42
0.0155	0.093	0.279	31	92	28330	29000	1.36
0.0155	0.124	0.372	41	91	28020	28700	1.39

*SOURCE: Reproduced from Reference[15]. Copyright 1998 Wiley-VCH Verlag GmbH

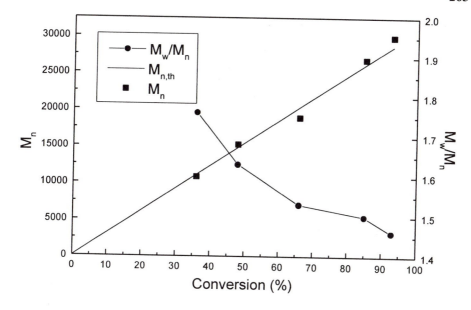

Figure 1. Molecular weight and molecular weight distribution dependence on monomer conversion for the bulk polymerization of MA initiated by AIBN/CuCl$_2$/bpy. [MA]=11.1M, [AIBN]=0.0155M, [CuCl$_2$]=0.031M, [bpy]=0.093M. (Reproduced from Reference 15. Copyright 1998 Wiley-VCH Verlag GmbH)

These experimental data indicated that the bulk polymerization of MA initiated by AIBN/Cu^{II}Cl$_2$/bpy system proceeded via a living/controlled mechanism. Furthermore, the measured molecular weight of the resulting polymer is in a good agreement with theory up to $M_n = 3 \times 10^4$, indicating the high initiation efficiency of AIBN. The molecular weight distribution of the products is rather narrow, $M_w/M_n \leq 1.42$ (Table 1), The reaction scheme is suggested as below:

Scheme 1

Initiation and termination at lower temperature:

$$I\text{–}I \xrightarrow{\Delta} 2\,I\cdot$$

$$I\cdot \;+\; M_t^{n+1}X \longrightarrow I\text{–}X \;+\; M_t^n$$

$$k_i' \Big| +M$$

$$I\text{–}P_1\cdot \;+\; M_t^{n+1}X \longrightarrow I\text{–}P_1\text{–}X \;+\; M_t^n$$

Initiation and propagation at higher temperature:

$$I-X + M_t^n \rightleftharpoons I\cdot + M_t^{n+1}X$$

$$k_i \downarrow +M$$

$$I-P_1-X + M_t^n \rightleftharpoons I-P_1\cdot + M_t^{n+1}X$$

$$P_n-X + M_t^n \rightleftharpoons P_n\cdot + M_t^{n+1}X$$

$$\left(+M\right)_{k_p}$$

It is known that $Cu^{II}Cl_2$ is a strong and efficient inhibitor of radical polymerization under usual conditions[19]. When the polymerization of MA proceeds at 65 ℃-70 ℃, it enables slower decomposition of AIBN which allows the dissolution of $Cu^{II}Cl_2$/bpy at the rate comparable to AIBN consumption. The oxidized transition–metal, $Cu^{II}Cl_2$, is a strong and efficient inhibitor of the polymerization of MA initiated by AIBN. $Cu^{II}Cl_2$ donates the halogen atom Cl to the initiating or propagating radicals, $I\cdot$ or $I-P\cdot$, forming the reduced transition–metal species, Cu^I, and the dormant species, I–Cl and P–Cl. At lower temperature, 65 ℃-70 ℃, the covalent species I–Cl and P–Cl are rather stable, therefore the reduced transition–metal species, Cu^I, can not promote ATRP process as it does in the living/controlled radical polymerization initiated by R–X/Mt/Lx at higher temperature. After ten hours at 65 ℃-70 ℃, most of AIBN was exhausted while the reduced transition–metal species (Cu^I) and the dormant species (I–Cl and P–Cl) were accumulated. During this period the inhibited reaction is predominant, and little polymer was generated. In order to carry out the fast ATRP, higher temperature is needed. When the system is heated to 100 ℃, however, a reversible activation of the resulting alkyl chlorides occurs in the presence of a coordinative ligand, then the polymerization of the system proceeds via the reverse ATRP mechanism.

Reverse ATRP of Styrene Initiated by BPO/Cu^{II}Cl$_2$/bpy

The polymerization of styrene (St) with BPO/Cu^{II}Cl$_2$/bpy as the initiating system can also result in a well-controlled radical polymerization with high initiation efficiency, 90%, and the polydispersity of resulting polymer is rather low, M_w/M_n = 1.32. Figure 2 shows that the measured molecular weight linearly increases with increasing monomer conversion, and matches the theoretical value calculated from eq. 1 for bulk polymerization of styrene initiated with BPO in the presence of 2 molar equiv. Cu^{II} Cl$_2$ and 4 molar equiv. bpy at 110 ℃.

Figure 3 shows the first-order kinetic plot, indicating that the concentration of the growing species remained constant. The first point in Figure 3 is beyond the experimental error from the linear plot, and the monomer conversion at this moment is apparently higher than that in a living polymerization. It seems that the polymerization occurs partially at low temperature.

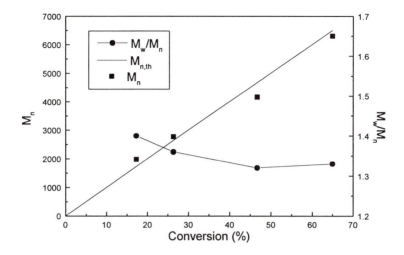

Figure 2. Molecular weight and molecular weight distribution dependence on monomer conversion for the heterogeneous reverse ATRP of styrene at 110 ℃ in bulk. [Styrene]:[BPO]:[CuCl₂]:[bpy]=100:1:2:4. (Reproduced from Reference 16. Copyright 1999 Wiley-VCH Verlag GmbH)

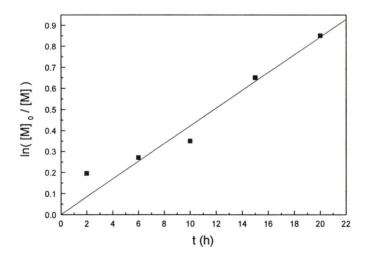

Figure 3. Semilogarithmic kinetic plot for the bulk reverse ATRP of styrene at 110 ℃. [Styrene]:[BPO]:[CuCl₂]:[bpy]=100:1:2:4. (Reproduced from Reference 16. Copyright 1999 Wiley-VCH Verlag GmbH)

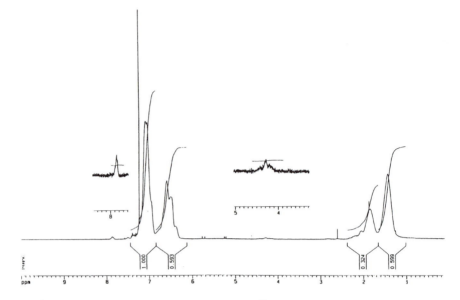

Figure 4. ^{1}*H-NMR of the polystyrene with BPO/Cu^{II}Cl$_2$/bpy. (Reproduced from Reference 16. Copyright 1999 Wiley-VCH Verlag GmbH)*

These experimental data indicate that the polymerization of styrene initiated by the BPO/Cu^{II}Cl$_2$/bpy system proceeds via a living/controlled radical mechanism. The molecular weight is essentially proportional to the reciprocal of the concentration of BPO. Furthermore, a good correlation between the calculated molecular weight and the experimental one was found, and the polydispersity is lower than 1.39, indicating a high initiation efficiency of BPO. The structure of resulting polystyrene was studied by ^{1}H-NMR. The ^{1}H-NMR spectrum of PS is illustrated in Figure 4. Signals were observed at 1.2~2.1ppm, originating from the methene and methine protons of the main chain. A signal at 7.9 ppm was assigned to the aromatic protons at ortho position of the benzoyl moiety attached to the polymer head group, The broad triplets at 4.4ppm was attributed to the end group chlorine. Moreover, comparison of the integration of the signals of end group with those of methene and methine in the main chain of PS gives a molecular weight, $M_{n,NMR}$ =12000, close to the one from GPC based on PS standards, $M_{n,SEC}$ =11000. The reaction scheme is suggested as below:

Scheme 2

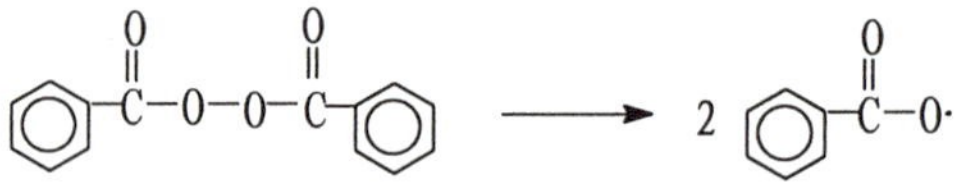

$$\text{Ph-C(=O)-O-O-C(=O)-Ph} \longrightarrow 2\ \text{Ph-C(=O)-O}\cdot \qquad (1)$$

$$\text{Ph-C(=O)-O}\cdot + nM \longrightarrow \text{Ph-C(=O)-O-Mn}\cdot \qquad (2)$$

$$\text{Ph–C(=O)–O–Mn·} + \text{CuCl}_2\text{·bpy} \longrightarrow \text{Ph–C(=O)–O–Mn–Cl} + \text{CuCl·bpy} \qquad (3)$$

$$\text{Ph–C(=O)–O–O–C(=O)–Ph} + \text{CuClbpy} \longrightarrow \text{Ph–C(=O)–O·} + \text{Ph–C(=O)–O}^- \cdot \text{Cu}^{2+}\text{Cl·bpy} \qquad (4)$$

At first, a BPO molecule decomposes into two primary radicals. The Cu^{II} cation can't take electron from primary radicals due to the strong oxidation of primary radicals resulting from BPO. Secondly, after a monomer adds to a primary radical, the oxidation of the radical decreases greatly. Thirdly, $Cu^{II}Cl_2$ can get rid of electron from this radical, forming dormant species and Cu^{I}. Finally, redox reaction between BPO and Cu^{I} takes place and produces one primary radical. This process continually repeats at lower temperature. From the analysis we can find that one mole BPO will result in one mole radicals. So the initiation mechanism of BPO $/Cu^{II}Cl_2$/bpy is somewhat different from that of AIBN/$Cu^{II}Cl_2$/bpy. After ten hours at 70℃, most of BPO was exhausted while the reduced transition–metal species, Cu^{I}, and the dormant species were accumulated. When the system is heated to 110℃, however, the polymerization under consideration proceeds via the ATRP mechanism which is the same as AIBN/$Cu^{II}Cl_2$/bpy system.

Reverse ATRP of MMA and St Initiated by BPO/Cu^{I}Cl/bpy

The "living"/controlled radical polymerization of MMA and St in butanone is also accessible using BPO/Cu^{I}Cl/bpy as the initiator. When a solution of methyl methacrylate, Cu^{I}Cl, bpy (2 molar equiv. relative to Cu^{I}Cl) and BPO, (1 molar equiv. relative to Cu^{I}Cl) was heated, the solution was homogeneous and gradually became viscous. As shown in Figure 5, the molecular weight measured, $M_{n,SEC}$, is higher than the theoretical one during the low monomer conversion period. With increasing monomer conversion, a linear increase of number average molecular weight, $M_{n,SEC}$, versus monomer conversion up to 90% is observed. It is interesting that the "living"/controlled radical polymerization is still valid when the number-average molecular weight of products approaches to as high as 100,000. This result indicates that BPO/Cu^{I}Cl/bpy acts as an efficient initiating system of "living"/controlled radical polymerization. The reaction temperature in this work is rather lower comparing with those of ATRP and reverse ATRP initiated by AIBN. The molecular weight distribution of products is rather narrow ($M_w/M_n < 1.6$). A linear plot of $\ln([M]_0/([M])$ versus polymerization time (Figure 6) demonstrates that the concentration of growing radicals remains constant during propagation, and termination is not significant. These experimental data suggest that the polymerization of MMA is a "living"/controlled radical polymerization process with a negligible amount of irreversible transfer and termination.

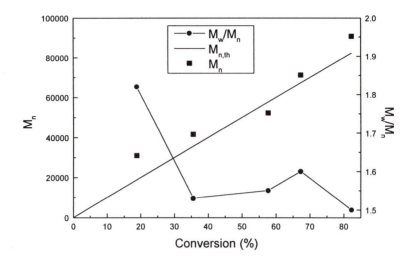

Figure 5. Molecular weight and molecular weight distribution dependence on monomer conversion in polymerization of MMA with BPO/CuCl/bpy. [MMA]=5M, [BPO] = [CuCl] = [bpy/2] = 0.005M, butanone, 40 ℃.

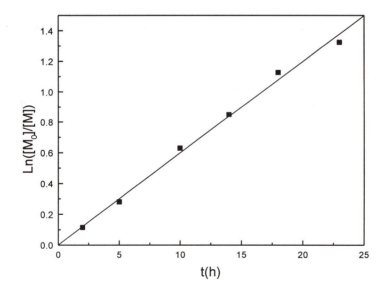

Figure 6. Semilogarithmic kinetic plot in polymerization of MMA with BPO/CuCl/bpy. [MMA] = 3M, [BPO] = [CuCl] = [bpy/2] = 0.01M, butanone, 40 ℃.

The reaction scheme is suggested as below:

Scheme 3

Initiation

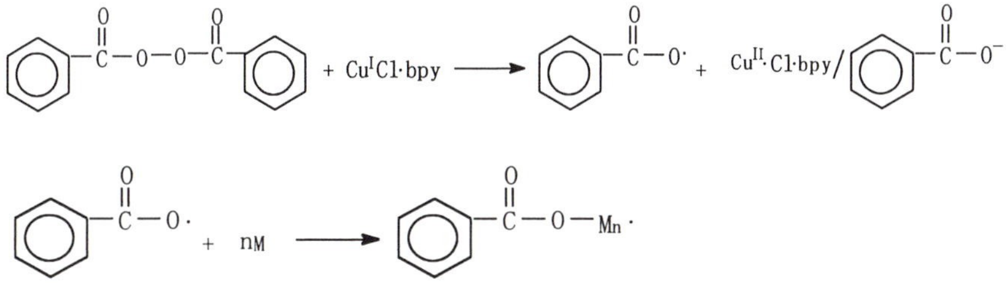

Propagation

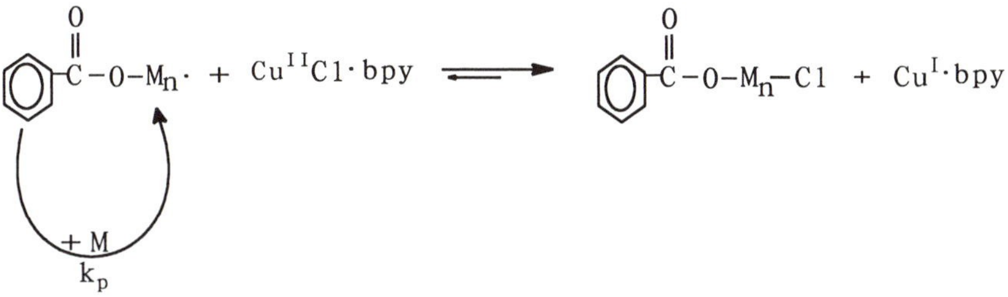

It is known that cuprous salts act as very efficient accelerators in the decomposition of peroxides[20, 21]. The coordination of the bidentate nitrogen ligand to Cu^I increases the solubility of the inorganic salt and facilitates the redox reaction between CuCl and BPO. First, radicals are generated through the electron transfer from Cu^I complexed by bpy to the peroxide (BPO), and initiate the radical polymerization of MMA. Then, Cu^{II} complexed by bpy reacts reversibly with growing radicals and gets rid of electron from these radicals, forming dormant species P–Cl and Cu^I. Finally in the presence of a coordinating ligand a reversible activation of the resulting alkyl chlorides occurs, and the polymerization proceeds via the ATRP mechanism. During the low monomer conversion period the molecular weight measured, $M_{n,SEC}$, is higher than the theoretical one, $M_{n,th}$. It seems that the initiation has not completed in the earlier stage of the polymerization at 40^0C. More probably, at the beginning of the polymerization a certain amount of primary radicals were generated, which start the general radical polymerization at first and then switched to ATRP due to the formation and presence of Cu^{II}.

Similarly, using BPO/Cu^ICl/bpy (1:1:2) initiating system, the reverse atom transfer radical polymerization of styrene can also afford polymers with the predetermined molecular weight and low polydispersities (≤ 1.5). The straight semilogarithmic kintic plot of $\ln([M]_0/[M])$ vs time, t, can be observed as well.

The block copolymer, PS-b-PMMA, was successfully synthesized by using BPO/CuCl/bpy as the initiating system, which further verifies the "living"/controlled nature of the polymerization system under consideration (Table II.)[17].

Table II. Characterization data of PS-b-PMMA

sample	Conv.	$M_{n,SEC}$	$M_{n,th}$	M_w/M_n
PMMA	98.6%	18600	19700	1.42
PMMA-PS	165.1%	31300	33000	1.76
PS	93.5%	16500	18700	1.50
PS-PMMA	178.8%	32900	35700	1.83

During the reverse ATRP of methyl methacrylate initiated by BPO/Cu^{I}Cl/bpy, a interesting color change has been observed. Before BPO was added into the reaction system, the color of the catalyst solution was brown, which showed the presence of Cu^I/bpy complex. After the initiator (BPO) was added, it turned from brown to blue immediately, which indicated the formation of a large quantity of Cu^{II}. During the polymerization, it turned into brown again and remained unchanged, which demonstrated the emergence of Cu^I/bpy complex once more. In order to verify these results, we monitored the polymerization process in-situ by means of UV spectrum. As shown in Figure 7, there was an absorption peak at 440 nm before BPO had been added into the reaction system, and this absorption peak disappeared completely as soon as the initiator was added, however, it appeared again after the reaction lasted for a certain period. UV analysis data conform to the color change during polymerization, which means that the preceding mechanism suggested by authors is reasonable.

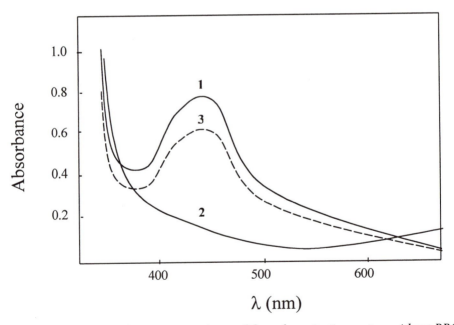

Figure 7. UV spectra: 1. copper complexes of the polymerization system without BPO; 2. copper complexes of the polymerization system at the moment of BPO (0.01M) being added; 3. copper complexes of the polymerization system after Polymerization for 5 hours. [MMA] = 3M, [CuCI] = [bpy/2] = 0.01M, butanone, 60 ℃.

Table III. Solvent effect on 'living'/controlled polymerization of MMA with
BPO/CuCl/bpy, [MMA] = 3M, [BPO] = [CuCl] = [bpy/2] = 0.01M, T = 60°C

Solvent	t (h)	Conv.%	M_n	M_w/M_n	$M_{n,th}$
Butanone	16	90.1	28700	1.42	27030
Cyclohexanone	16	83.3	23900	1.48	24990
THF	16	23.1	22800	2.07	6930
Toluene	16	43.2	54800	3.38	12960
Acetonitrile	16	no polymer			
Bulk	24 (30°C)	no polymer			

As shown in Table III, solvent has significant effect on the reverse atom transfer radical polymerization initiated by this redox system. The reverse ATRP of methyl methacrylate can be conducted in butanone or cyclohexanone. However, the controlled polymerization process of methyl metharcylate can't be attained in bulk, and in the solvents such as toluene, THF and acetonitrile. The polarity of solvent not only considerably affects the solubility of catalysts and dormant species, but also influences the process of initiation and propagation as well as the equilibrium between the active species and dormant species, which gives rise to the experimental results mentioned. Among these selected solvents, non-polar solvents such as toluene and bulk do not dissolve CuCl/bpy complex, so those solvents result in the poor controlled polymerization process. However, it is not clear: why polar solvents such as acetonitrile and THF hinder the controlled polymerization of MMA.

Conclusion

Reverse atom transfer radical polymerization of methyl acrylate in the presence of a conventional radical initiator (AIBN) in bulk was successfully implemented via a new polymerization procedure. The system first reacts at 65°C-70°C for ten hours, then polymerizes at 100°C. Various molar ratios of AIBN to $Cu^{II}Cl_2$ were used in this work, all of which result in a well-controlled radical polymerization with high initiation efficiency and narrow molecular weight distribution, i.e., the polydispersity is as low as $M_w/M_n = 1.36$.

Reverse atom transfer radical polymerization of styrene in the presence of a conventional radical initiator (BPO) in bulk was also successfully carried out by the means of the same polymerization procedure as AIBN. However, the initiation mechanism of BPO is different from that of AIBN because there is the redox reaction between BPO and Cu^I generated from the reaction of radicals with Cu^{II}. The molecular weight of the resultant polymer is in agreement with the theoretical value calculated in accordance with the suggested mechanism.

The "living"/controlled radical polymerization of MMA and St was implemented using BPO/Cu^{I}Cl/bpy as the initiating system in solvents at ambient temperature (40 °C). The mechanism is suggested that the initiation is a redox reaction, and the polymerization proceeds by the way of reverse ATRP, which affords controlled polymerization of MMA and St with predetermined molecular weight and narrower molecular weight distribution, $M_w/M_n < 1.6$.

Experiment Part

Reagents. Methyl acrylate, Methyl methacrylate, styrene, butanone, cyclohexanone and THF were refluxed from CaH$_2$, distilled in vacuum. CuCl was washed repeatedly by acetic acid and ethanol, and then dried in vacuum. Azo-bis-isobutyronitrile and Benzoyl peroxide was purified by the general method[22]. CuCl$_2$ (Aldrich) and 2.2′ -bipyridine (Aldrich) and acetonitrile (GC) were used as received.

Reverse ATRP of methyl acrylate initiated by AIBN/Cu^{II}Cl$_2$/bpy in bulk. Cu^{II}Cl$_2$, bpy, AIBN, Methyl acrylate were added to a flask with stirrer. The heterogeneous mixture was first degassed three times by freeze-pump-thaw cycles and sealed under vacuum, secondly immersed in an oil bath, heated at 65°C for 10 hours, then reacted at 100°C. After a given time, the flask was opened and the sample was dissolved with THF and filtered, then precipitated in methanol, filtered, and dried.

Reverse ATRP of styrene initiated by BPO/Cu^{II}Cl$_2$/bpy in bulk. Cu^{II}Cl$_2$, bpy, BPO, Styrene were added to a flask with stirrer. The heterogeneous mixture was first degassed three times by freeze-pump-thaw cycles and sealed under vacuum, secondly immersed in an oil bath , heated at 70°C℃ for 10 hours, then reacted at 110°C. After a given time, the flask was opened and the sample was dissolved with THF and filtered, then precipitated in methanol, filtered, and dried.

Reverse ATRP of methyl methacrylate and styrene initiated by BPO/Cu^{I}Cl/bpy. A typical polymerization was carried out by heating a reddish brown butanone solution of methyl methacrylate or styrene, Cu^{I}Cl, and bpy in a glass sealed under vacuum after freeze-pump-thaw cycles at 40 °C. The reaction solution was diluted with THF and filtered, then precipitated in methanol, filtered, and dried.

Polymer characterization. The monomer conversion was determined by gravimetry or GC. Molecular weight and molecular weight distribution were obtained by gel permeation chromatography (GPC) that was carried out with a Waters 208 instrument, using tetrahydrofuran as eluent, and calibration with polystyrene or poly(methyl methacrylate) standards. UV analysis was carried out using a PE-UV/VIS spectrometer. ^{1}HNMR spectrum was taken on 300 MHz spectrometers at room temperature in CDCl$_3$.

Acknowledgment: This work was sponsored by W. L. Gore & Associates, Inc. and the National Natural Science Foundation of China. Partial experimental work of this article had been finished with the help of Dr. Ping Xia, Miss Shenmin Zhu, Mister

Wenping Tu. The authors are much obliged to Prof. Müller A. H. E. for his kind offer of PMMA standards, and thank Prof. Guojian Wang very much for the characterization in Prof. Müller's laboratories. The authors are also grateful to Professor Matyjaszewski for his advice to our manuscript.

References

1. Otsu, T.; Yoshida, M. *Makromol. Chem., Rapid Commun.,* **1982**, 3, 127, 133.
2. Georges, M. K.; Veregin, R. P. N.; Kazmaier, P.M.; Hamer, G. K. *Macromolecules,* **1993**, 26, 5316.
3. Wayland, B. B.; Poszmik, G.; Mukerjee, S. L.; Fryd, M. *J. Am. Chem. Soc.,* **1994**, 116, 7943.
4. Greszta, D.; Mardare, D.; Matyjaszewski, K. *Macromolecules,* **1994**, 27, 638.
5. Matyjaszewski, K.; Gaynor, S.; Wang, J. S. *Macromolecules,* **1995**, 28, 2093.
6. Wang, J. S.; Matyjaszewski, K. *J. Am. Chem. Soc.,* **1995**, 117, 5614.
7. Asscher, M.; Vofsi, D. *J. Chem. Soc.,* **1963**, 1887.
8. Wang, J. S.; Matyjaszewski, K. *Macromolecules,* **1995**, 28, 7901.
9. Grannel, C.; Dubois, P.; Jerome, R.; and Teyssie, P. M*acromolecules,* **1996**, 29, 8576.
10. Kato, M.; Kamigaito, M.; Sawamoto, M. and Higashimura, T. *Macromolecules,* **1995**, 28, 1721.
11. Percec, V.and Barboiu B. *Macromolecules,* **1995**, 28, 7970.
12. Wang, J. S.; Matyjaszewski K. *Macromolecules,* **1995**, 28, 7572.
13. Xia, J. H.; Matyjaszewski K. *Macromolecules,* **1997**, 30, 7692.
14. Moineau, G.; Dubios, Ph.; Jerome, R.; Senninger T., Teyssie, Ph. *Macromolecules*, **1998**, 31, 545.
15. Wang, W.; Dong, Z.; Xia P.; Yan, D. *Macromol. Rapid Commun.,* **1998**, 19, 647.
16. Zhu, S.; Wang W.; Tu, W.; Yan, D. *Acta Polym.,* **1999**, 50, 267.
17. Wang, W.; Yan, D.; Chen X.; Wu, H. World Patent Pending.
18. Xia, J.; Matyjaszewki, K. *Macromolecules,* **1999**, 32, 5199.
19. Bamford, H. In *Comprehensive Polymer Science (First Supplement)*, Allen, G.; Aggarwal, S. L.; Russo, S. Eds., Pergamon: Oxford, 1992, P1.
20. Moad, G.; Solomon, D. H. *The Chemistry of Free Radical Polymerization*, Elsevier: Oxford, U.K., 1995
21. Barton, J.; Borsig, E. *Complexes in Free Radical Polymerization*, Elsevier: New York, 1998
22. *Polymer Experimental Technology*, Polymer Science Department of Fu Dan University, Fu Dan Publishing Company: Shanghai, 1996, p361

OTHER METHODS FOR CONTROLLED
RADICAL POLYMERIZATION

Chapter 20

Synthesis of Defined Polymers by Reversible Addition–Fragmentation Chain Transfer: The RAFT Process

Ezio Rizzardo, John Chiefari, Roshan T. A. Mayadunne, Graeme Moad, and San H. Thang

CSIRO Molecular Science, Bag 10, Clayton South MDC, Victoria 3169, Australia

Free radical polymerization in the presence of thiocarbonylthio compounds of general structure Z-C(=S)S-R provides living polymers of predetermined molecular weight and narrow molecular weight distribution by a process of reversible addition-fragmentation chain transfer. A rationale for selecting the most appropriate thiocarbonylthio compounds for a particular monomer type is presented with reference to the polymerization of methacrylates, styrenes, acrylates, acrylamides and vinyl acetate. The efficacy of the process is further demonstrated by the synthesis of narrow polydispersity polystyrene-*block*-poly(methyl acrylate)-*block*-polystyrene and 4-armed star polystyrene.

Introduction

Living radical polymerization allows the preparation of polymers with predetermined molecular weight, narrow molecular weight distribution and tailored architecture (e.g. end-functional, block, star) and thereby offers a vast range of new and advanced materials (*1,2*). With applications ranging from surfactants, dispersants, coatings and adhesives, to biomaterials, membranes, drug delivery, and microelectronics; the new materials have the potential of revolutionizing a large part of the polymer industry.

Recently, we communicated an effective and versatile living radical polymerization process that functions by reversible addition-fragmentation chain transfer (RAFT process) (*3-7*). The process is performed simply by adding a suitable thiocarbonylthio compound **1** to an otherwise conventional free radical polymerization. In this paper, we report in detail on the polymerization of methacrylates, styrenes, acrylates, acrylamides and vinyl acetate, and offer a rationale for selecting preferred thiocarbonylthio compounds for particular monomer types. Further, we demonstrate the efficacy of the process by reporting on the synthesis of an ABA triblock in two steps and a star polymer.

1

a   Z = Ph, R = C(CH$_3$)$_2$CN

b   Z = Ph, R = C(CH$_3$)$_2$Ph

c   Z = CH$_3$, R = C(CH$_3$)$_2$Ph

d   Z = Ph, R = C(CH$_3$)$_2$COOEt

e   Z = Ph, R = C(CH$_3$)$_2$CH$_2$C(CH$_3$)$_3$

f   Z = Ph, R = C(CH$_3$)(CN)CH$_2$CH$_2$CH$_2$OH

g   Z = SCH$_3$, R = C(CH$_3$)$_2$CN

h   Z = Ph, R = C(CH$_3$)(CN)CH$_2$CH$_2$COOH

i   Z = Ph, R = C(CH$_3$)(CN)CH$_2$CH$_2$COO$^-$Na$^+$

j   Z = N(CH$_2$CH$_3$)$_2$, R = CH$_2$Ph

k   Z =   R = CH$_2$Ph

l   Z = SCH$_3$, R = CH(Ph)COOH

m   Z = Ph, R = CH$_2$Ph

n   Z =   R = CH$_2$Ph

o   Z = OEt, R = C(CH$_3$)$_2$CN

p   Z = OEt, R = CH$_2$CN

q   Z = OEt, R = CH(CH$_3$)COOEt

r   Z = N(CH$_2$CH$_3$)$_2$, R = C(CH$_3$)$_2$COOEt

s   Z = N(Ph)(CH$_3$), R = CH$_2$CN

t   Z = Ph, R = C(CH$_3$)$_2$Cl

u   Z = SCH$_2$Ph, R = CH$_2$Ph

v   Z =   R = CH$_2$CN

Experimental

General

Meth(acrylates),styrene, *N,N*-dimethylacrylamide and vinyl acetate were purified by filtration through neutral activity I alumina followed by fractional distillation. The purified monomers were stored in a freezer and flash distilled immediately prior to use. Acrylic acid was fractionated from copper wire (to inhibit polymerization). *N*-Isopropylacrylamide was recrystallized from a mixture of benzene and *n*-hexane.

Solvents were AR grade and redistilled. Benzene (AR grade) was purified to remove thiophene and redistilled.

The syntheses of many of the thiocarbonylthio compounds **1** (RAFT agents) mentioned in this paper are described elsewhere (*3,8-10*).

Gel permeation chromatography (gpc) was carried out on a Waters Associates liquid chromatograph equipped with 6 Ultrastyragel columns (10^6, 10^5, 10^4, 10^3, 500 and 100 Å) and differential refractometer using tetrahydrofuran as the eluent at a flow rate of 1.0 mL/min. The gpc of polyacrylamides was performed at 80 °C on a set of 3 Styragel HT columns (10^5, 10^3, and 500 Å) using dimethylformamide/lithium bromide (0.05 M) as the eluent at a flow rate of 1.0 mL/min. In both cases, the instruments were calibrated with polystyrene standards. The number average molecular weight $M_n(gpc)$ quoted in Tables and Figures are not corrected for the type of polymer analyzed (*11*).

RAFT polymerizations

Polymerizations were carried out on a 2-5 mL scale in flame sealed ampoules. Prior to sealing, the contents were carefully degassed by three freeze-evacuate-thaw cycles under high vacuum. The concentrations of all reagents and experimental conditions are presented in the Tables and legends to Figures. Conversions were determined gravimetrically after exhaustive removal of volatiles under high vacuum. The volatile-free samples were used to acquire gpc data, i.e., the polymer products were not precipitated/fractionated prior to analysis.

Results and Discussion

Mechanism of RAFT polymerization and general remarks

The mechanism of the RAFT process is believed to involve a series of reversible addition-fragmentation steps as shown in Scheme 1. Addition of a propagating radical $P_n^\bullet$ to the thiocarbonylthio compound **1** gives the adduct radical **2** which fragments to a polymeric thiocarbonylthio compound **3** and a new radical R$^\bullet$. The radical R$^\bullet$ then re-initiates polymerization to give a new propagating radical $P_m^\bullet$. Subsequent addition-fragmentation steps set up an equilibrium between the propagating radicals $P_n^\bullet$ and $P_m^\bullet$ and the dormant polymeric thiocarbonylthio compounds **3** and **4** by way of the intermediate radical **5**. Equilibration of the growing chains gives rise to a narrow molecular weight distribution. Throughout the polymerization (and at the end) the vast majority of the polymer chains are end capped by a thiocarbonylthio group (dormant chains). The overall process is outlined in Scheme 2.

Compelling evidence for this mechanism has been obtained by direct observation of the intermediate radical **5** with ESR (*12*) and by the end group analysis of the polymer products **6** by NMR and UV/visible spectroscopy (4) Additional evidence is provided by MALDI-TOF mass spectroscopy analysis of a poly(methyl methacrylate) prepared with the aid of thiocarbonylthio compound **1b**. Although a variety of spectra were obtained under different operating conditions, the main populations could be clearly attributed to polymer chains capped by the $C(CH_3)_2Ph$ and $PhC(S)S$ moieties (13).

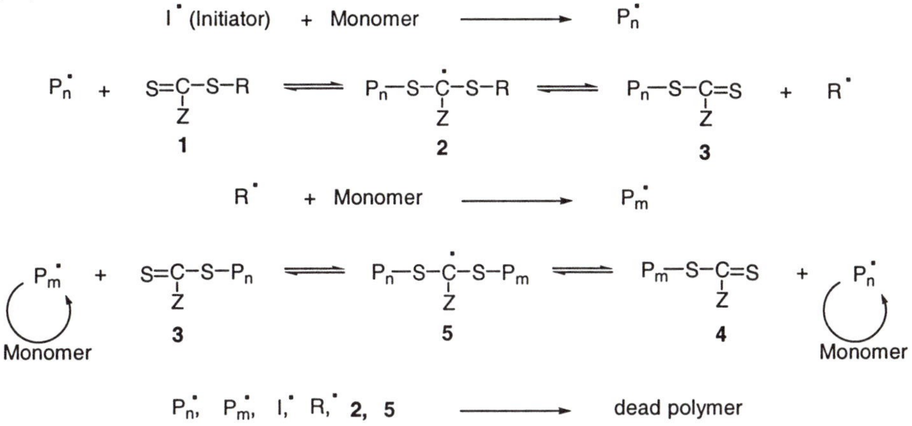

Scheme 1. Mechanism of RAFT polymerization

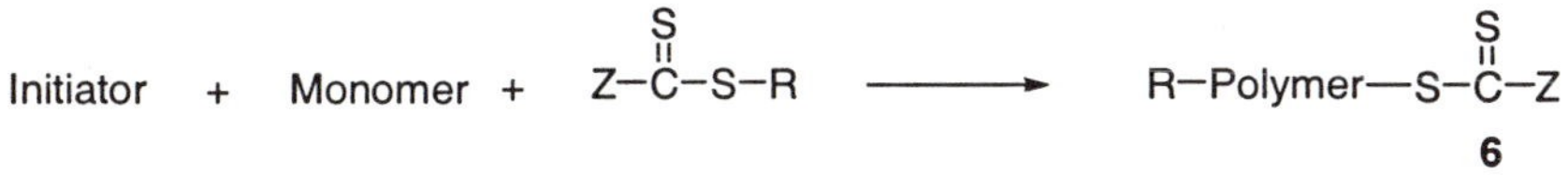

Scheme 2. Overall process for RAFT polymerization

For RAFT polymerization to function effectively, the choice of the thiocarbonylthio reagent **1** is extremely important. The requirements (see Scheme 1) are high rate constants for both the addition of propagating radicals to the thiocarbonylthio species **1**, **3** and **4** and fragmentation of the intermediate radicals **2** and **5**, relative to the rate constant for propagation. In other words, the species **1**, **3** and **4** must have high chain transfer constants (14).

It is generally accepted that rate constants for reactions involving radicals depend on a complex interplay between stability effects, steric effects, and polarity effects in the substrate, the radical and the products (15). A clear definition of all these factors for RAFT polymerization is not possible at this time. Nevertheless, the results obtained thus far have allowed us to identify the main drivers of the process and these are discussed below with reference to the polymerizations of methacrylates, acrylates, styrenes, acrylamides and vinyl acetate in the presence of thiocarbonylthio compounds selected from **1a – 1v**

In general terms, the rates of addition of radicals to thiocarbonylthio compounds **1**, **3** and **4** (Scheme 1) are strongly influenced by the substituent Z. It appears that radical stabilizing groups such as phenyl enhance the rate of addition. On the other hand, when Z is a nitrogen or oxygen substituent (e.g., **1j** or **1o**), the reagents are practically inert in the polymerization of acrylate, methacrylate and styrene monomers. This is attributed to a lowering of the double bond character of the C=S bond by conjugation with the lone pair of electrons on the heteroatom substituent (Scheme 3). This effect is largely eliminated when the nitrogen lone pair is part of an aromatic system as in **1k** or when the nitrogen is conjugated to a carbonyl group as in **1n**. The substituent Z is also expected to influence the rate of fragmentation of the intermediate radicals **2** and **5**. It may be reasoned that substituents which stabilize these radicals would retard their rate of fragmentation and *vice versa*.

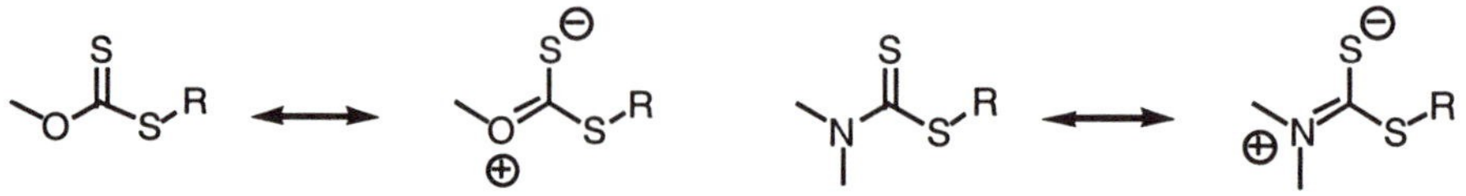

Scheme 3. Resonance forms of xanthates and dithiocarbamates

A high rate constant for fragmentation of radical **2** to the polymeric thiocarbonylthio species **3** (Scheme 1) is achieved by selecting a substituent R that is a good homolytic leaving group relative to the polymer chain P_n. The homolytic leaving ability of a group R appears to increase with increasing radical stability of the expelled radical R$^{\bullet}$ and increasing steric bulk of R. The cumyl group (C(CH$_3$)$_2$Ph), for example, is a much better leaving group than benzyl (CH$_2$Ph). While polarity factors in both the substituents Z and R are expected to play a role in the reactivity of the RAFT agents **1**, clear evidence is not as yet available.

It should also be noted that the radical R$^{\bullet}$ expelled from **1** (Scheme 1) must be sufficiently reactive to initiate polymerization efficiently. This means that a substituent R with high homolytic leaving ability need not be suitable for all polymerizations. For example, a RAFT agent **1** where R is cumyl would be a poor choice for the RAFT polymerization of vinyl acetate.

The key to understanding RAFT polymerization is to note that the process is conducted at low radical concentrations and using high ratios of RAFT agent (**1**) to initiator consumed during the experiment. The low radical concentration (determined by the choice of initiator and conditions) will minimize radical-radical termination and thus formation of "dead" chains. It should also be realised that, as a first approximation (neglecting, for example, chain length dependence for the rate of termination) the proportion of dead chains produced by radical-radical termination is expected to be the same in all current forms of living radical polymerization. In RAFT polymerization, unlike self-initiating system such as NMRP or ATRP, this loss of radicals is compensated for by a continuous low rate of initiation.

As can be gleaned from Scheme 1, the number of polymer chains initiated in RAFT polymerization will be equal to the total number of initiating radicals (I$^{\bullet}$+ R$^{\bullet}$). The number of dormant polymer chains (**3** and **4**) will be equal to the number of RAFT agent molecules (**1**) (the same as the number of radicals R$^{\bullet}$). The number of dead polymer chains will therefore be equal to the number of initiating radicals I$^{\bullet}$, if termination is by disproportionation or one half of this if termination is by combination. Since RAFT polymerization can be successfully conducted at ratios of initiating radicals I* to RAFT agent of 0.1 or less, the proportion of dead chains can be kept quite low.

The low ratios of initiating radicals I* to RAFT agent also serve to explain the low polydispersities of the polymers produced to high conversion, even though continuous initiation is employed. The chains initiated at high conversion will remain short but are relatively very few in number. These short chain radicals will react with the dormant chains (3 and 4) and in so doing will liberate long chain radicals which in turn will activate other dormant chains. As a result, most of the remaining monomer will be consumed in lengthening the dormant chains by the equilibrium depicted in Scheme 1. The few short chains remaining in the product are unlikely to be detected by current analytical methods.

RAFT polymerizations can be performed under conditions where it can be estimated that the proportion of dead chains is less than 10%. In these cases the molecular weight of the polymer ($M_n(calc)$) can be estimated by the equation:

$M_n(calc)$ = ([Monomer]/[RAFT agent]) x fractional conversion x MW of monomer.

RAFT polymerization of methacrylates

The free radical polymerization of methacrylates involves relatively bulky propagating radicals of moderate reactivity. The results in Table 1, in keeping with our earlier discussion, demonstrate that efficient RAFT polymerization of these monomers requires thiocarbonylthio compounds with bulky substituents (R) (and R* is stabilized) and radical stabilizing substituents (Z) that enhance the rate of addition of radicals to the C=S double bond. The most effective RAFT agents, in terms of producing polymers of low polydispersity and predetermined molecular weight, are 1a, 1b, 1f and 1t (Table 1, Entries 1, 3, 8 and 10 respectively).

Entries 2 and 4 show that RAFT agent 1b has a higher chain transfer constant than 1c as the polymer reaches a lower polydispersity with the former (M_w/M_n = 1.19 versus 1.33) at similar monomer conversions. This is attributed to a higher rate of addition of the methacrylate propagating radical to the RAFT agent 1. The higher chain transfer constant of 1b versus 1c is also evident by comparing the discrepancies between $M_n(gpc)$ and $M_n(calc)$ in Entries 2 and 4. The $M_n(calc)$ in Entry 4 is significantly lower than $M_n(gpc)$ indicating that the RAFT agent 1c is not all consumed at 33% conversion of monomer.

The effect of the substituent R on the reactivity of the RAFT agent is shown by examining the results obtained with 1a, 1b, 1d, 1e and 1f in Entries 1,3,6,7 and 8 respectively. Both 1a and 1b provide PMMA with narrow polydispersity and good correspondence between $M_n(gpc)$ and $M_n(calc)$ at high conversion. RAFT agent 1d also provides good molecular weight control but M_w/M_n is only 1.48. This indicates a relatively low chain transfer constant likely due to the poorer homolytic leaving ability of the ethyl isobutyrate group compared to cyanoisopropyl (in 1a) or cumyl (in 1b).

Table 1. RAFT polymerization of methacrylates[a]

Entry	Monomer (M)	RAFT Agent(10^2 M)	Time (h)	% Conv.	M_n (gpc)	M_n (calc)	M_w/M_n
1	MMA (7.01)	1a (1.11)	16	95	52 300	59 995	1.16
2	MMA (7.01)	1b (1.11)	4	27	18 000	17 050	1.19
3	MMA (7.01)	1b (1.11)	16	95	56 200	59 995	1.12
4	MMA (7.01)	1c (1.11)	4	33	26 200	20 840	1.33
5	MMA (7.01)	1c (1.11)	16	95	53 500	59 995	1.18
6	MMA (7.01)	1d (1.11)	16	95	52 900	59 995	1.48
7	MMA (7.01)	1e (1.11)	16	96	130 500	60 630	1.89
8	MMA (7.01)	1f (1.13)	16	92	55 300	57 100	1.05
9	MMA (7.01)	1g (1.11)	8	95	59 300	59 995	1.13
10	MMA (7.01)	1t (1.11)	16	95	52 800	59 995	1.14
11	MMA (7.01)	1b (36.72)	64	80	3 200	1 530	1.17
12	MMA (7.01)	1b (14.69)	24	95	6 600	4 530	1.21
13	MMA (9.34)	1a (2.92)	4	99	29 100	31 665	1.27
14	DMAEMA (4.45)	1b (1.25)	16	92	28 100	51 420	1.21
15	MMA/HEMA[b] (4.62)	1b (1.11)	16	75	28 000	32 100	1.21

Abbreviations: MMA: methyl methacrylate; DMAEMA: *N,N*-(dimethylamino)ethyl methacrylate; HEMA: 2-hydroxyethyl methacrylate.

a) Unless otherwise stated, all polymerizations were carried out at 60 °C with benzene as solvent and AIBN as initiator at a concentration of 6.1×10^{-3} M. Entry 9: 1,1'-azobis(1-cyclohexanecarbonitrile) (7.4×10^{-3} M) as initiator, 80 °C. Entries 11 and 12: AIBN (0.03 M) as initiator. Entry 13: 2,2'-azobis(methyl isobutyrate) (4.3×10^{-2} M) as initiator. Entry 15: ethyl acetate as solvent.

b) Mole ratio of monomers 91/9.

This in turn, may be the result of a lower radical stabilizing effect of COOEt compared to CN or Ph. The RAFT agent 1e (Entry 7) gives neither molecular weight control nor narrow polydispersity, indicating the need for both steric bulk and radical stabilization in the leaving group R. By contrast, the best results are obtained with

RAFT agent **1f** (Entry 8) in which steric bulk is greater than in **1a** or **1b** and radical stabilization (by CN) is greater than in **1e**.

Two other effective RAFT agents for methacrylate polymerization are **1g** and **1t** (Entries 9 and 10) respectively. The evolution of molecular weight and polydispersity with conversion for the polymerization of MMA in the presence of **1t** under the conditions of Entry 10 (Table 1) is shown in Figure 1.

Entries 11 and 12 (Table 1) show that good control can be maintained in the preparation of low molecular weight PMMA whilst Entry 13 demonstrates that RAFT polymerization can be performed at a relatively fast rate (high conversion in 4 hrs) with retention of both narrow polydispersity and molecular weight control. Finally, Entries 14 and 15 show that the RAFT process is compatible with tertiary amino and hydroxyl functionalities.

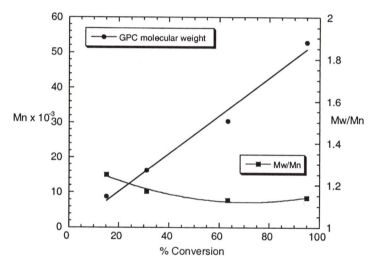

Figure 1. Evolution of molecular weight and polydispersity with conversion for the polymerization of methyl methacrylate in the presence of p-chlorocumyl dithiobenzoate 1t. For conditions see Entry 10, Table 1.

RAFT polymerization of styrenes

Table 2 shows a series of results obtained in the RAFT polymerization of styrenes. Compared to methacrylates, the less bulky propagating radical and lower propagation rate constant provides more freedom in the choice of the RAFT agent **1**, particularly in the substituent R. Entry 10 demonstrates that even benzyl has sufficiently high leaving group ability to promote the equilibrium needed for the RAFT process to operate (Scheme 1).

Figure 2 shows a plot of molecular weight and polydispersity versus conversion for the thermal polymerization of styrene at 100 °C under the conditions of Entry 2 (Table 2). To be noted is the close correspondence between $M_n(gpc)$ and $M_n(calc)$ and the fact that narrow polydispersities are reached at very low conversions. This indicates a very high transfer constant for **1b** in the polymerization of styrene.

Table 2 shows that either a conventional azo initiator or thermal initiation can be successfully utilized in the RAFT polymerization of styrene. Entries 3, 4 and 5 show that control over molecular weight is progressively lost as the concentration of RAFT agent is reduced, indicating that the normal events of radical polymerization (radical-radical termination) are gaining importance relative to the RAFT process. Nevertheless, the polydispersity in Entry 5 is substantially lower than that of a control experiment ($M_w/M_n = 1.74$).

Table 2. RAFT polymerization of styrenes[a]

Entry	Monomer (M)	RAFT Agent($10^2 M$)	Time (h)	Temp (°C)	% Con.	M_n (gpc)	M_n(calc)	M_w/M_n
1	St (8.73)	**1b** (0.84)	50	60	38	40 000	41 070	1.17
2	St (8.73)	**1b** (2.93)	64	100	81	25 200	25 100	1.12
3	St (8.73)	**1b** (0.98)	16	110	55	48 100	50 955	1.07
4	St (8.73)	**1b** (0.49)	16	110	57	88 200	105 615	1.16
5	St (8.73)	**1b** (0.25)	16	110	60	137 100	217 900	1.30
6	St (8.73)	**1h** (5.87)	64	100	61	8 900	9 435	1.05
7	St (8.73)	**1g** (2.95)	48	110	91	27 800	28 000	1.09
8	St (8.73)	**1l** (2.93)	48	110	92	29 200	28 500	1.07
9	St (8.73)	**1j** (2.97)	6	100	15	317 100	4 590	1.86
10	St (8.73)	**1k** (2.97)	30	100	60	15 600	18 340	1.20
11	St-SO₃Na (1.21)	**1i** (1.66)	14	70	84	10 500[b]	12 640	1.20

Abbreviations: St: styrene; St-SO₃Na: p-styrenesulfonic acid sodium salt.
a) Unless otherwise stated, all polymerizations were carried out in bulk without added initiator. Entry 1: AIBN (1.7×10^{-3} M) as initiator. Entry 11: 4,4'-azobis(4-cyanopentanoic acid) (4.2×10^{-3} M) as initiator in water as solvent.
b) Molecular weights determined with reference to poly(p-styrenesulfonic acid sodium salt) standards. 0.1 M aqueous NaNO₃/acetonitrile (80/20) as eluent and flow rate 0.8 mL/min.

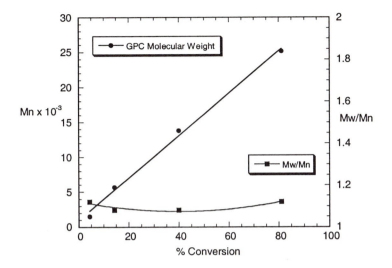

Figure 2. Evolution of molecular weight and polydispersity with conversion for the thermal polymerization of bulk styrene at 100 °C in the presence of cumyl dithiobenzoate 1b (0.029 M).

RAFT polymerization affords end-functional polymers when functional RAFT agents are employed. Examples are found in Entries 6 and 8 of Table 2. The structure of the polystyrene **7** prepared with **1l** (Entry 8) is shown in Scheme 4. It should be noted that the trithiocarbonate end group in **7** can be chemically transformed to other functional groups. A particularly easy transformation is that to thiol (—SH) by treatment with nucleophiles such as R_2NH or OH^- (7,16).

$$CH_3S-\overset{\overset{S}{\|}}{C}-S-CH(Ph)COOH \quad \xrightarrow[110\ ^{\circ}C]{St} \quad CH_3S-\overset{\overset{S}{\|}}{C}-S\left(\overset{|}{\underset{Ph}{CH}}-CH_2\right)_n CH(Ph)COOH$$

7

Scheme 4. End - functional polystyrene prepared with RAFT agent **1l**

Entry 9 shows that *N,N*-dialkyl dithiocarbamates are ineffective in the RAFT polymerization of styrene because of the reduced reactivity of the C=S double bond to radical addition in these compounds (see Scheme 3 and earlier discussion). The polymer of Entry 9 is of similar M_n and polydispersity to that obtained in the absence of a RAFT agent under the same conditions (the same applies to methacrylates and acrylates, but not vinyl acetate - see later). However, when the lone pair on the nitrogen is part of an aromatic system (e.g. **1k**) and no longer available for conjugation with the C=S bond (see Scheme 3) good activity in the RAFT process is observed, as in Entry 10.

RAFT polymerizations can be performed in an aqueous medium. This is exemplified by the controlled polymerization of the sodium salt of styrenesulfonic acid using water soluble RAFT agent **1i** and initiator as shown in Entry 11.

Polymerization of acrylates and acrylic acid

The free radical polymerization of acrylates is characterized by propagating radicals of relatively low steric bulk and high reactivity. These factors facilitate both the addition of the propagating radicals to the C=S double bond and fragmentation of the R group in the RAFT agent (see Scheme 1) allowing a wider choice in the Z and R substituents. The easily prepared RAFT agents **1m** and **1k** provide excellent control and have been used to make polymers of relatively high and low molecular weight respectively (Table 3 Entries 1 and 2). Of all the RAFT agents listed above, only the dithiocarbamates **1j**, **1r**, and **1s** show little or no activity in the polymerization of acrylates. Xanthates exhibit moderate activity and provide good molecular weight control but the polydispersities remain high. An example is the experiment listed in Entry 7 of Table 3.

Table 3. RAFT polymerization of acrylates/acrylic Acid[a]

Entry	Monomer (M)	RAFT Agent(10^2 M)	AIBN (10^4M)	Time (h)	% Conv.	M_n (gpc)	$M_{n(calc)}$	M_w/M_n
1	nBA (2.79)	**1m** (0.16)	3.3	8	40	91 700	89 300	1.14
2	MA (2.22)	**1k** (1.48)	7.3	16	74	8 800	9 500	1.17
3	MA (2.22)	**1g** (0.17)	7.3	4	55	65 400	61 800	1.06
4	MA (2.22)	**1g** (0.17)	7.3	16	89	97 600	99 900	1.25[b]
5	MA (4.44)	**1n** (0.36)	3.3	24	97	108 200	102 900	1.18
6	MA (4.44)	**1n** (0.18)	3.4	16	90	210 700	190 900	1.24
7	tBA (2.33)	**1o** (2.12)	5.4	1	72	11 000	10 100	1.77
8	MA (4.44)	**1a** (0.36)	3.3	48	62	72 200	65 700	1.15
9	AA (5.83)	**1a** (0.36)	3.3	16	53	66 800[c]	73 800[c]	1.13

Abbreviations: nBA: *n*-butyl acrylate; MA: methyl acrylate; tBA: *tert*-butyl acrylate; AA: acrylic acid

a) Unless otherwise stated, all polymerizations were carried out at 60 °C in benzene. Entry 7: ethyl acetate as solvent. Entry 9: methanol as solvent.
b) Bimodal
c) GPC molecular weight on a PAA sample after methylation.

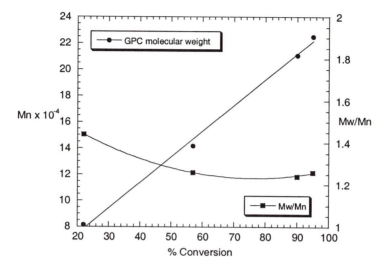

*Figure 3a. Evolution of molecular weight and polydispersity with conversion for the polymerization of methyl acrylate in the presence of **1n**. For conditions see Entry 6, Table 3.*

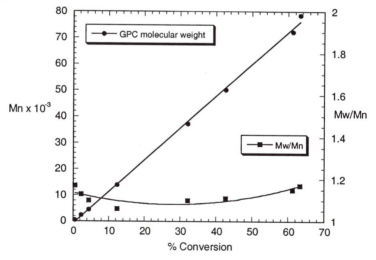

*Figure 3b. Evolution of molecular weight and polydispersity with conversion for the polymerization of methyl acrylate in the presence of cyanoisopropyl dithiobenzoate **1a**. For conditions see Entry 8, Table 3.*

A common and negative feature of the RAFT polymerization of acrylates is a yet to be defined side reaction which manifests as a bimodal distribution in the gpc chromatograms of the polymers, particularly at high conversions and for $M_n(gpc)$ over approximately 50 000. An example of this is the polymerization of methyl acrylate using **1g** as the RAFT agent shown in Entries 3 and 4 of Table 3. At 55% conversion, the gpc of the polymer exhibited a symmetrical peak of M_w/M_n = 1.06 but at 89% conversion, a small second peak appeared and M_w/M_n broadened to 1.25. The second peak is relatively small and has peak molecular weight (M_p) approximately double that of the main peak. For this reason, it may be due to the expected combination of propagating radicals, although some combination with the intermediate radicals **5** (Scheme 1) (~3xMp) cannot be excluded from the gpc data.

Interestingly, bimodality in the molecular weight distribution is not observed when **1n** is used as the RAFT agent. Entries 5 and 6 (Table 3) show that the polymerization of methyl acrylate can be taken to high conversion and high M_n with retention of narrow polydispersity and molecular weight control. The evolution of M_n and M_w/M_n with conversion for a polymerization of methyl acrylate in the presence of **1n** is shown in Figure 3a. To be noted is the relatively high polydispersity at low conversion attributed to the relatively low chain transfer constant of this RAFT agent (see Scheme 3 and relevant discussion). By contrast, the high transfer constant of RAFT agent **1a** gives rise to polymer of very low polydispersity at very low conversion. This is shown by the relevant plot in Figure 3b.

RAFT polymerization can also be applied to acid monomers. Entry 9 shows the results of a polymerization of acrylic acid in methanol using **1a** as the RAFT agent.

RAFT polymerization of acrylamides

The free radical polymerization of acrylamides is also characterized by propagating radicals of low steric bulk and high reactivity. Consequently, it is anticipated that a wide range of RAFT agents (**1**) will be useful in their RAFT polymerization. To date we have investigated *N,N*-dimethylacrylamide (DMA) and *N*-isopropylacrylamide (NIPAM) with the RAFT agent **1b**. The results of these experiments are shown in Table 4. A plot of M_n and M_w/M_n versus conversion (Figure 4), demonstrates the effectiveness of the RAFT process with these monomers.

Table 4. RAFT polymerization of acrylamides[a]

Entry	Monomer (M)	RAFT Agent(10^2 M)	AIBN (10^3M)	Time (h)	% Conv	M_n (gpc)	M_n(calc)	M_w/M_n
1	DMA (3.88)	**1b** (0.36)	1.5	2	8.5	12 000	9 100	1.13
2	DMA (3.88)	**1b** (0.36)	1.5	8	64	59 500	68 300	1.05
3	DMA (3.88)	**1b** (0.36)	1.5	16	76	88 700	81 100	1.08
4	NIPAM (1.77)	**1b** (0.39)	1.2	24	56	24 500	28 700	1.15

Abbreviations: DMA: *N,N*-dimethylacrylamide; NIPAM: *N*-isopropylacrylamide.
a)   At 60 °C, DMA in benzene, NIPAM in 1,4-dioxane

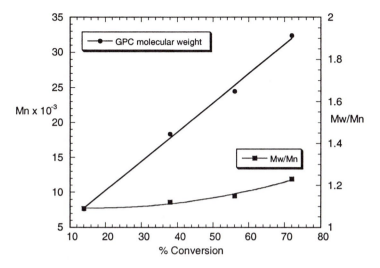

Figure 4. Evolution of molecular weight and polydispersity with conversion for the polymerization of N-isopropyl acrylamide in the presence of 1b. For conditions see Entry 4, Table 4.

RAFT polymerization of vinyl acetate

The polymerization of vinyl acetate is completely inhibited in the presence of the preferred RAFT agents for (meth) acrylates, acryamides and styrenes, i.e. dithioesters, trithiocarbonates and aromatic dithiocarbamates. On the other hand, xanthates and dithiocarbamates which show little or no activity with these monomers function effectively in the RAFT polymerization of vinyl acetate.

These findings may be rationalized by considering that the vinyl acetate propagating radical is poorly stabilized and of low steric bulk and hence is both highly reactive and poor as a homolytic leaving group. As a consequence, we attribute the inhibition mentioned above to slow fragmentation of the intermediate radical 5 (Scheme 1) when P_n and P_m are poly(vinyl acetate) chains. We surmize, however, that when the substituent Z in 5 is OR or NR_2 the radical is destabilized (higher energy), relative to Z = Ph or CH_3 for example, and its fragmentation becomes more facile. An additional driving force for the fragmentation of radical 5 may be provided by the higher stability of the product 3 and 4 brought about by conjugation of the OR or NR_2 substituents with the C=S double bond (see Scheme 3). It is clear from our results, however, that the reduction in double bond character of the C=S bond by conjugation with the heteroatoms is not sufficient to inhibit the addition of the highly reactive vinyl acetate propagating radical to 1, 3 and 4 in contrast to the behaviour of the less reactive propagating radicals from (meth)acrylate, acrylamide and styrene monomers.

Table 5. RAFT polymerization of vinyl acetate[a]

Entry	VAc (M)	RAFT Agent(10^2 M)	Initiator[b] (10^3M)	Time (h)	Temp (°C)	% Conv.	M_n (gpc)	M_n (calc)	M_w/M_n
1	5.43	**1p** (4.99)	8.5	2	100	41	4 600	3 840	1.16
2	5.43	**1p** (4.99)	8.5	16	100	92	9 100	8 610	1.37
3	10.86	**1p** (9.96)	8.7	4	100	66	7 000	6 190	1.18
4	10.86	**1q** (2.51)	3.9	16	100	63	32 000	23 440	1.33
5	10.86	**1p** (1.65)	5.9	18.5	80	88	47 000	49 810	1.62
6	7.24	**1r** (4.99)	28.0	24	75	95	15 000	11 850	1.50
7	10.86	**1s** (4.98)	61.0	16	60	96	22 700	18 000	1.24
8	7.24	**1s** (5.06)	28.0	24	75	93	13 400	11 440	1.29
9	7.24	**1s** (10.06)	28.0	24	75	95	7 100	5 880	1.25

a) Ethyl acetate as solvent for Entries 1,2,6,8 and 9, bulk monomer for Entries 3,4,5 and 7.

b) Initiator: 1,1'-Azobis(1-cyclohexanecarbonitrile) (Vazo-88) for Entries 1,2,3,5,6,8 and 9, 2,2'-azobis(2,4,4-trimethylpentane) (VR-110) for Entry 4, and AIBN for Entry 7.

The addition of radicals to xanthates followed by β-scission (fragmentation) is well documented in synthetic organic chemistry (*17*) and their use as agents for living radical polymerization has also been reported recently by others (*18*).

Selected experimental results for the RAFT polymerization of vinyl acetate in the presence of xanthates and dithiocarbamates are reported in Table 5. This shows that narrow polydispersities and good control of molecular weight are readily achieved for polymers of $M_n < 30\ 000$ and particularly at moderate conversions (Entries 1 and 3). As the target molecular weight is increased by using lower concentrations of RAFT agent the polydispersities deteriorate (Entries 4 and 5).

It is also evident that the *N*-aryl, *N*-alkyl dithiocarbamate **1s** provides better results (Entries 7,8 and 9) than the *N,N*-dialkyl dithiocarbamate **1r** (Entry 6). This is attributed to the higher transfer constant for **1s** brought about by the phenyl substituent reducing the electron density on nitrogen (see Scheme 3 and earlier discussion). Further reduction of the electron density on nitrogen however, as in RAFT agent **1v**, results in inhibition of polymerization.

The evolution of M_n and M_w/M_n with conversion for the polymerization of vinyl acetate in the presence of xanthate **1p** is shown in Figure 5. The significant increase in M_w/M_n at higher conversions is attributed to the greater propensity for the vinyl acetate propagating radical to undergo side reactions (e.g. chain transfer to monomer and polymer) when compared to the propagating radicals of the monomers discussed previously.

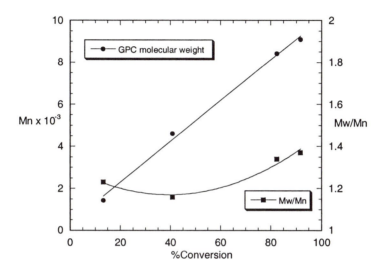

*Figure 5. Evolution of molecular weight and polydispersity with conversion for the polymerization of vinyl acetate at 100 °C in the presence of xanthate **1p**. For conditions see Entry 2, Table 5.*

Synthesis of block and star polymers

The living nature of RAFT polymerization allows the synthesis of polymers of complex architecture. Some examples of narrow polydispersity AB, ABA and ABC block copolymers, star polymers and gradient copolymers have been reported recently (*3,5-7*). An additional interesting example is provided by the use of the symmetrical trithiocarbonate **1u** to prepare ABA block copolymers in two steps, as shown in Scheme 5 (*19*).

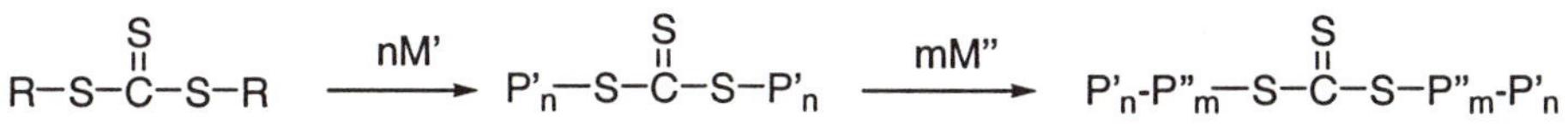

$$R-S-\overset{\overset{S}{\|}}{C}-S-R \xrightarrow{nM'} P'_n-S-\overset{\overset{S}{\|}}{C}-S-P'_n \xrightarrow{mM''} P'_n-P''_m-S-\overset{\overset{S}{\|}}{C}-S-P''_m-P'_n$$

Scheme 5: ABA block copolymer synthesis using symmetrical trithiocarbonate

294

This approach was used to synthesize polystyrene-*block*-poly(methyl acrylate)-*block*-polystyrene by first polymerizing styrene (M') to M_n 13 200 and following with methyl acrylate (M") to M_n 53 300. The GPC traces of the polymers from this experiment are shown in Figure 6.

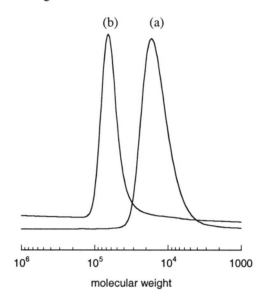

*Figure 6. GPC traces for (a) polystyrene of M_n 13 200 and M_w/M_n 1.22 by heating styrene and **1u** (2.93x10^{-2} M) at 100 °C for 20 hrs, and (b) polystyrene-block-poly(methyl acrylate)-block-polystyrene of M_n 53 300 and M_w/M_n 1.19 prepared by heating the above polystyrene (8.5x10^{-3} M) and 2,2'-azobis(2-cyanopropane) (3.6x10^{-3} M) in methyl acrylate at 60 °C for 4 hrs.*

Star polymers are readily prepared by utilizing RAFT agents containing multiple thiocarbonylthio moeities. The RAFT agents can be designed to operate in two distinct ways: (a) those in which the propagating arms are attached to the core and (b) those in which the arms propagate away from the core and become part of the star polymer by addition to the core. The outstanding advantage of the second approach is that complications due to star-star radical coupling are avoided. An example of this is provided by the tetrafunctional trithiocarbonate **8** as outlined in Scheme 6. In **8**, fragmentation will expel the benzyl radical in preference to the primary alkyl radical. Likewise, the propagating radical will be preferentially expelled from the dormant chains. As a result, the arms of the star polymer will always be in the dormant form. Termination of the propagating radicals would give rise to linear polymer of lower molecular weight than the star. This is clearly demonstrated in the gpc traces for a four-armed polystyrene star prepared with **8** (Figure 7). To be noted are the very narrow molecular weight distributions of the star polymers and the low level of dead chains even at high conversion.

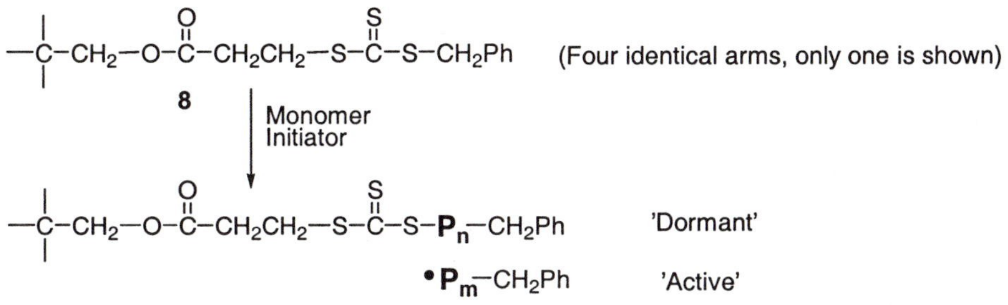

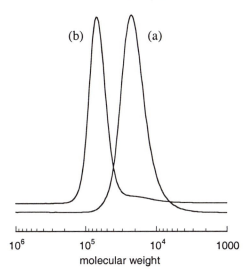

$-\overset{|}{\underset{|}{C}}-CH_2-O-\overset{O}{\overset{\|}{C}}-CH_2CH_2-S-\overset{S}{\overset{\|}{C}}-S-CH_2Ph$ (Four identical arms, only one is shown)

8 | Monomer
Initiator

$-\overset{|}{\underset{|}{C}}-CH_2-O-\overset{O}{\overset{\|}{C}}-CH_2CH_2-S-\overset{S}{\overset{\|}{C}}-S-\mathbf{P_n}-CH_2Ph$ 'Dormant'

$\bullet\mathbf{P_m}-CH_2Ph$ 'Active'

Scheme 6: Synthesis of star polymers by propagation of arms away from core

Figure 7. GPC traces for 4-armed star polystyrene prepared by thermal polymerization of bulk styrene in the presence of trithiocarbonate 8 ($7.3x10^{-3}$ M) at 110 °C for (a) 6 hrs, 25% conversion (M_n 25 600, M_w/M_n 1.18) and (b) 48 hrs, 96% conversion (M_n 92 000, M_w/M_n 1.07 excluding the lower molecular weight peak)

Conclusion

With appropriate choice of thiocarbonylthio reagents, RAFT polymerization yields well defined polymers from methacrylate, styrene, acrylate, acrylamide and vinyl acetate monomers. Any of a wide range of solvents, initiators and temperatures can be employed. The living nature of the process is indicated by a linear increase in molecular weights with conversion, a correspondence between calculated molecular weights and those measured by gpc, and the narrow polydispersity of the polymers produced. Thiocarbonylthio compounds can be designed to generate ABA triblocks in two steps or to avoid star-star coupling in the synthesis of star polymers.

Acknowledgement. We are grateful to Drs C. Berge, M. Fryd and R. Matheson of DuPont Perfomance Coatings for their support of this work.

References
1.   Rizzardo, E.; Moad, G. in *The Polymeric Materials Encyclopaedia: Synthesis, Properties and Applications*; Salamone, J.C. Ed.; CRC Press: Boca Raton, Florida, 1996; vol. 5, p 3834-3840.
2.   Matyjaszewski, K. *ACS Symp. Ser.,* **1998**, *685*, 2-30.
3.   Le, T.P.; Moad, G.; Rizzardo, E.; Thang, S.H. PCT Int. Appl. WO 9801478 A1 980115 [*Chem. Abstr.* **(1998)** *128*: 115390].
4.   Chiefari, J.; Chong, Y.K.; Ercole, F.; Krstina, J.; Jeffery, J.; Le, T.P.T.; Mayadunne, R.T.A.; Meijs, G.F.; Moad, C.L.; Moad, G.; Rizzardo, E.; Thang, S.H. *Macromolecules* **1998**, *31*, 5559-5562.
5.   Rizzardo, E.; Chiefari, J.; Chong, Y.K.; Ercole, F.; Krstina, J.; Jeffery, J.; Le, T.P.T.; Mayadunne, R.T.A.; Meijs, G.F.; Moad, C.L.; Moad, G.; Thang, S.H. *Macromol. Symp.* **1999**, 143, 291-308.
6.   Chong, Y.K.; Le, T.P.T.; Moad, G.; Rizzardo, E.; Thang, S.H. *Macromolecules* **1999**, *32*, 2071-2074.
7.   Mayadunne, R.T.A.; Rizzardo, E.; Chiefari, J.; Chong, Y.K.; Moad, G.; Thang, S.H. *Macromolecules* **1999**, *32*, 6977-6980.
8.   Moad, G.; Rizzardo, E.; Thang, S.H. PCT Int. Appl. WO 99/05099.
9.   Thang, S.H.; Chong, (Bill) Y.K.; Mayadunne, R.T.A.; Moad, G.; Rizzardo, E. *Tetrahedron lett.* **1999**, *40*, 2435-2438.
10. Chiefari, J.; Mayadunne, R.T.A.; Moad, G.; Rizzardo, E.; Thang, S.H. PCT Int. Appl. WO 99/31144.
11. Hutchinson, R.A.; Paquet, D.A. Jr.; McMinn, J.H. *Macromolecules* **1995**, *28*, 5655-5663. Hutchinson, R.A.; Paquet, D.A. Jr.; McMinn, J.H.; Beuermann, S.; Fuller, R.E.; Jackson, C. *DECHEMA Monogr.* **1995**, *131*, 467-492.
12. Hawthorne, D.G.; Moad, G.; Rizzardo, E.; Thang, S.H. *Macromolecules* **1999**, *32*, 5457-5459.
13. Moad, G.; Rizzardo, E.; Thang, S.H. unpublished results.
14. Moad, G.; Chiefari, J.; Chong, Y.K.; Krstina, J.; Mayadunne, R.T.A.; Postma, A.; Rizzardo, E.; Thang, S.H. *Polym. International* **2000**, in press.
15. Ghosez-Giese A.; Giese, B. *ACS Symp. Ser.,* **1998**, *685*, 50-61.
16. Kato, S.; Ishida, M. *Sulfur Reports*, **1988**, *8*, 155; Mayer, R.; Scheithauer, S. In *Houben - Weyl Methods of Organic Chemistry*; Buechel, K.H.; Falbe, J.; Hagemann, H.; Hanack, M. Eds.; Thieme: Stuttgart, 1985; vol E, p 891.
17. Zard, S.Z. *Angew. Chem. Int. Ed. Engl.* **1997**, 36, 672-685.
18. Corpart, P.; Charmot, D.; Biadatti, T.; Zard, S.; Michelet, D. PCT Int. Appl. WO 9858974 [*Chem. Abstr.* **(1999)** 130: 82018].
19. Mayadunne, R.T.A.; Rizzardo, E.; Chiefari, J.; Krstina, J.; Moad, G.; Postma, A.; Thang, S.H. *Macromolecules* **2000**, 33, 243-245.

Chapter 21

Preparation of Macromonomers via Chain Transfer with and without Added Chain Transfer Agent

John Chiefari, Justine Jeffery, Roshan T. A. Mayadunne,
Graeme Moad, Ezio Rizzardo, and San H. Thang

CSIRO Molecular Science, Bag 10, Clayton South MDC, Victoria 3169, Australia

Two methods are described for the preparation of macromonomers based predominantly on monosubstituted monomers. The first method involves the use of cobalt mediated catalytic chain transfer in copolymerizations of functional acrylic and styrene monomers with α-methylvinyl monomers to prepare macromonomers that are predominantly composed of the monosubstituted monomer. We examine the influence of the cobalt concentration, cobalt type, monomer/comonomer ratio, comonomer type, and reaction temperature on the relative amount of the desired 1,1-disubstituted alkene end group (derived from the α-methylvinyl monomer) to the 1,2-disubstituted alkene end group (derived from the monosubstituted monomer). An alternative procedure for macromonomer preparation uses no added chain transfer agent or α-methylvinyl comonomer. This procedure involves heating a mixture of the monosubstituted monomer in an appropriate solvent with an azo or peroxy initiator at temperatures between 80°C to 240°C. The effect of reaction temperature and monomer concentration on macromonomer molecular weight and the degree of branching are discussed.

Macromonomers, possessing a 1,1-disubstituted alkene end group (see structure **1**), are versatile intermediates which have applications as transfer agents for molecular weight control (*1-3*) and have also seen widespread use as precursors to block (*1,4*) and graft copolymers (*3,5*).

1

2. Y = Ph, X = Ph
3. Y = CO$_2$Bu, Z = CO$_2$Me

Two effective and well known methods for the preparation of macromonomers are based on the use of radical addition-fragmentation chain transfer agents (*e.g.* allyl sulfides) (*6,7*) and catalytic chain transfer (CCT) with cobalt complexes (*8-10*). The effectiveness of the CCT process for molecular weight control of polymers and for preparing macromonomers based on α-methylvinyl monomers *e.g.* methacrylates has been well documented (*8-10*). In the past few years there has been increasing interest in the cobalt mediated CCT polymerization of monosubstituted monomers (*11*) and the cobalt mediated CCT copolymerization of monosubstituted monomers with α-methylvinyl monomers as highlighted in a number of papers (*12-14*), patents (*15*) and more recently a kinetic analysis of the process (*16,17*).

We have, for sometime, been interested in the preparation of macromonomers that are based on monosubstituted monomers. We have been successful in utilizing the cobalt mediated CCT copolymerization of, among others, styrene and acrylates, with α-methylvinyl monomers *e.g.* α-methylstyrene (AMS) and methyl methacrylate (MMA), to prepare macromonomers of general structure **2** and **3** (*18*). In this paper, we extend the scope of CCT copolymerization with cobalt complexes, to the preparation of functionalized macromonomers **4**, **5** and **6**, while minimizing the formation of the corresponding undesired oligomers **7**, **8** and **9**.

4   **7**

5   **8**

6   **9.** R = Butyl and 2-Hydroxyethyl

As part of our continuing efforts to find new methods to prepare macromonomers, we also report on a method of macromonomer synthesis that is clean, simple and economical to perform. It involves heating a mixture of either acrylate, styrene or vinyl benzoate monomer in an appropriate solvent with an azo or peroxy initiator at elevated temperatures, typically >150 °C (*19,20*). There are two major advantages in this new process over the cobalt mediated CCT process. Firstly, no added chain transfer agent is needed, thereby overcoming purification steps for removal of cobalt impurities. Secondly, no α-methylvinyl comonomer is required thus insuring a homopolymer and 1,1-disubstituted alkene end group exclusivity.

Macromonomer Synthesis via Cobalt Mediated CCT Polymerization.

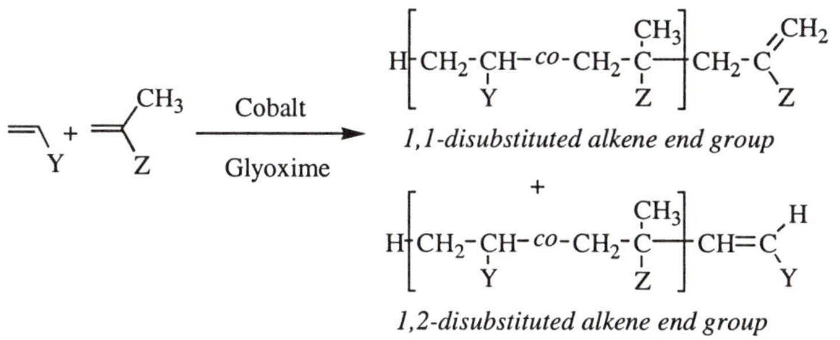

1,1-disubstituted alkene end group

1,2-disubstituted alkene end group

Scheme 1: Products from CCT copolymerization of a monosubstituted monomer (e.g.
Y=CO₂R or Ph) with an α-methylvinyl monomer (e.g. Z=CO₂R or Ph).

The CCT copolymerization of monosubstituted monomers with α-methylvinyl monomers in the presence of a cobalt glyoxime complex, as depicted in scheme 1, produces two distinct copolymer products. The two products are distinguished by their different vinyl end groups. The macromonomer, which is the desired product, possesses a 1,1-disubstituted alkene end group. The second product, which possesses the 1,2-disubstituted alkene end group, is regarded as an undesired by-product since this oligomer shows reduced reactivity towards further free radical polymerization. The macromonomer purity (% macromonomer) is a measure of the ratio of the 1,1-disubstituted alkene end group over the total alkene end groups. The two types of vinyl end groups can be readily identified by ^{1}H-nmr spectroscopy.

The mechanism of CCT copolymerization, as shown in scheme 2, describes the origins of the alkene end groups. They can form by disproportionation involving the propagating radical and the Co(II) complex or by collapse of a polymer-Co(III) complex. This latter pathway may be responsible for the lower polymerization rates observed with CCT (see Table I). The identity of the alkene end group arises from the terminal unit on the propagating chain. Hence, the need for the α-methylvinyl comonomer as this gives rise to the desired 1,1-disubstituted alkene end group. The reaction of propagating radicals with the Co(II) species also produces Co(III)-H. This species reacts with monomer to produce a new H-terminated propagating radical.

In contrast to the first reported cobalt complexes, which were based on porhyrins (*11*), the cobalt complexes we have used in this study are based on glyoximes *i.e.* cobalt(II)diphenylglyoxime **10**, isopropylcobalt(III)dimethylglyoxime **11** and methylcobalt(III)diethylglyoxime **12** (*21*). The organocobalt(III) complexes, which possess a weak Co-C bond, are converted *in situ* into the active cobalt(II) species by homolytic cleavage of the Co-C bond.

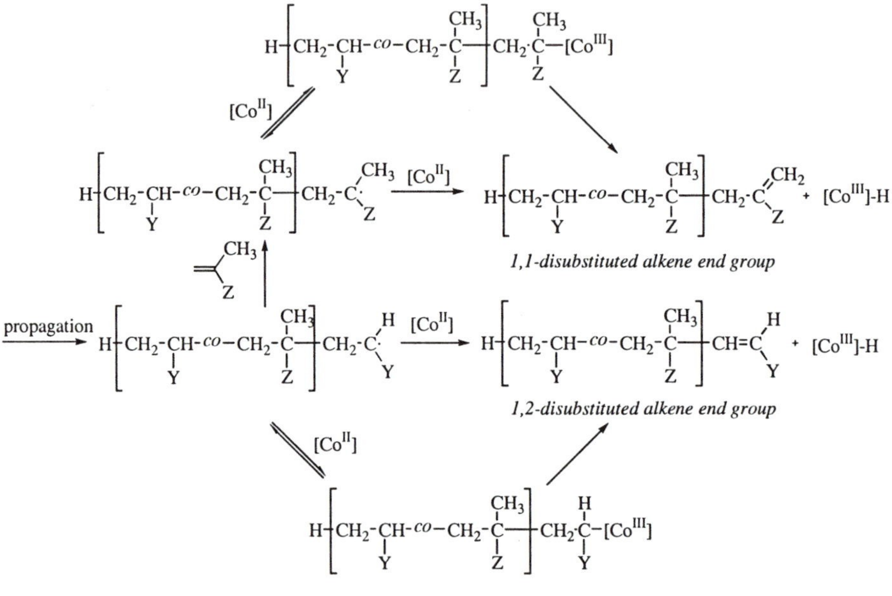

$$H\left[CH_2\text{-}CH\text{-}co\text{-}CH_2\text{-}\underset{\underset{Y}{|}}{\overset{\overset{CH_3}{|}}{C}}\right]CH_2\text{-}\underset{\underset{Z}{|}}{\overset{\overset{CH_3}{|}}{C}}\text{-}[Co^{III}]$$

$$[Co^{II}]$$

$$H\left[CH_2\text{-}CH\text{-}co\text{-}CH_2\text{-}\underset{\underset{Y}{|}}{\overset{\overset{CH_3}{|}}{C}}\right]CH_2\text{-}\overset{\overset{CH_3}{|}}{C}\cdot_Z \quad [Co^{II}] \longrightarrow H\left[CH_2\text{-}CH\text{-}co\text{-}CH_2\text{-}\underset{\underset{Y}{|}}{\overset{\overset{CH_3}{|}}{C}}\right]CH_2\text{-}C\overset{CH_2}{\underset{Z}{}} + [Co^{III}]\text{-}H$$

1,1-disubstituted alkene end group

$$\overset{CH_3}{\underset{Z}{}}$$

propagation $\longrightarrow H\left[CH_2\text{-}CH\text{-}co\text{-}CH_2\text{-}\underset{\underset{Y}{|}}{\overset{\overset{CH_3}{|}}{C}}\right]CH_2\text{-}\overset{\overset{H}{|}}{C}\cdot_Y \quad [Co^{II}] \longrightarrow H\left[CH_2\text{-}CH\text{-}co\text{-}CH_2\text{-}\underset{\underset{Y}{|}}{\overset{\overset{CH_3}{|}}{C}}\right]CH=C\overset{H}{\underset{Y}{}} + [Co^{III}]\text{-}H$

1,2-disubstituted alkene end group

$$[Co^{II}]$$

$$H\left[CH_2\text{-}CH\text{-}co\text{-}CH_2\text{-}\underset{\underset{Y}{|}}{\overset{\overset{CH_3}{|}}{C}}\right]CH_2\overset{\overset{H}{|}}{\underset{Y}{C}}\text{-}[Co^{III}]$$

Scheme 2: *Mechanism of CCT copolymerization and origin of 1,1-disubstituted and 1,2-disubstituted alkene end groups.*

The nature of the R group determines the mechanism of homolytic cleavage. It can be thermally induced, it can involve an S_H2 mechanism, or by a combination of both mechanisms.

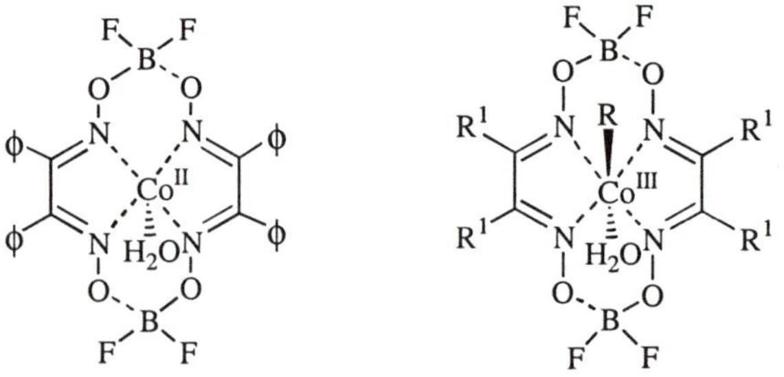

10. Co(II)DPG

11. R = Isopropyl R^1 = Me
12. R = Me R^1 = Et

The study conducted by Greuel and Harwood on the CCT copolymerization of styrene and MMA showed that the monomer/comonomer ratio influenced the relative proportion of the two different alkene end groups and the amount of comonomer incorporated in the copolymer (*14*). These observations were later confirmed by an extensive kinetic analysis for the CCT copolymerization of styrene and MMA (*16*) and more recently for the CCT copolymerization of styrene and AMS (*17*). We have confirmed the observations of Greuel and Harwood but, in addition, we have also shown that macromonomer purity is influenced by; (a) the cobalt complex concentration, (b) the type of cobalt complex, (c) the comonomer type and (d) the reaction temperature (*18,22*). In our studies on the preparation of polyacrylate and polystyrene macromonomers we found that lowering the concentration of the cobalt complex increases the proportion of the 1,1-disubstituted alkene end group over the 1,2-disubstituted alkene end group *i.e.* higher %macromonomer.

In addition, Table I shows that in the CCT copolymerization of styrene and AMS (scheme 1, Y=Ph, Z=Ph), the isopropylcobalt(III) complex **11** produces marginally lower macromonomer purity than either cobalt(II)diphenylglyoxime or methylcobalt(III)diethylglyoxime complexes **10** and **12** respectively (*23*). This may reflect the different steric requirements for the three different complexes.

Table I: The Effect of Different Cobalt Complexes on the Purity of Macromonomer 2.[a]

Co complex	$[Co]$ $x10^5$ M	$\bar{M}_n$	$\bar{M}_w/\bar{M}_n$	%Conv	%Macromonomer
10	6.1	1,455	1.70	12	90
	0	71,900	1.70	23	0
11	6.1	1,065	1.60	13	70
	0	58,300	1.80	19	0
12	6.1	1,390	1.70	15	85
	0	72,300	1.70	19	0

[a] Styrene (1.0g), AMS (0.12g), n-butyl acetate (2g), $3x10^{-4}$M 2,2'-azobis(2,4,4-trimethyl-pentane) and $6.1x10^{-5}$ M cobalt complex was heated at 125°C for 2h.

To prepare macromonomers with a high proportion of the 1,1-disubstituted alkene end group and with a low comonomer content in the polymer backbone, the choice of the α-methylvinyl comonomer is important. It must be chosen such that the rate of reaction of the cobalt complex with the α-methylvinyl derived propagating species is substantially greater than for the propagating species derived from the monosubstituted monomer (see scheme 2). There are two factors which influence this: (i) rate constants for chain transfer and (ii) relative concentrations of the propagating species (determined by monomer concentrations, propagation rate constants and cross propagation rate constants as reflected in the monomer reactivity ratios). We have experimentally shown that AMS is superior to MMA as a comonomer for preparing polystyrene macromonomers (*18,22*). This was recently confirmed by a kinetic model on the cobalt mediated CCT copolymerization of styrene with MMA (*16*) and styrene with AMS (*17*). In contrast, we have shown that both AMS and methacrylates are excellent α-methylvinyl comonomers for preparing polyacrylate macromonomers which have a low content of the α-methylvinyl comonomer in the backbone (*18*).

Previous studies of styrene (*11,24*) and acrylate (*12*) polymerizations with CCT at or below 70°C have produced polymers with multimodal distributions (arising from side reactions), and the reported conditions led to catalyst poisoning and inhibition of polymerization. The reported inhibition phenomenon observed at lower temperatures is attributed to the formation of a polymer-cobalt species (*12,25,26*). To avoid complications arising from the formation of stable intermediates we have prepared polyacrylate and polystyrene macromonomers at temperatures of 80°C and above. Interestingly, we have observed that for the CCT copolymerization of acrylates with α-methylvinyl comonomers, a higher proportion of the 1,1-disubstituted alkene end group is obtained at 80°C than at 125°C.

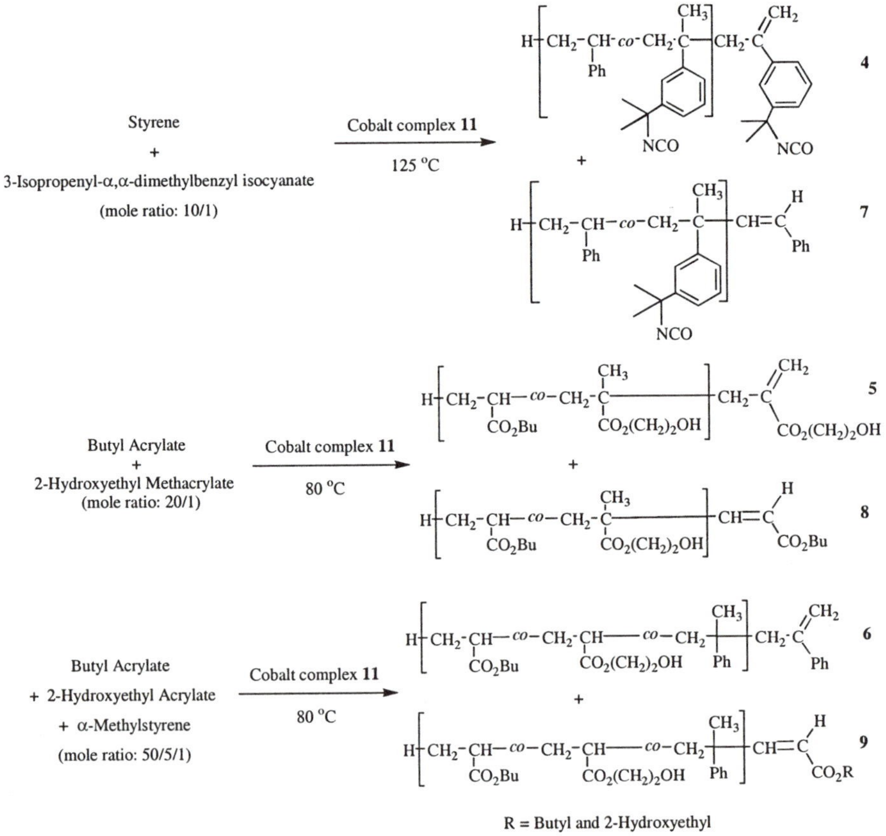

Scheme 3

In contrast, the CCT copolymerization of styrene with α-methylvinyl comonomers, a higher proportion of the 1,1-disubstituted alkene end group is obtained at 125°C than at 80°C (*18,22*). We have used these guidelines to extend the scope of the cobalt mediated CCT copolymerization to prepare macromonomers **4**, **5** and **6** according to scheme 3. The results of these experiments are summarized in Table II.

The ^{1}H-nmr spectra of macromonomers **5** and **6** clearly show the 1,1-disubstituted alkene end groups (Figure 1). The vinyl end group for macromonomer **5** is at ~δ6.3 and ~δ5.55, which is indicative of a methacrylate derived vinyl end group. The vinyl end group for macromonomer **6** is at ~δ5.3 and ~δ5.15, which is characteristic for an AMS derived vinyl end group. The ^{1}H-nmr spectrum of macromonomers **5** and **6** also provides an estimate of comonomer incorporation. For example, macromonomer **5** has about 10% HEMA incorporated (Figure 1(a)), whereas macromonomer **6** has 10-15% HEA incorporated in addition to about 10% AMS (Figure 1(b)).

Table II: Preparation of Macromonomers shown in scheme 3.

Macromonomer	[Co] $x10^5$ M	$\bar{M}_n$	$\bar{M}_w/\bar{M}_n$	% Conv	% Macromonomer
4 [a]	20.0	640	1.40	d	53
4 [a]	5.1	1,275	1.75	d	75
5 [b]	39.6	1,700	1.80	6	80
5 [b]	9.9	12,900	2.00	23	100
6 [c]	37.6	1,260	1.55	16	78
6 [c]	9.4	1,840	1.80	22	100

[a] A mixture of styrene (1g), TMI (0.2g), n-butyl acetate (2g), $3x10^{-4}$M 2,2'-azobis(2,4,4-trimethylpentane) and cobalt complex **11** was heated at 125°C for 2h. ^{1}H-nmr(d$_6$-acetone): δ4.9 and δ5.20 (1,1-disubstituted alkene end group); δ6.0-6.2, (1,2-disubstituted alkene end group); δ6.6-7.4, ArH. [b] A mixture of butyl acrylate (1.3g), 2-hydroxyethyl methacrylate (65mg), n-butyl acetate (2g), $6x10^{-4}$M 2,2'-azobis(isobutyronitrile) and cobalt complex **11** was heated at 80°C for 1h. [c] A mixture of butyl acrylate (1.3g), 2-hydroxyethyl acrylate (116mg), AMS (26mg), n-butyl acetate (2g), $6x10^{-4}$M 2,2'-azobis(isobutyronitrile) and cobalt complex **11** was heated at 80°C for 2h. [d] Accurate conversions could not be determined due to residual TMI.

304

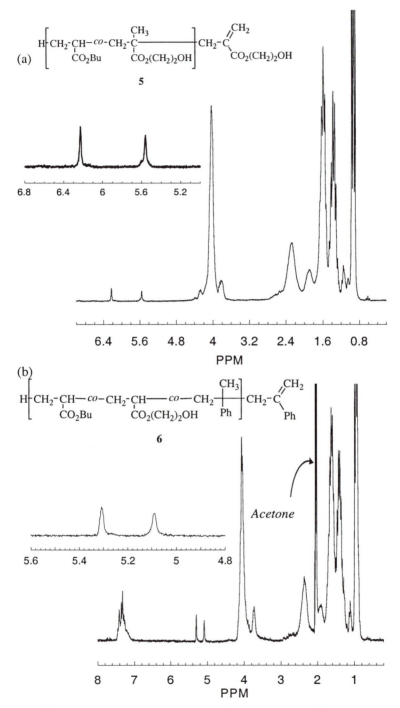

Figure 1: ¹H-nmr spectra identifying the 1,1-disubstituted alkene end groups for (a) poly(acrylate) macromonomer **5** with signals at ~δ6.3 and ~δ5.55. (b) poly(acrylate) macromonomer **6** with signals at ~δ5.3 and ~δ5.15.

The 1,2-disubstituted alkene end group *e.g.* in structures **8** and **9**, is also readily identifiable by ¹H-nmr spectroscopy at ~δ6.8 and ~δ5.85. Figure 2 shows the ¹H-nmr spectra in the vinyl end group region of macromonomer **5**, a mixture of macromonomer **5** and oligomer **8** (see Table II) and poly(butyl acrylate) oligomer **13** (prepared by polymerizing butyl acrylate with cobalt complex **11**).

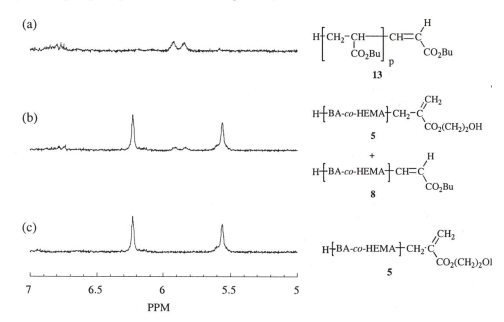

Figure 2: *The vinyl end group region for (a) poly(butyl acrylate) oligomer **13**, (b) 80% macromonomer **5** with oligomer **8** (see table III) and (c) 100% macromonomer **5***

Macromonomer Synthesis without added Chain Transfer Agent.

We have recently reported that macromonomers **14-16** can be prepared without added chain transfer agent or added α-methylvinyl comonomer (scheme 4) (*19,20,23*).

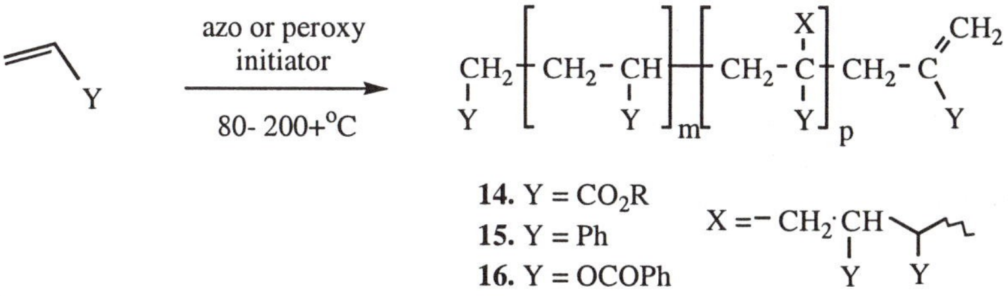

Scheme 4

The method is simple and efficient producing macromonomers with a high proportion of the 1,1-disubstituted alkene end group. In general, the requirements for the process are to heat monomer in an appropriate solvent with an azo or peroxy initiator at temperatures of 80°C to 240°C. Ideally, a low radical flux is required to minimize radical termination reactions. Furthermore, the solvent should have a low transfer constant. Under these conditions, branched macromonomers **14-16** have been prepared in high purity (Table III).

The 'purity' of macromonomers is a measure of the relative amount of macromonomer compared to non-macromonomer products (*i.e.* products from radical-radical termination or chain transfer). This ratio can be determined by calculating the molecular weight of the macromonomers from their ^{1}H-nmr spectra and dividing by the molecular weight obtained from GPC chromatogram.

Table III: Preparation of Macromonomers 14, 15 and 16 under various conditions.

Macromonomer	[M] wt%	Solvent	Temp. (°C)	M_n	M_w/M_n	%Conv	Macromonomer purity (%)
14[a]	10	n-BuAc	150	1,560	1.80	79	>90
15[b]	5	n-AmAc	170	3,090	1.90	99	>90
16[c]	10	DMF	220	910	1.40	14	>50[d]

[a] Butyl acrylate (1g), n-butyl acetate (9g) and 6.3×10^{-5} M 2,2'-azobis(2-methylpropane) was heated at 150°C for 8 hours. [b] Styrene (0.5g), n-amyl acetate (9g) and 3×10^{-4} M 2,2'-azobis(2-methylpropane)was heated at 170°C for 7 hr. [c] Vinyl benzoate (10g), N,N-dimethylformamide (90g) and 6.5×10^{-4}M cumene hydroperoxide was heated at 220°C under microwave irradiation for 30min. ^{1}H-nmr(d$_6$-acetone): δ5.5-5.4 and δ5.35-5.1 multiplets (1,1-disubstituted alkene end group and backbone methines); δ8.0-7.5 and δ7.4-6.9 multiplets, ArH. [d] Overlap of vinyl end group signal with backbone methines prevented accurate determination of purity by NMR spectroscopy.

The proposed mechanism for macromonomer formation is depicted in Scheme 5. The crucial step involves β-scission of radical **17** to give macromonomer and a propagating radical. The formation of radical **17** occurs by chain transfer to polymer backbone. There are two plausible mechanisms by which this can occur; (a) by a [1,5-H] shift *i.e.* backbiting via a six membered transition state, or (b) intermolecular chain transfer involving a second propagating radical. Our results suggest that backbiting is the predominant mechanism. Due to its bimolecular nature, the intermolecular mechanism is likely to lead to 'dead' polymer.

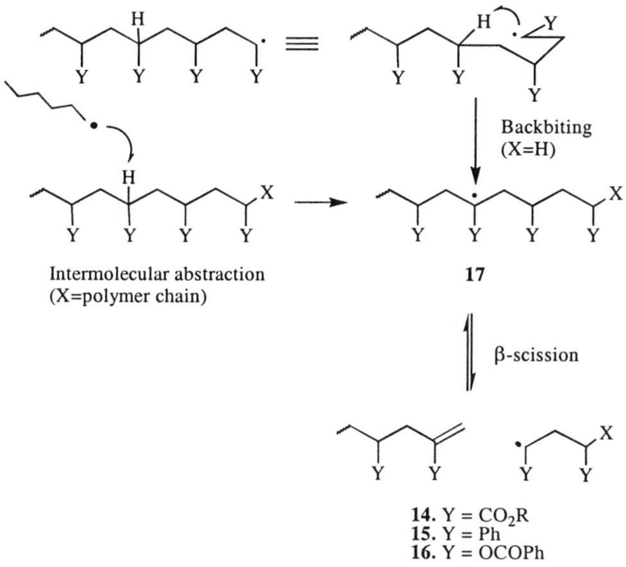

Scheme 5: Proposed mechanism of macromonomer formation.

The 1,1-disubstituted alkene end group structure for macromonomers **14** and **15** has been verified by comparison of their ^{1}H-nmr spectra with macromonomers **5** and **6**, derived by cobalt mediated CCT.

For example, macromonomer **14** possesses an alkene end group derived from a methacrylate with vinyl H signals at $\sim\delta6.2$ and $\delta5.5$ (Figure 3(a)). This compares to the vinyl end group for macromonomer **5** at $\sim\delta6.3$ and $\delta5.55$ (see Figure 1(a)). Similarly, macromonomer **15** possesses an alkene end group derived from AMS with vinyl H signals at $\sim\delta5.1$ and $\delta4.8$ (Figure 3(b)). This compares with vinyl H signals for macromonomer **6** at $\sim\delta5.3$ and $\delta5.1$ (Figure 1(b)). The slight upfield shift observed for macromonomer **15** is likely to be due to the shielding effects of the phenyl groups from the penultimate monomer units.

As a consequence of generating a radical on a polymer backbone *viz.* **17**, propagation from this radical can occur resulting in the formation of branches. Branching can arise by two routes: (a) copolymerization with the macromonomer and/or (b) monomer addition to radical **17**. We have detected branchpoints, and measured the level of branching using ^{13}C-nmr spectroscopy. The branchpoints of poly(acrylate) macromonomers of type **14** being clearly visible at $\sim\delta35-40$ ppm (*27*).

The macromonomer molecular weight and the degree of branching can be influenced by reaction temperature and monomer concentration. In a butyl acrylate polymerization, the results in Table IV show that as reaction temperature is increased the molecular weight of macromonomer **14** decreases. We attribute this to a higher incidence of backbiting and β-scission of radical **17** (see scheme 5) resulting in a lower molecular weight.

308

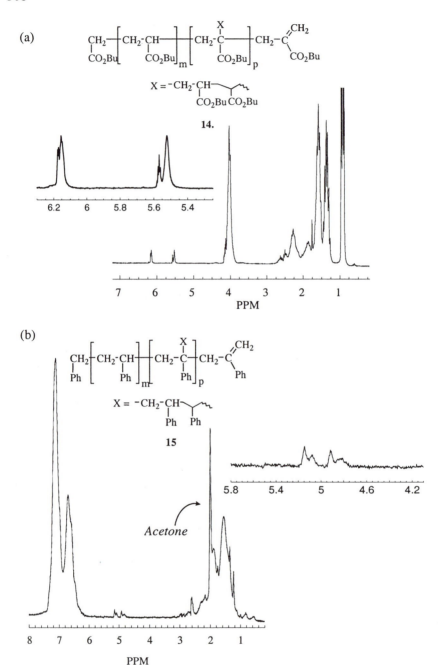

Figure 3: The ¹H-nmr spectra of (a) poly(butyl acrylate) macromonomer **14** ($\bar{M}_n$ ~1560, M_w/M_n ~1.80), run in deuterated chloroform, showing the alkene end group at ~δ6.2 and δ5.5, and (b) polystyrene macromonomer **15** ($\bar{M}_n$ ~3090, M_w/M_n ~1.90), run in deuterated acetone, showing the alkene end group at ~δ5.1 and δ4.8.

Table IV: Effect of Reaction Temperature on Molecular Weight of Macromonomer 14.

Temp. °C	M_n	M_w/M_n	% Conv.	Macromonomer purity %
150 [a]	3,680	2.60	70	>90
200 [b]	1,810	2.50	94	>90
220 [b]	1,460	2.70	87	>90

[a] Butyl acrylate (1g) in n-butyl acetate (3g) and 2.8×10^{-4} M 2,2'-azobis(2,4,4-trimethylpentane) was heated for 20min. under microwave irradiation. [b] Butyl acrylate (15g) in N,N-dimethylformamide (44.6g) and 1.17×10^{-4} M cumene hydroperoxide was heated for 20min. under microwave irradiation.

Reaction temperature can also be used to control the degree of branching in macromonomers prepared by this technique. The results in Table V show that increasing the reaction temperature decreases the degree of branching. This is likely to be due to an increase in the incidence of β-scission, which reduces the mole fraction of radical **17**. In addition, a higher reaction temperature may favor fragmentation of the addition adduct of radical **17** with monomer/macromonomer.

Table V: The Effect of Reaction Temperature on Degree of Branching in Acrylate Polymerization.

Monomer	Temp. °C	M_n	M_w/M_n	% Conv.	Branch points/ macromonomer
BA [a]	150	4,090	2.55	50	1.4
MA [b]	240	1,090	2.70	58	0.3

[a] Butyl acrylate (1g) in n-butyl acetate (3g) and 2.4×10^{-4} M 2,2'-azobis(2,4,4-trimethylpentane) was heated for 20min. [b] Methyl acrylate (15g) in N,N-dimethylformamide (44.6g) and 1.17×10^{-4} M cumene hydroperoxide was heated for 20min. under microwave irradiation.

The influence of monomer concentration on macromonomer molecular weight is summarized in Table VI. In general, increasing the monomer concentration increases the molecular weight of the macromonomer. For example, an increases in monomer concentration from 13wt% to 100wt% increases the molecular weight of the macromonomer from 1,800 to 16,300. This is due simply to an increase in the propagation rate with increasing monomer concentration. The macromonomer purity appears to decrease with increasing monomer concentration. This is particularly evident at monomer concentrations of 80wt% and above (see Table IV). This may reflect the increasing significance of the intermolecular mechanism. At high monomer concentration and at high monomer conversion (*i.e.* a high polymer concentration), the intermolecular mechanism may be more predominant resulting in higher proportions of 'dead' (*i.e.* H terminated) polymer (see scheme 5). A second contributing factor may also be the effect that high levels of branching have on the GPC elution time and hence the measured molecular weight. In particular, increasing levels of branching will tend to 'lower' the molecular weight of the polymer since the polymer elutes slower. This will manifest itself as an apparent lower macromonomer purity.

Table VI: Effect of Monomer and Macromonomer Concentration on the Degree of Branching and Molecular Weight of Poly(Butyl Acrylate) Macromonomer.

[BA] wt%	M_n	M_w/M_n	% Conv.	% Macromonomer purity	Branch points/ macromonomer
13 [a]	1,800	1.70	8	>90	0.9
25 [b]	3,540	2.75	82	>90	1.8
40 [c]	4,500	2.95	82	>90	2.0
60 [c]	10,500	2.50	90	>90	3.8
80 [c]	13,900	3.30	96	80	4.2
100 [c]	16,300	4.00	67	70	4.8

[a] Butyl acrylate (1.28g) in n-butyl acetate (8.87g) and 7.8×10^{-4} M 2,2'-azobis(2-methylpropane) was heated at 150°C for 6 hrs. [b] Butyl acrylate (1g) in n-butyl acetate (3g) and 2.4×10^{-4} M 2,2'-azobis(2,4,4-trimethylpentane) was heated at 150°C for 40min. [c] A mixture of butyl acrylate and n-butyl acetate (at the specified monomer concentration) was heated at 150°C for 20min with 1×10^{-4} M 2,2'-azobis(2,4,4-trimethylpentane).

Conclusion

We have reported the use of two methods that effectively broadens the range of macromonomers accessible. In the first method, we have extended the applicability of the cobalt complex mediated CCT process by preparing functional poly(acrylate) and poly(styrene) macromonomers that are based predominantly on the monosubstituted monomers. In addition, we outlined an alternative method for accessing branched macromonomers. This method relies on a mechanism of chain transfer to polymer followed by β-scission. The procedure is clean and efficient, and has advantages over the current preparative methods, including the cobalt complex mediated CCT polymerization, since no added chain transfer agent is required. Furthermore, no α-methylvinyl comonomer is needed for the desired 1,1-disubstituted alkene end group.

Acknowledgments. We are grateful to Drs. C. Berge, M. Fryd, and R. Matheson of E.I. DuPont de Nemours and Co. (Performance Coatings) for their support of this work and for valuable discussions.

Literature Cited

1   Krstina, J.; Moad, C.L.; Moad, G.; Rizzardo, E.; Berge, C.T.; Fryd, M.; *Macromol. Symp.*, **1996**, *111*, 13.

2   Rizzardo, E.; Chong, Y.K.; Evans, R.A.; Moad, G.; Thang, S.H. *Macromol. Symp.* **1996**, *111*, 1

3   Cacioli, P.; Hawthorne, D.G.; Laslett, L.; Rizzardo, E.; Solomon, D.H.; *J. Macromol. Sci. Chem.*, **1986,** *A23*, 839.

4   Krstina, J.; Moad, G.; Rizzardo, E.; Winzor, C.L.; Berge, C.T.; Fryd, M.; *Macromolecules*, **1995**, *28*, 5381.

5   Darmon, M.J.; Berge, C.T.; Antonelli, J.A.; US 5 362 826   1992, (*Chem. Abstr.*, **1993**, *120,* 299554)

6   Rizzardo, E.; Meijs, G. F.; Thang, S.H.; *Macromol. Symp.*, **1995**, *98*, 101.

7   Nair, C.P.R.; Chaumont, P.; Charmot, D.; *J. Polym. Chem.*, **1995**, *33*, 2773.

8   Burczyk, A. F.; O'Driscoll, K.F.; Rempel, G.L. *J. Polym. Sci., Polym. Chem. Ed.*, **1984**, *22*, 3255.

9   Davis, T.P.; Haddleton, D.M.; Richards, S.N.; *J. Macromol. Sci., Rev. Macromol. Chem. Phys.*, **1994**, *C34*, 274.

10   Davis, T.P.; Kukulj, D.; Haddleton, D.M.; Maloney, D.R.; *Trends Polym. Sci.,* **1995**, *3*, 365.

11   Martchenko, A.; Bremner, T.; O'Driscoll, K.F.; *Eur. Polym. J.,* **1997**, *33,* 713.

12   Smirnov, B.R.; Plotnikov, V.D.; Ozerkovskii, B.V.; Roshchupkin, V.P.; Yenikolopyan, N.S.; *Polym. Sci. USSR,* **1981**, *23,* 2807.

13   Organova, A.G.; Smirnov, B.R.; Ioffe, N.T.; Kim, I.P.; *Izvestiya Akademii Nauk SSSR, Ser. Khim.,* **1984**, *6*, 1258.

14   Greuel, M.P.; Harwood, H.J.; *Polym. Prep. (Am. Chem. Soc., Div. Polym. Chem.),* **1990**, *32,* 545.

15   Lin, J-C.; Abbey, K.J.; *US 4,680,354,* **1987**, Janowicz, A.H.; Melby, L.R.;*US 4,680,352,* **1987**, Janowicz, A.H.; *US 4,694,054,* **1987**, Janowicz, A.H.; *US 4,722,984,* **1988**, Janowicz, A.H.; *US 4,746,713,* **1988**, Janowicz, A.H.; *US 4,886,861,* **1989**, Janowicz, A.H.; Melby, L.R.; Ittel, S.D.; *EP0 199 436,* **1996**, Janowicz, A.H.; Melby, L.R.; Ittel, S.D.; *EP0 196 783,* **1996**.

16   Heuts, J.P.A.; Kukulj, D.; Forster, D.J.; Davis, T.P.; *Macromolecules,* **1998**, *31,* 2894.

17   Heuts, J.P.A.; Kukulj, D.; Davis, T.P.; *Macromolecules,* **1998**, *31,* 6034.

18   Ittel, S.D.; Gridnev, A.A.; Moad, C.L.; Moad, G.; Rizzardo, E.; Wilczek, L.; *WO 97 31,030,* (*Chem. Abstr.,* **1997**, *128,* 234752).

19   Chiefari, J; Moad, G; Rizzardo, E; Gridnev, A. A.; WO 9847927, **1998**, (Chem. Abstr. **1998**:709104).

20   Chiefari, J.; Jeffery, J.; Mayadunne, R.T.A.; Moad, G.; Rizzardo, E.; Thang, S.H.; *Macromolecules,* **1999**, *32,* 7700.

21   The use of organocobalt(III) complexes as CCTA has been described before (see reference 1). These complexes have advantages over the corresponding cobalt(II) complexes in stability and ease of handling. See; Hawthorne, D.G.; US 5324 879, **1991,** (*Chem. Abstr.,* **1987**, *107,* 237504), Gridnev, A.A.; *Polym. Sci. USSR,* **1989**, *31,* 2369, Gridnev, A.A.; *Polym. J. (Tokyo),* **1992**, *24,* 613.

22   Chiefari, J.; Moad, C.L.; Moad, G.; Rizzardo, E.; Thang, S.H., *Macromolecules* in preparation. Chiefari, J.; Jeffery, J.; Moad, G.; Rizzardo, E.; Thang, S.H.; *Macromolecules,* in preparation.

23   Chiefari, J.; Jeffery, J.; Moad, G.; Rizzardo, E.; Thang, S.H.; *Polym. Prep. (Am. Chem. Soc., Div. Polym. Chem.),* **1999**, *40,* 344.

24   Yenikolopyan, N. S.; Smirnov, B.R.; Ponomarev, G.V.; Belgovskii, I.M.; *J. Polym. Sci., Polym. Chem. Ed.,* **1981**, *19,* 879.

25   Wayland, B.B.; Poszmik, G.; Mukerjee, S.L.; Fryd, M.; *J. Am. Chem. Soc.,* **1994**, *116,* 7943.

26   Heuts, J.P.A.; Forster, D.J.; Davis, T.P.; Yamada, B.; Yamazoe, H.; Azukizawa, M.; *Macromolecules,* **1998**, *31,* 2894.

27   Ahmad, N.M.; Heatley, F.; Lovell, P.A.; *Macromolecules,* **1998**, *31,* 2822.

Chapter 22

End-Group Control in Catalytic Chain Transfer Polymerization

Johan P. A. Heuts[1], David A. Morrison, and Thomas P. Davis[1]

Centre for Advanced Macromolecular Design, School of Chemical Engineering
and Industrial Chemistry, The University of New South Wales,
Sydney, New South Wales 2052, Australia

The application of catalytic chain transfer polymerization as a technique to produce end-functionalized low-molecular weight polymers is discussed in terms of simple kinetic models and illustrated by practical examples. Since end-functionalities other than vinyl groups need to be introduced via a functional monomer it is shown to be necessary and possible to control the endgroup by careful selection of a comonomer. Simple kinetic modeling reveals the required properties of this comonomer to be a high affinity of its derived radical to react with Co(II) rather than with other monomers, and that the other present monomers need to possess the opposite properties. The use of α-methyl styrene as a monomer to selectively introduce end-functionalities is discussed in detail. Furthermore it is shown that the use of α-hydroxy methyl functionalized monomers in catalytic chain transfer polymerization lead to the introduction of aldehyde endgroups, which are very versatile building blocks for post-polymerization modifications. The α-methyl styrene analogue, i.e., 2-phenyl allyl alcohol, has great promise for the selective introduction of these aldehyde endgroups.

In certain applications of (free-radical) polymers low molecular weight materials are required, which may or may not be subsequently modified. Good examples of this can be found in the development of high-solids organic coatings, self-reinforced hydrogel materials for contact lenses or even the production of surfactants. In the first example, the reduction of organic solvent content, required by environmental legislation causes an increasing viscosity of the resin, which can be compensated for by using a resin of a lower molecular weight; this resin is subsequently cross-linked. The second example shows the use of hydrophobic macromonomers, which are copolymerized with more hydrophilic comonomers to produce high performance hydrogel materials, and in the last example, a hydrophobic

---

[1]Corresponding authors.

oligomer is extended by a hydrophilic block, or vice versa, to produce a surface active molecule.

It is clear from the above that there is an industrial need for low molecular weight materials and the chosen examples further illustrate the fact that these materials often need to contain functional groups. Common practices for producing low molecular weight materials are the use of very high initiator concentrations or the use of added chain transfer agents such as thiols. These practices are based on the fact that both a higher initiator concentration and the addition of a chain transfer agent cause an increase in the rate of chain stopping relative to propagation. These effects can be quantitatively described by the Mayo equation (Eq 1), which expresses the reciprocal of the number average degree of polymerization, DP_n, as the ratio of the rates of all chain stopping events to the rate of propagation:([1,2])

$$\frac{1}{DP_n} = (1+\lambda)\frac{\langle k_t\rangle[R^\bullet]}{k_p[M]} + C_M + C_S\frac{[S]}{[M]} \tag{1}$$

In this expression, λ is the fraction of termination by disproportionation, $\langle k_t\rangle$ the average termination rate coefficient, $[R^\bullet]$ the overall radical concentration, k_p the propagation rate coefficient, $[M]$ the monomer concentration, C_M the chain transfer constant to monomer and C_S the chain transfer constant to chain transfer agent S, where the latter two chain transfer constants are defined as the ratios of the chain transfer rate coefficients to the respective molecules and the propagation rate coefficient. It is clear upon examination of Eq 1 that an increase in initiator concentration will increase the total radical concentration and hence increase the first term on the RHS of Eq 1. Similarly, the addition of a chain transfer agent leads to the appearance of the last term on the RHS of Eq 1. Both effects thus increase DP_n^{-1}, and hence decrease DP_n.

Since the use of additional amounts of initiator causes a change in polymerization rate and the associated reaction heat (2), chain transfer agents are more generally used for molecular weight control. These chain transfer agents are often thiols which have chain transfer constants in the order of 10^{-1} to 10 for most common monomer systems (3). They act through a hydrogen transfer reaction with the growing radical chain, which abstracts the hydrogen from the –SH group, thus creating a dead polymer chain and a thiyl radical which can subsequently initiate a new chain (4). A major drawback of thiols is that the chain transfer constants are relatively low and that they are consumed during the reaction. This implies that relatively large amounts of thiols, which are odorous and toxic, are required in the reaction mixture, and probably end up in the polymer product and hence need to be removed.

Catalytic Chain Transfer Polymerization.

An interesting alternative to the use of thiols has emerged over the past two decades in the shape of catalytic chain transfer agents, i.e., certain low-spin Co(II) complexes (for example, see Scheme 1) which were found to catalyze the chain transfer to monomer reaction and hence provide a means for molecular weight control (5-7).

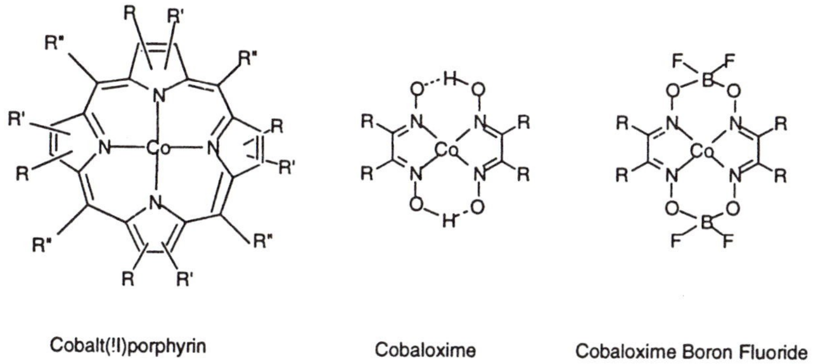

Cobalt(II)porphyrin   Cobaloxime   Cobaloxime Boron Fluoride

Scheme 1. Low-spin cobalt catalysts

These Co(II) complexes act by facilitating the hydrogen transfer reaction from the growing polymeric radical to the monomer, very likely by forming an intermediate Co(III)H complex (5-8). The overall reactions are clearly illustrated by Scheme 2, which shows that if an α-methyl group is present in the radical, the hydrogen is abstracted from this α-methyl group, whereas if such group is not present, it will be abstracted from the α-carbon in the polymer backbone. Furthermore, it is clear that chains terminated by this process are characterized by a vinyl endgroup and that those chains initiated by this process are characterized by a hydrogen "initiator" fragment. It is thus possible to produce macromonomers consisting solely of monomer units (9,10).

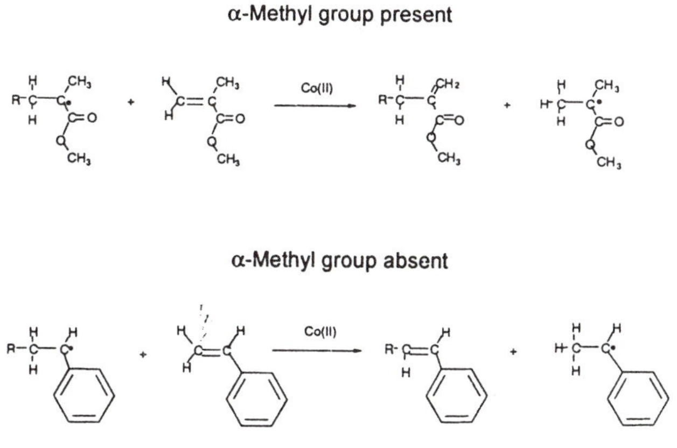

α-Methyl group present

α-Methyl group absent

Scheme 2. Catalytic process

The catalytic chain transfer process is highly efficient, displaying chain transfer constants of up to about 10^5 for the most active catalyst found to date, i.e., the cobaloximes (see Scheme 1). Currently the most commonly used cobaloximes are those that contain BF_2-bridges, which provide greater hydrolytic stability and protection against oxidation (11). A great advantage of this technique compared to the use of thiols is that the process is catalytic, which means that the (catalytic) chain transfer agent is not consumed in the process, which, in combination with the very high chain transfer constants, results in the use of only ppm amounts of the catalyst for very large molecular weight reductions (5-7). Some of the most important relevant features of free-radical polymerizations in the presence of the two discussed types of chain transfer agent are summarized in Table 1.

For the present purpose it is important to compare the way that possible end-functionalities can be introduced into the polymer. In both cases one of the two end-groups consists of a hydrogen atom: in the case of the thiol as chain transfer agent this is the terminal endgroup; in the Co(II) case it is the initiator group. The real functionality hence lies in the other group. In the thiol case, the initiator group is the thiyl moiety and hence any possible functionality to be introduced as an endgroup needs to be present in the chain transfer agent. However, in the Co(II) case, no fragment of the chain transfer agent remains in the polymer chain, but the functional end-group is simply the terminal monomer unit, now containing an unsaturated carbon-carbon bond. Hence, if any particular end-functionality needs to be built in via catalytic chain transfer polymerization, this functional group needs to be present in the monomer.

Table 1. Comparison of Typical Features of Conventional and Catalytic Chain Transfer in Free-Radical Polymerization

	Conventional - Thiol	Catalytic – Co(II)
chain transfer constant, C_s^a	10^{-1} - 10	10^3 - 10^4
action of chain transfer agent	reactant	catalyst
required for $M_n = 1,000^a$	~ 10 wt%	~ 10 ppm
environmental & health	odorous, toxic	no specific problems
α functionality of polymer	thiyl fragment	hydrogen atom
ω functionality of polymer	hydrogen atom	repeat unit with vinyl group

[a] based on methyl methacrylate polymerizations

In what follows we will make an attempt to describe our current efforts to control the nature of the endgroups in catalytic chain transfer polymerization including some of our recent efforts to introduce the aldehyde endgroup, a well-known synthon in synthetic organic chemistry.

Experimental

Materials

The bis(methanol) complexes of the BF_2-bridged cobaloximes with R = CH_3 (i.e., bis[(difluoroboryl)dimethyl-glyoximato]cobalt(II) – COBF in the remainder of this chapter) and R = phenyl (i.e., bis[(difluoroboryl)diphenyyl-glyoximato]cobalt(II) – COPhBF) are prepared according to the method described by Bakac et al.(*12*) Monomers (commercial monomers typically: Aldrich, 99%), passed through a column of activated basic alumina (ACROS, 50-200 micron) prior to use, and solvents (typically of analytical grade) are purged with high purity nitrogen (BOC) for 1.5 hours prior to use. AIBN (DuPont) is recrystallized twice from methanol and used as initiatior.

Chain Transfer Constant Measurements

The procedures employed to measure chain transfer constants have been described extensively in previous publications (*13-19*), and can been found in the references for the experimental data presented in this chapter.

Polymerizations

Catalytic chain transfer polymerizations in our laboratory are typically perfomed at temperatures between 40 and 80°C, either in bulk or solvent (~30 – 50% solids), with [Co(II)] = $1 \cdot 10^{-6}$ - $1 \cdot 10^{-4}$ mol·dm^{-3} and [AIBN] = $1 \cdot 10^{-4}$ - $1 \cdot 10^{-2}$ mol·dm^{-3}. At all times it is ensured that oxygen-free conditions are maintained, using common Schlenk and syringe techniques. For more specific details we refer to the references cited for the reported experimental data (*14,16,20*).

Molecular Weight Analysis

Molecular weight analyses were performed using size-exclusion chromatography. The equipment has been described in detail in previous references (*14-19*).

NMR Analysis

^{1}H-NMR analyses of the polymer samples were performed on a 300MHz Bruker ACF 300 spectrometer using $CDCl_3$ (Aldrich, 99.8 atom% D) as a solvent.

Endgroup Control

As stated in the introduction, catalytic chain transfer polymerization yields polymer chains consisting solely of monomer units with α and ω functionalities of a hydrogen atom and a vinyl group respectively (ignoring the contribution of the chains that contain initiator fragments or those that were terminated by radical-radical termination). In other words, the great majority of chains will not contain any "foreign" functional groups and in order to introduce a particular functional group, this functional group needs to be present in the reacting monomer. If this functionality is only to appear as a functional endgroup in a chain consisting solely of monomer units of a different nature, it is clear that a functional comonomer should be used with particular reactive properties. Basically, propagating radicals with the functional monomer as terminal units should mainly undergo chain transfer reactions, whereas those with "main chain" monomer(s) as terminal units should mainly propagate.

It is clear that in order to predict whether a particular (co)monomer can be used to introduce a functional endgroup, it is imperative that both the copolymerization and the chain transfer behavior of the comonomer combination is understood. In the following two sub-sections of this section we will briefly outline the most relevant theoretical concepts of conventional and Co(II)-mediated free-radical copolymerization. This section concludes with actual examples of endgroup control by catalytic chain transfer polymerization.

Free-Radical Copolymerization

In contrast to homopolymerizations, the kinetics in free-radical copolymerization are governed by more than a single type of radical and monomer. At the most basic and simplest level one may consider that in a copolymerization of two monomers A and B, the chain growth is caused by four different propagation steps, i.e., the additions of monomers A and B to propagating chains with terminal units derived from A and B (see Scheme 3). The four corresponding propagation rate coefficients can be conveniently grouped into two single kinetic parameters, i.e., the monomer reactivity ratios, r_A and r_B, which express the relative rate coefficients of homo-propagation to cross-propagation for radicals $\sim A^{\bullet}$ and $\sim B^{\bullet}$, respectively (2,21).

$$\left. \begin{array}{l} \sim A^{\bullet} + A \xrightarrow{k_{AA}} \sim AA^{\bullet} \\ \sim A^{\bullet} + B \xrightarrow{k_{AB}} \sim AB^{\bullet} \end{array} \right\} \quad r_A = \frac{k_{AA}}{k_{AB}}$$

$$\left. \begin{array}{l} \sim B^{\bullet} + B \xrightarrow{k_{BB}} \sim BB^{\bullet} \\ \sim B^{\bullet} + A \xrightarrow{k_{BA}} \sim BA^{\bullet} \end{array} \right\} \quad r_B = \frac{k_{BB}}{k_{BA}}$$

Scheme 3. Basis for terminal copolymerization kinetics

A useful property of the monomer reactivity ratios is that they can adequately describe the instantaneous composition of the copolymer chains formed if the monomer feed composition is known; the instantaneous fraction of monomer A in the copolymer, F_A, as a function of the monomer reactivity ratios and the faction of monomers A and B, f_A and f_B respectively, is given by Eq 2 *(2,21)*.

$$F_A = \frac{r_A f_A^2 + f_A f_B}{r_A f_A^2 + 2 f_A f_B + r_B f_B^2} \qquad (2)$$

Furthermore, it is found that the overall propagation rate coefficient, $\langle k_p \rangle$, based on overall radical and monomer concentrations, is a function of the same parameters and the homopropagation rate coeffcients, k_{AA} and k_{BB} (see Eq 3) *(21)*.

$$\langle k_p \rangle = \frac{r_A f_A^2 + 2 f_A f_B + r_B f_B^2}{\dfrac{r_A f_A}{k_{AA}} + \dfrac{r_B f_B}{k_{BB}}} \qquad (3)$$

Both the instantaneous copolymer composition and average propagation rate coefficient implicitly contain the ratio of the two propagating radical concentrations, A_{AB}, of which an expression (i.e., Eq 4) can be derived from the steady state assumption in all propagating radical concentrations. This ratio is important as it is basically a weighting factor for the contributions of the reactions of either radical to the overall observed kinetics and copolymer composition. From this ratio the fractions of the different propagating radicals is easily derived and given by Eq 5 *(2)*.

$$A_{AB} = \frac{[\sim A^\bullet]}{[\sim B^\bullet]} = \frac{k_{BB} r_A f_A}{k_{AA} r_B f_B} \qquad (4)$$

$$\phi_A = \frac{A_{AB}}{1 + A_{AB}} \qquad (5)$$

So far, we treated free-radical copolymerization to consist of four basic propagation steps and we presented the corresponding kinetic expressions. This simple model was derived in the early 1940s and is known as the Mayo-Lewis or terminal model *(2,21)*. It has been clear from the onset of free-radical copolymerization that copolymer compositions were adequately modelled by Eq 2, and that overall trends in radical and monomer reactivities derived from this scheme were sufficiently reliable. However, kinetic predictions of polymerization rates were faulty from the very beginning and after many years of accounting for this defect by the introduction of a cross-termination parameter (*2*), it was unambiguously shown by Fukuda and co-workers (*22*) that this defect was caused by an oversimplification of the propagation kinetics. Several alternative corrections on the terminal model have since been suggested (*22*), and although it is not the aim of the current chapter to go too much into detail of this discussion, we would like to point out that many experimental and recent theoretical results suggest that the nature of the group preceding the terminal radical unit, i.e., the penultimate unit, can and will affect the radical reactivity (*21,23*). Since every single terminal unit can have two different penultimate units, the kinetic scheme presented in Scheme 3 should not contain two, but four different radicals, each with a corresponding homo- and a cross-propagation reaction. It is clear that the mathematics to describe the kinetics of this model become more complicated, and the resulting expressions for copolymer composition and average propagation rate coefficient can be obtained by making the substitutions shown in Eqs 6 and 7, in Eqs 2-4 (*21,23*).

$$r_i \to \bar{r}_i = \frac{r_i''(r_i' f_i + f_j)}{r_i'' f_i + f_j} \text{ with } r_i' = \frac{k_{iii}}{k_{iij}} \text{ and } r_i'' = \frac{k_{jii}}{k_{jij}} \quad (6)$$

$$k_{ii} \to \bar{k}_{ii} = \frac{k_{iii}(r_i' f_i + f_j)}{r_i' f_i + \dfrac{f_j}{s_i}} \text{ with } s_i = \frac{k_{jii}}{k_{iii}} \quad (7)$$

In these expressions i and j denote A or B ($i \neq j$), and $k_{(x)yx}$ denote the rate coefficient of the addition of monomer Z to a propagating radical with penultimate unit X and terminal unit Y.

In Figure 1 we have shown the copolymerization behavior of styrene (S) and methyl methacrylate (MMA) in terms of the average propagation rate coefficient and the fraction of propagating styrene radicals, ϕ_S. It is clear from this figure that the simple terminal model and the more complicated penultimate model give qualitatively the same trend for ϕ_S, but that the $\langle k_p \rangle$ behavior predicted from these models is very different (with the penultimate model giving the correct prediction) (*24,25*). Similar discrepancies for the average propagation rate coefficient are generally observed in the free-radical copolymerization of other systems. Similar data are shown for the copolymerization of α-methyl styrene (AMS) and styrene, for which it was found that the terminal model was sufficiently adequate to describe the propagation kinetics (which is an exception rather than a rule!) (*26*).

Comparison of Figures 1 and 2 shows that in both cases the fractions of the radicals derived from the slower monomer (in terms of homopropagation rate coefficients) are always larger than the corresponding fractions in the monomer feed. This basically means that these radicals "hang around" longer before they propagate again, and in the meantime are more prone to possible side reactions. It is indeed this property that we wish to make use of in endgroup control, as will be discussed later (vide infra).

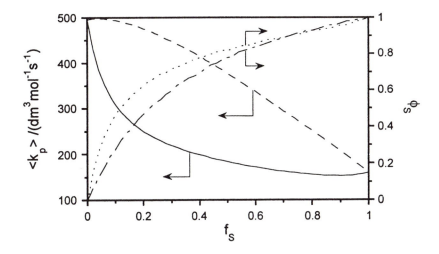

Figure 1. Average propagation rate coefficient and fraction of propagating styrene radicals in the copolymerization of styrene and methyl methacrylate at 40°C. Penultimate predictions: $\langle k_p \rangle$ (——), ϕ_S (•••). Terminal predictions: $\langle k_p \rangle$ (---), ϕ_S (- • -). Used model parameters: $k_{SS} = 160$ $dm^3 mol^{-1} s^{-1}$, $r_S = 0.48$, $s_S = 0.37$, $k_{MMAMMA} = 496$ $dm^3 mol^{-1} s^{-1}$, $r_{MMA} = 0.42$, $s_{SMMA} = 2$.

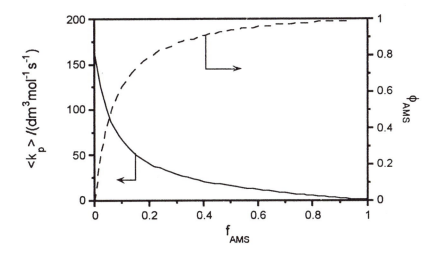

Figure 2. Average propagation rate coefficient and fraction of propagating α-methyl styrene radicals in the copolymerization of α-methyl styrene and styrene at 40°C. Used model parameters: $k_{SS} = 160$ $dm^3 mol^{-1} s^{-1}$, $r_S = 0.95$, $k_{AMSAMS} = 1.12$ $dm^3 mol^{-1} s^{-1}$. $r_{AMS} = 0.10$.(26)

In summary we wish to note that the use of a more complicated model, such as the penultimate model for which we have given the expressions here, will generally be required for a more quantitative description of the kinetics. However, (qualitative) information about propagating radical concentrations, which is of most importance for endgroup control, can mostly be adequately provided by the simple terminal model. For (mathematical) simplicity we will therefore only use the terminal description in the remainder of this chapter, unless otherwise stated.

Catalytic Chain Transfer Copolymerization

In the previous section we discussed the consequences of having different propagating radicals on the propagation kinetics. It is clear that a similar effect should exist for chain transfer. The reactions shown in Eqs 8a and 8b are those that should account for the catalytic chain transfer process in a catalytic chain transfer copolymerization and both reactions have an associated chain transfer rate coefficient, $k_{tr,A}$ and $k_{tr,B}$.

$$\sim A^{\bullet} + Co(II) \xrightarrow{k_{tr,A}} \sim A + Co(III)H \tag{8a}$$

$$\sim B^{\bullet} + Co(II) \xrightarrow{k_{tr,A}} \sim B + Co(III)H \tag{8b}$$

These two chain transfer reactions will contribute to an observed, chain transfer rate coefficient $\langle k_{tr} \rangle$, which can be expressed as an average of the two individual rate coefficients, weighted by the radical fractions (Eq 9). The experimentally observed average chain transfer constant, $\langle C_S \rangle$, can then be expressed as the ratio of the average chain transfer rate coefficient and the average propagation rate coefficient (Eq 10).

$$\left\langle k_{tr,Co} \right\rangle = \phi_A k_{tr,A} + (1 - \phi_A) k_{tr,B} \tag{9}$$

$$\left\langle C_S \right\rangle = \frac{\left\langle k_{tr,Co} \right\rangle}{\left\langle k_p \right\rangle} \tag{10}$$

The applicability of this model for the description of catalytic chain transfer copolymerization was previously tested experimentally for the systems S/MMA (19) and AMS/S (15), and the average chain transfer constants of these systems are shown in Figures 3 and 4, respectively.

It is clear from these figures that the model provides an adequate description for both systems, but that the S/MMA system requires the more complicated penultimate model for an accurate description. Comparison of the two figures shows that in the case of the S/MMA system the more slowly propagating radical (i.e., the

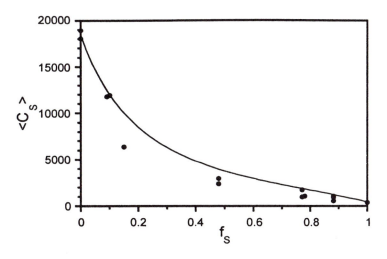

Figure 3. Average chain transfer constant in the COPhBF-mediated copolymerization of S and MMA at 40˚C. (●) Experimental data. (—) Prediction using parameters listed in Figure 1 and $C_S(S)=400$, $C_S(MMA)=18500$.(19)

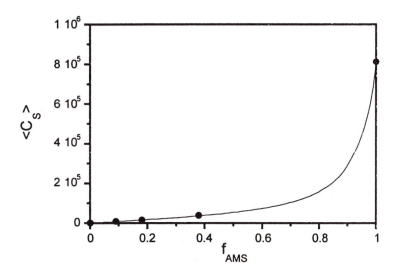

Figure 4. Average chain transfer constant in the COBF-mediated copolymerization of AMS and S at 40˚C. (●) Experimental data. (—) Prediction using parameters listed in Figure 2 and $C_S(AMS) = 9.1\cdot10^5$, $C_S(S)=1000$.(15)

~S• radical) is also less reactive towards the Co(II) catalyst, and that the net effect is a decrease in average chain transfer constant with increasing fraction of styrene in the monomer feed. In contrast, Figure 3 shows that the more slowly propagating radical is the more reactive one towards the Co(II), leading to a very large increase in average chain transfer constant. We will show in the following section that this latter behavior is exactly what is required for an effective endgroup control.

Endgroup Control

In the previous sections we have outlined some of the most important aspects of free-radical copolymerization with and without the addition of a (catalytic) chain transfer agent. Here we will combine these concepts for the prediction of the fraction of a particular endgroup. It is clear that for a particular endgroup to form, we need radicals of the desired nature and a large reactivity of these radicals towards the Co(II) complex. So, we can define the probability of the formation of an endgroup of type A as the product of the probability of finding a radical of type A and the probability of ~A• reacting with Co(II). The former probability is simply given by the fraction of ~A• radicals, ϕ_A, as given by Eq 5. The probability of the ~A• radicals reacting with the Co(II) complexes, P(A|Co), is simply given as the ratio of the rate of reaction with Co(II) and all possible reaction rates involving ~A• radicals (Eqs 11-13) (15,19,27,28).

$$P(A \mid Co) = \frac{\text{rate of } \sim A^{\bullet} \text{ reacting with Co(II)}}{\sum_{\text{all possible x}} \text{rate of } \sim A^{\bullet} \text{ reacting with x}} \qquad (11)$$

$$P(A \mid Co) = \frac{k_{tr,A}[\sim A^{\bullet}][Co(II)]}{[\sim A^{\bullet}]\left(k_{tr,A}[Co(II)] + k_{AA}[A] + k_{AB}[B]\right)} \qquad (12)$$

$$P(A \mid Co) = \frac{C_{S,A}[Co(II)]}{C_{S,A}[Co(II)] + [A] + r_A^{-1}[B]} \qquad (13)$$

It is clear from Eq 13, that the prediction of P(A|Co) only involves the knowledge of readily obtainable kinetic parameters, i.e., all those required for modelling catalytic chain transfer copolymerization (vide supra). If we now want to derive an expression for the fraction of A endgroups, Φ_A, we need to realize that the probability of forming an A endgroup, i.e., $\phi_A \times P(A|Co)$, needs to be normalized by the overall probability of forming an endgroup altogether. If this is taken into account, the fraction of A endgroups is simply given by Eq 14 (15,19,27,28).

$$\Phi_A = \frac{\phi_A \times P(A \mid Co)}{\phi_A \times P(A \mid Co) + (1 - \phi_A) \times P(B \mid Co)} \tag{14}$$

Let us now consider our two previous case studies, the catalytic chain transfer copolymerizations of styrene with methyl methacrylate and α-methyl styrene with styrene. In both systems, the first monomer, i.e., styrene and α-methyl styrene, respectively, are the slow monomers, and in both cases, the fraction of their derived radicals is larger than the fraction of these monomers in the monomer feed (see Figures 1 and 2). However, as is clear from Figures 3 and 4, the behavior of the average chain transfer constant is very different in both cases. Whereas the addition of the slow monomer styrene in the S/MMA case leads to a decrease in $\langle C_S \rangle$, the addition of the slow monomer α-methyl styrene in AMS/S leads to an increase. This means that in the case of S/MMA we have two opposing effects for the production of a certain endgroup; addition of the slower monomer leads to an increase in ϕ_S, but the overall chain transfer probability decreases. Overall, we can therefore not give an a priori prediction of the nature of the dominant endgroups in this system. In the case of AMS/S the addition of more α-methyl styrene leads to an increase in ϕ_{AMS} and an increase in overall chain transfer probability; hence we find a synergistic effect in this case. We can therefore make the a priori prediction that at some monomer feed composition with $f_{AMS} \ll 1$, the nature of the dominant endgroups will be that of α-methyl styrene, i.e., $\Phi_{AMS} \approx 1$.

In Figure 5, we have shown the predictions made by Eq 14 for the S/MMA and AMS/S systems, and our qualitative expectations are clearly recognized. The two opposing effects in S/MMA cause the fraction of styrene endgroups to be proportional to the fraction of styrene in the monomer feed. In contrast, the synergistic effects in AMS/S lead to a situation in which nearly all chains have an α-methyl styrene endgroup at α-methyl styrene feed fractions of about 5%.

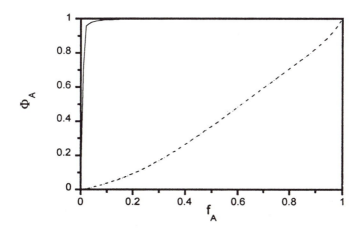

Figure 5. Fraction of A endgroups as a function of monomer A in the the monomer feed. (—) A = AMS in an AMS/S copolymerization at 40°C, model parameters as given in Figures 2 and 4. (---) A = S in S/MMA copolymerization at 40°C, model parameters as given in Figures 1 and 3. In all calculations: $[Co(II)] = 1 \cdot 10^5 M$

The predictions shown in Figure 5 have been confirmed experimentally, albeit at higher temperatures, by Harwood and co-workers (29) for the S/MMA system, and by the CSIRO/DuPont team (10) for the AMS/S system. The latter group has used the principle outlined above for the production of macromonomers of styrenes, acrylates and certain methacrylates, especially at higher temperatures (10). They have also shown that the fraction of AMS endgroups in poly(styrene)-end-α-methyl styrene depends on the nature and the amount of Co(II) complex (30). In Figure 6, we have shown the effect of the amount of Co(II) complex on Φ_{AMS}. It is clear that the fraction of α-methyl styrene endgroups increases with a decreasing Co(II) concentration, an effect that was also found experimentally by the CSIRO group at higher temperatures [10]. Finally, we wish to note that, in mathematical terms (see Eq 13), we also cover the effect of changing chain transfer constants (e.g., caused by changing catalyst nature).

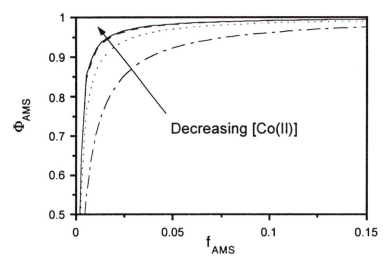

Figure 6. Effect of cobalt(II) concentration on the fraction of AMS endgroups in a catalytic chain transfer copolymerization of AMS and S. Parameters as in Figure 5, [Co(II)] = $1 \cdot 10^{-6}$ M, $1 \cdot 10^{-5}$ M, $1 \cdot 10^{-4}$ M and $5 \cdot 10^{-4}$ M.

Aldehyde End-Functionalities

In the previous section the general principles were shown for endgroup control in catalytic chain transfer polymerization, and it was shown that the endgroup functionalities in general consist of a vinyl group and whatever functional group is present in the ω monomer unit. The vinyl group is, in principle, an intrinsic result of the catalytic chain transfer process, and although this group is reactive for post-polymerization modification, a more versatile functional group would often be desirable. One such versatile group that is accessible via catalytic chain transfer polymerization is the aldehyde functionality, which can undergo a multitude of reactions (see Scheme 4) (31).

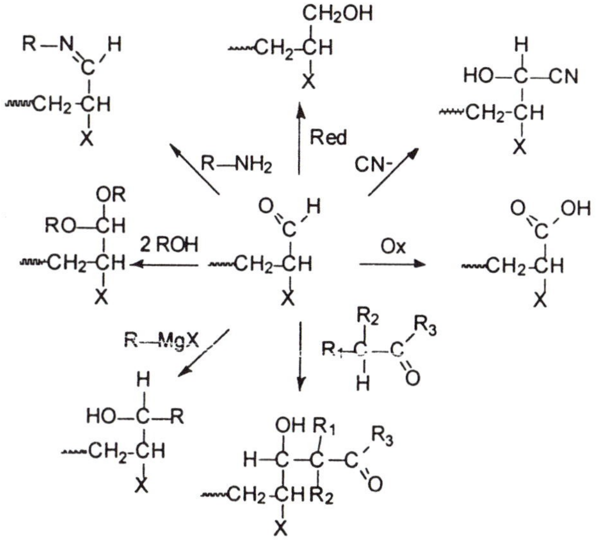

Scheme 4. Versatility of aldehyde chemistry in crosslinking reactions

The aldehyde end-functionality is introduced when a monomer containing an α-hydroxymethyl group is used in catalytic chain transfer polymerization. When a propagating radical containing a terminal unit of this nature undergoes the chain transfer reaction, the primary endgroup will be an enol, which can subsequently tautomerize to an aldehyde (see Scheme 5). This principle has been demonstrated by us previously using the monomer ethyl α-hydroxymethacrylate (EHMA) (*32*). In Figure 7, the relevant region in the ¹H-NMR spectrum of the formed polymer is enlarged, and the signals arising both from the enolic and the aldehydic proton are clearly visible.

Since the polymerization behavior of EHMA is likely to be similar to that of the other methacrylates, its use to selectively introduce aldehyde endgroups under ordinary polymerization conditions will in many cases not be conceivable. It is likely that in copolymerizations EHMA will behave similarly to MMA, and may therefore not become the dominating radical species. Preliminary studies in our laboratory on the copolymerization of EHMA and styrene indeed failed to show the presence of a significant amount of aldehyde or enol endgroups. Our earlier discussions on α-methyl styrene (vide supra), however, immediately identify a more suitable candidate to selectively introduce aldehyde endgroups, i.e., the α-hydroxy methyl analog of α-methyl styrene. This monomer, i.e., 2-phenyl allyl alcohol, should also display a very low propagation rate coefficient and a very large chain transfer constant with the cobaloximes. Preliminary studies in our laboratory confirmed these expectations; attempts to polymerize 2-phenyl allyl alcohol in bulk at 50°C using 0.144 M AIBN did not lead to any detectable amount of polymer formation, even after 9 days of heating. Similar attempts using pulsed laser polymerization with total pulsing times of up to 13 hours did not lead to polymer formation. However, heating a 50% (v/v) solution of 2-phenyl allyl alcohol in toluene for 200 hours at 45°C in the presence of

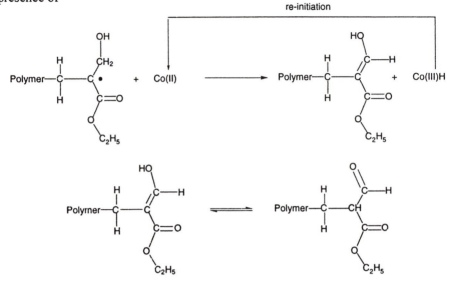

Scheme 5. Catalytic chain transfer isomerism reaction

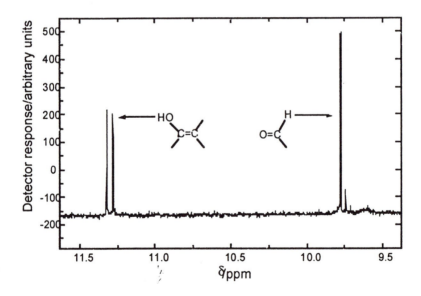

Figure 7. Expansion of ^{1}H-NMR spectrum of polyEHMA clearly indicating the enol and aldehyde signals.(32)

1.7·10^{-4} M cobaloxime boron fluoride and 1·10^{-2} M AIBN clearly resulted in the formation of aldehyde groups (see Figure 8). This result indicates that the catalytic chain transfer process indeed takes place, but that the formed monomeric radicals do not propagate, but immediately undergo the chain transfer reaction. Investigations into the copolymerization behavior of this very promising monomer are currently underway.

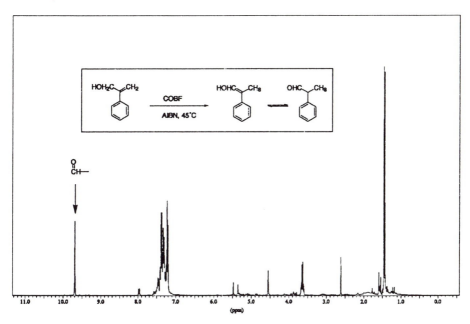

Figure 8. ^{1}H-NMR spectrum of the product of 2-phenyl allyl alcohol after heating at 45°C for 200 hours in the presence of AIBN and COBF. The aldehyde signal is clearly present.

Conclusions

Catalytic chain transfer polymerization using low-spin Co(II) complexes is an efficient, relatively simple and clean procedure for the preparation of low-molecular weight end-functionalized polymers. By default, the procedure introduces a vinyl endgroup into the polymer chain, and any other desired end-functionalities need to be introduced by the use of monomers containing these functional groups. Using conventional free-radical (co)polymerization kinetics, the requirements for the selective introduction of a particular (functional) monomer as an endgroup were discussed; in a comonomer mixture, the monomer required as the endgroup needs to display a very low reactivity towards addition to another monomer, but a very high reactivity towards the Co(II) complex. The monomers making up the main chain of the polymer should be highly reactive towards addition to monomers (including the "endgroup" monomer) and be relatively unreactive towards the Co(II) complex. A (semi-)quantitative prediction of whether a particular monomer mixture is suitable for this purpose is simple and straightforward, and only reqires knowledge of some basic, and often readily available, kinetic parameters.

It was also shown that α-hydroxy methyl substituted monomers yield the very versatile aldehyde building block when used in catalytic chain transfer polymerization. This was clearly demonstrated for ethyl α-hydroxymethacrylate, which shows a polymerization behavior largely similar to other methacrylates, and for 2-phenyl allyl alcohol, which shows similar polymerization characteristics to α-methyl styrene. This latter functional monomer shows great promise for the selective introduction of an aldehyde endgroup in a wide range of free-radical polymers.

Acknowledgments

Discussions relating to this work with Michael Gallagher, Alexei Gridnev, Dave Haddleton, Dax Kukulj, Graeme Moad and Keith Moody, experimental input by Michael Zammit and Darren Forster, and financial support by ICI Plc and the Australian Research Council are all gratefully acknowledged.

References

1. Mayo, F. R. *J. Am. Chem. Soc.* **1943**, *65*, 2324.
2. Odian, G. *Principles of Polymerization*; 2nd ed.; Wiley: New York, 1981.
3. Brandrup, A.; Immergut, E. H. *Polymer Handbook*; 3rd ed.; Brandrup, A.; Immergut, E. H., Ed.; Wiley: New York, 1989.
4. Moad, G.; Solomon, D. H. *The Chemistry of Free Radical Polymerization*; Pergamon: Oxford, 1995.
5. Karmilova, L. V.; Ponomarev, G. V.; Smirnov, B. R.; Belgovskii, I. M. *Russ. Chem. Rev.* **1984**, *53*, 132.
6. Davis, T. P.; Haddleton, D. M.; Richards, S. N. *J. Macromol. Sci., Rev. Macromol. Chem. Phys.* **1994**, *C34*, 234.
7. Davis, T. P.; Kukulj, D.; Haddleton, D. M.; Maloney, D. R. *Trends Polym. Sci.* **1995**, *3*, 365.
8. Heuts, J. P. A.; Forster, D. J.; Davis, T. P. In *Transition Metal Catalysis in Macromolecular Design*; Boffa, L. S. and Novak, B. M., Eds., ACS Symposium Series Vol. 760; American Chemical Society: Washington DC, 2000.
9. Krstina, J.; Moad, C. L.; Moad, G.; Rizzardo, E.; Berge, C. T.; Fryd, M. *Macromol. Symp.* **1996**, *111*, 13.
10. (a) Chiefari, J.; Jeffery, J.; Moad, G.; Rizzardo, E.; Thang, S. H. *Polym. Prepr. (Am. Chem. Soc., Div. Polym. Chem.)* **1999**, *40(2)*, 344. (b) Ittel, S.D.; Gridnev, A.A.; Moad, C.L.; Moad, G.; Rizzardo, E.; Wilczek, L.; WO 97 31,030, (Chem. Abstr.,1997, 128, 234752).
11. Haddleton, D. M.; Maloney, D. R.; Suddaby, K. G.; Muir, A. V. G.; Richards, S. N. *Macromol. Symp.* **1996**, *111*, 37.
12. Bakac, A.; Brynildson, M. E.; Espenson, J. H. *Inorg. Chem.* **1986**, *25*, 4108.

331

13. Suddaby, K. G.; Maloney, D. R.; Haddleton, D. M. *Macromolecules* **1997**, *30*, 702.
14. Kukulj, D.; Davis, T. P. *Macromol. Chem. Phys.* **1998**, *199*, 1697.
15. Kukulj, D.; Heuts, J. P. A.; Davis, T. P. *Macromolecules* **1998**, *31*, 6034.
16. Heuts, J. P. A.; Forster, D. J.; Davis, T. P.; Yamada, B.; Yamazoe, H.; Azukizawa, M. *Macromolecules* **1999**, *32*, 2511.
17. Heuts, J. P. A.; Forster, D. J.; Davis, T. P. *Macromolecules* **1999**, *32*, 3907.
18. Heuts, J. P. A.; Forster, D. J.; Davis, T. P. *Macromol. Rapid Commun.* **1999**, *20*, 299.
19. Heuts, J. P. A.; Kukulj, D.; Forster, D. J.; Davis, T. P. *Macromolecules* **1998**, *31*, 2894.
20. Heuts, J. P. A.; Muratore, L. M.; Davis, T. P. *Macromol. Chem. Phys.* **1999**, *In press*.
21. Fukuda, T.; Kubo, K.; Ma, Y.-D. *Prog. Polym. Sci* **1992**, *17*, 875.
22. Fukuda, T.; Ma, Y.-D.; Inagaki, H. *Macromolecules* **1985**, *18*, 17.
23. See, for example, Heuts, J. P. A.; Gilbert, R. G.; Maxwell, I. A. *Macromolecules* **1997**, *30*, 726.
24. Coote, M. L.; Zammit, M. D.; Davis, T. P.; Willett, G. D. *Macromolecules* **1997**, *30*, 8182.
25. Coote, M. L.; Johnston, L. P. M.; Davis, T. P. *Macromolecules* **1997**, *30*, 8191.
26. Kukulj, D.; Davis, T. P. *Macromolecules* **1998**, *31*, 5668.
27. Galbraith, M. N.; Moad, G.; Solomon, D. H.; Spurling, T. H. *Macromolecules* **1987**, *20*, 675.
28. Heuts, J. P. A.; Coote, M. L.; Davis, T. P.; Johnston, L. P. M. In *Controlled Radical Polymerization*; Matyjaszewski, K., Ed.; ACS Symposium Series, Vol. 685; American Chemical Society: Washington, DC, 1998; p 120.
29. Greuel, M. P.; Harwood, H. J. Polym. Prepr. (Am. Chem. Soc., Div. Polym. Chem.) **1990**, 32(1), 545.
30. Chiefari, J.; Moad, G.; Rizzardo, E. Unpublished results
31. Morrison, R. T.; Boyd, R. N. *Organic Chemistry*; 5th ed.; Allyn and Bacon, Inc.: Boston, 1987.
32. Davis, T. P.; Zammit, M. D.; Heuts, J. P. A.; Moody, K. *Chem. Commun.* **1998**, 2383.

New Materials by Controlled Radical Polymerization

Chapter 23

Synthesis of Polystyrene–Polyacrylate Block Copolymers by Nitroxide-Mediated Radical Polymerization

Sophie Robin and Yves Gnanou[1]

Laboratoire de Chimie des Polymères Organiques, ENSCPB-CNRS-Université Bourdeaux 1, Avenue Pey Berland, BP 108, 33402 Talence Cedex, France

This paper discusses the conditions the best suited to the preparation of well-defined polystyrene (PS) / poly(n-butyl acrylate) (PBuA), using a β-hydrogen containing phosphonylated nitroxide. Using kinetic data to compute the rates of cross-addition and propagation of the second monomer, it is demonstrated that polymerizing first n-butyl acrylate (BuA) should give rise to well-defined PBuA-b-PS copolymer samples. This prediction was experimentally confirmed subsequently.
When styrene was polymerized first, the copolymer formed was contaminated with a substantial amount of residual PS macroinitiator : the difference between the rates of cross-addition and propagation of BuA resulted in a fast growth of those of the PBuA blocks that were initiated, causing the medium to partition in mesophases with the residual PS precursor entrapped in the monomer-poor phase.

Introduction

Nitroxide-Mediated Radical Polymerization currently enjoys renewed impetus after it was discovered that several families of vinylic monomers can be polymerized under living/controlled conditions, provided the nitroxide used to control propagation is adequately chosen (*1,2,3,4,5*). For instance, (N-*tert*-butyl-N-(1-diethylphosphono-2,2-dimethyl)propylnitroxyl) (DEPN) is a β-hydrogen bearing nitroxide that was found to bring about the controlled polymerization of both styrene and alkyl acrylates when associated with AIBN as initiator (*6*). Other β-hydrogen containing nitroxides were recently shown suitable for the controlled radical polymerization of various vinylic monomers, including styrene, alkyl systems lies in the frequency of the deactivation/activation process that depends both on the rate of coupling of propagating

[1]Corresponding author.

radicals with nitroxides and on the acrylates, methacrylates, and 1,3 dienes (7,8,9). The key to the control of such rate of homolysis of the C—ON bond formed. When used to polymerize styrene or alkyl acrylates, the AIBN/DEPN system not only provides high equilibrium constants ($K=k_d/k_{rec}$) –reflected in a fast propagation of the two monomers- but it also sets three decade difference between the rate constants of recombination and propagation, affording samples of particularly narrow polydispersity. There lies the beauty of polymerizations mediated by DEPN.

Notwithstanding the fact that the same initiator (AIBN)/nitroxide (DEPN) system induces totally different kinetics with styrene and n-butyl acrylate, the controlled character of their respective polymerization prompted us to investigate their block copolymerization by sequential addition (10).

This paper thoroughly discusses the constraints associated with such synthesis and discloses the conditions the best suited to obtain well-defined PS/PBuA block copolymer samples. In particular, it demonstrates that the order of addition of monomers matters to achieve high blocking efficiency.

Results and Discussion

Kinetic Predictions as to the Order of Polymerization of Monomers

For a successful synthesis of diblock copolymers by controlled radical polymerization, it is essential to retain the chain end functionality of the first block and achieve an efficient cross-propagation. With respect to the first factor, that concerns the production of precursors containing a minimal amount of dead chains, all precautions should be taken to optimize propagation over termination or transfer reactions.

Since the ratio of the probability of termination to that of propagation ($k_t[P^\bullet]/k_p[M]$) increases with the monomer conversion, it is crucial to discontinue the polymerization of the first monomer before its total consumption so as to prevent too high a concentration of dead "precursor" chains.

As to the blocking efficiency, it not only depends on the rate constants of cross-addition and on the propagation rate constants of the two monomers, but it is also subordinated to the equilibrium constants ($K=k_d/k_{rec}$) between their respective dormant and active species (Scheme 1). In contrast to K, the reactivity ratios of these two monomers and their rate constants of propagation can be found in the literature (11,12,13). Though important, the knowledge of these values has only an indicative value that can well be used to choose the monomer to polymerize first.

336

Indeed, polymerizing the monomers in the order of decreasing rate constants of propagation does not necessarily entail a fast initiation of the second block by the first one. To bring about a complete cross-addition and thus achieve high blocking efficiency, one has to also consider the equilibrium constants for the two polymerizing systems and the concentration of polymeric radicals [P*] they imply for a given concentration of dormant species.

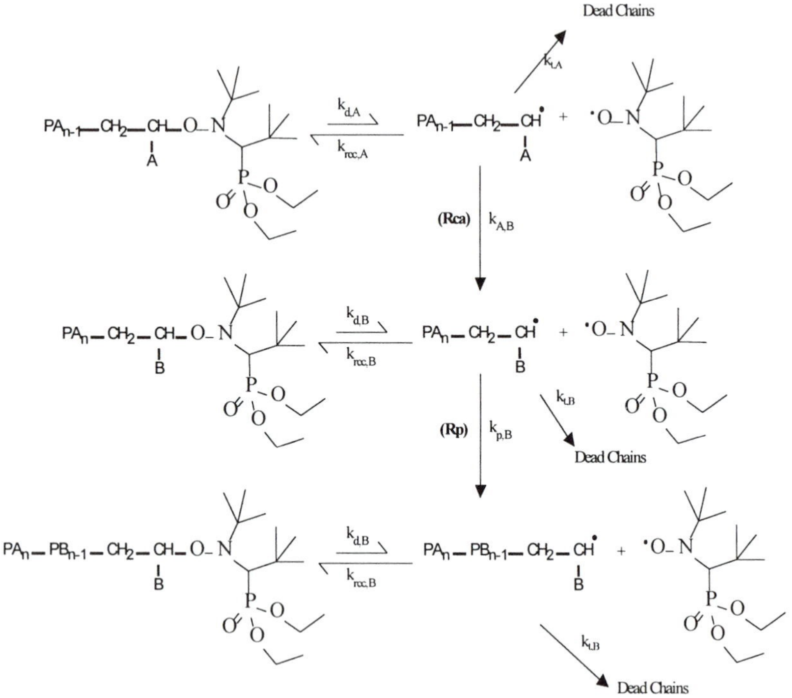

Scheme 1 : Reactions occurring while synthesizing block copolymers by nitroxide-mediated polymerization (A,B= Ph or CO2Bu).

From the knowledge of [P*] as a function [P-DEPN], the initial rates of cross-addition and propagation of the 2nd block can be easily deduced. At the very onset of the polymerization of the second monomer, the concentration of the two kinds of radicals should be quite contrasted, so that initiation should be initially favored over propagation. However, this situation cannot prevail for ever ; as conversion increases, the concentration of the first block is bound to continuously decrease –and alternately that of the 2nd block to progressively increase-, entailing a constant variation of the concentration of the two kinds of polymeric radicals ([PS*], [PBuA*]). Given the profiles of the concentration of the two species, the rate of cross-addition is bound to be overtaken by the rate of propagation but the question is whether this outrunning occurs soon or late in the conversion of the 2nd monomer. Should the rate of propagation overshadows that of cross-addition while much of the first block is not yet to decompose, the resulting copolymers are likely to be contaminated with a substantial amount of residual precursor chains.

To construct the [P-DEPN] versus [P•] curve, we referred to our previous work on the kinetics of DEPN-mediated polymerizations of styrene and n-butyl acrylate (14). Using PS and PBuA oligomers fitted with end-standing DEPN-based alkoxyamines, the polymerization of styrene and n-butyl acrylate was triggered at 120°C and their conversion followed as a function of time. From the $\ln([M]_0/[M])$ versus time plot, the corresponding [P•] could be easily deduced, using k_p values found in the literature (11) and the relation proposed by Fischer (15) for the slope of the linear variation. In that work, K values could also be determined upon concomitantly following the conversion in monomer and the evolution of free [DEPN], K being equal to $K=[P•]_t*[DEPN]_t/[P-DEPN]_0$ as demonstrated by Fischer (15). Based on this methodology, the concentration of polymeric radicals ([PS•], [PBuA•]) were computed as a function of the corresponding alkoxyamines and the curves subsequently drawn for the two systems. Because K dramatically varies from one monomer to another one ($K_S=3.5\cdot10^{-9}$ mol.l^{-1}, $K_{BuA}=1.2\cdot10^{-10}$ mol.l^{-1} (16)), a same concentration of dormant species is found to afford quite different values of [PS•] and [PBuA•] as expected (Figure 1).

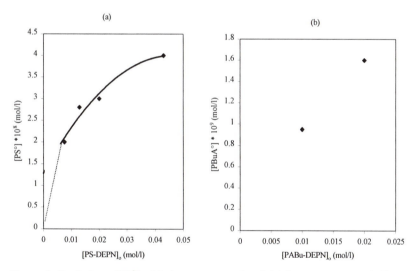

Figure 1: Evolution of [P•] with the corresponding initial concentration of alkoxyamine [P-DEPN]$_0$, in the case of PS-DEPN (a) and PBuA-DEPN (b) as macroinitiators.

All data are thus available to compute the rate of cross-addition (R_{CA}) as a function of [P-DEPN] and compare the value obtained with that of propagation (R_p) for a same experiment (Figure 2).

$$R_{CA} = k_{A,B} \cdot [P_A^\bullet] \cdot [B]$$

$$R_p = k_{p,B} \cdot [P_B^\bullet] \cdot [B]$$

Obviously, the two rates are set to vary oppositely and thus to cross each other, this particular point being not only determined by the respective K and [P•] values of the two systems, but by the initial concentration of dormant species [P-DEPN] as well.

Figure 2 and Figure 3 show the evolution of the rates of cross-addition and propagation with the concentration of residual precursor when PBuA and PS-based alkoxyamines were used to polymerize styrene and *n*-butyl acrylate, respectively.

Several remarks can be inferred from these two figures. To prepare diblock copolymers based on PS and PBuA, it is recommended first to polymerize *n*BuA and subsequently grow the PS block from the formed PBuA. For instance, the use of an initial concentration of $7.8 \cdot 10^{-3}$ mol.l^{-1} for the [PBuA-DEPN] precursor implies that only 4% of this macroinitiator will be present in the medium at the time the rate of propagation outruns that of cross-addition (Figure 2). Not only is this order of polymerization of monomers particularly favorable to the obtainment of copolymers with low compositional heterogeneity, but it also allows to grow rather short PS blocks from PBuA blocks.

Quite different is the situation that prevails when [PS-DEPN] is used to grow PBuA as second block (Figure 3). Notwithstanding a much higher concentration of polystyryl radicals (Figure 1), the sluggishness of the latter species to add *n*-butyl acrylate is the major cause for the slow rate of initiation observed. Whatever the initial concentration of PS-DEPN macroinitiator, the rate of propagation of *n*-butyl acrylate indeed outruns that of cross-addition while about 50% of PS precursor chains are still present in the reaction medium. This order of polymerization of monomers with PS as first block therefore appears not suited to the production of samples that are homogeneous in size and composition.

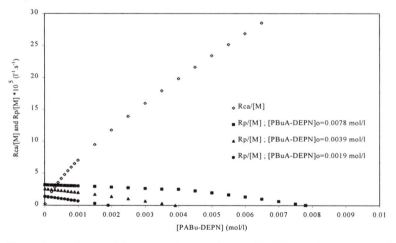

Figure 2 : Evolution of the rates of cross-addition ($R_{CA}/[M]$) and propagation ($R_{p,S}/[M]$) of styrene as a function of $[PBuA\text{-}DEPN]_0$, used as macroinitiator.

However, it can be seen from Figure 3 that the differences between the rates of cross-addition and propagation tend to narrow with low initial concentrations of PS macroinitiator, meaning that PBuA with a large size might be grown with little amount of PS-DEPN precursor left aside.

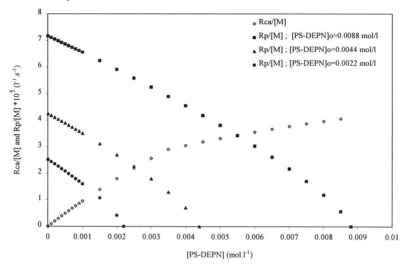

Figure 3 : Evolution of the rates of cross-addition (R_{CA}/[M]) and propagation ($R_{p,BuA}$/[M]) of n-butyl acrylate as a function of [PS-DEPN]$_0$, used as macroinitiator

To our knowledge, it is the first time that such an attempt is made to predict the outcome of a block copolymer synthesis mediated by nitroxides. However, one cannot disregard the fact that the model used to describe the kinetics of initiation and propagation may be oversimplified. In our reasoning, we have assumed the two operating systems -PS$^\bullet$, PBuA$^\bullet$ and DEPN in equilibrium with their corresponding dormant species- to behave independently one from the other which might not be the reality. Since each system sets its own level of nitroxides in excess, the active species that is known to release more free nitroxide –here PS$^\bullet$- may force the other kind of radicals to recede to lower concentrations than they would have taken in simple homopolymerizations. Since the direct measurement of the concentration of the two kinds of radicals in such copolymer synthesis is not possible, we had no other option than to rely on our previously described reasoning to account for the results experimentally obtained.

Comparison of Experimental Results with Kinetic Predictions

Synthesis of PBuA-b-PS Copolymer with PBuA as First Block
The synthesis of the first PBuA block (PBuA-DEPN) was obtained under conditions previously described (*10,14*) ; a sample of 50,000g.mol^{-1} molar mass was targeted, AIBN being used as initiator and DEPN as stable radical in a 1:2.5 ratio. As previously

mentioned, the polymerization was discontinued before the total consumption of monomer and a polymer exhibiting a polydispersity of 1.2 and 41,000g.mol⁻¹ as molar mass was eventually obtained after 9.5h of reaction at 120°C (Table I). The efficiency of the initiation step was found equal to 0.85 ; only a larger excess of DEPN could have afforded an efficiency close to 1 as previously demonstrated by Benoit *et al.* (*14*), but it would have unnecessarily slowed down the rate of polymerization.

Another sample of PBuA-DEPN was prepared from the following alkoxyamine $CH_3CH(COOCH_3)$-DEPN : polymerization proceeded at much faster rate in this case because a smaller excess of free DEPN (0.07 instead of 0.25 in the case of AIBN as an initiator) was used as compared to the previous case. The same molar mass of 50,000 g.mol⁻¹ was targeted and like previously the polymerization was discontinued at 70% conversion. Before adding styrene to grow the 2ⁿᵈ block, residual *n*BuA was stripped off in the reaction medium containing the PBuA-DEPN and the mixture was heated to 120°C.

Table I : Characteristics of PBuA(40,000g/mol)-PS(50,000g/mol) Diblock Copolymers

Exp.	Step	$M_{n,th}^{(1)}$ (g/mol)	Time (h)	$\rho^{(2)}$ (%)	$M_{n,exp}^{(3)}$ (g/mol)	$I_p^{(4)}$	$f^{(5)}$
1.1	PBuA⁽⁶⁾	35,500	9.5	71	41,700	1.2	0.85
1.2	PBuAPS	53,700	0.75	19	55,500	1.3	0.97
1.3	PBuA⁽⁷⁾	34,000	3	68	31,500	1.2	1.0
1.4	PBuAPS	85,000	2	35	87,600	1.2	0.97

⁽¹⁾Calculated from conversion.⁽²⁾Conversion measured by gravimetry. ⁽³⁾Determined by SEC using RI detection and PS standards. ⁽⁴⁾ Polydispersity index obtained from SEC.⁽⁵⁾Initiation efficiency calculated as $M_{n,th}/M_{n,exp}$. ⁽⁶⁾ Initiation by AIBN with [DEPN]/[AIBN]=2.5. ⁽⁷⁾ Initiation by the alkoxyamine mentioned in the text with [DEPN]/[Alkoxyamine]=0.07.

The polymerization was purposely discontinued at low conversion in styrene to check whether PS block of short size could be grown in a homogeneous way (Table I). As shown in Figure 4, the size exclusion chromatography traces of the two copolymers obtained are found to be entirely shifted towards the higher molar mass region upon superimposing them with that of the precursor.

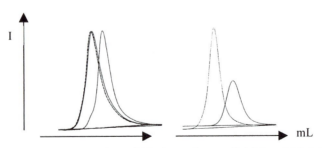

Figure 4 : SEC traces of PBuA-b-PS diblock copolymers (1.1,1.2 and 1.3,1.4, Table I).

No tailing could be detected in either SEC traces, confirming that all PBuA precursor chains efficiently initiated the polymerization of styrene. For instance, a PS block of 20,000 g.mol⁻¹ could be attached with perfect efficiency of the initiation step. Another piece of evidence of the excellent definition of the PBuA-b-PS block copolymer formed is the perfect fit between UV and RI detections : this bears out the fact that these samples exhibit high homogeneity in composition.

Synthesis of PS-b-PBuA Copolymer with PS as First Block

As we were mainly interested in PS-b-PBuA copolymers with a PS block of short size, a sample of PS-DEPN 20,000g.mol⁻¹ molar mass was synthesized. Using [DEPN] and [AIBN] in a 2.5 molar ratio, this macroinitiator (PS-DEPN) could be obtained with an excellent efficiency (f=1.0) and a narrow molar mass distribution (I_p=1.2) (*14*).

Table II : Characteristics of PS(20,000g/mol)-PBuA(100,000g/mol) Diblock Copolymers

Exp.	*Step*	$M_{n,th}$[(1)] (g/mol)	*Time* (h)	ρ[(2)] (%)	$M_{n,exp}$[(3)] (g/mol)	I_p[(4)]	f[(5)]
2.1	PS[(6)]	18,800	5.5	94	18,900	1.2	1.0
2.2	PS-PBuA	104,900	6	86	65,900	2.5	-

[(1)]Calculated from conversion.[(2)]Conversion measured by gravimetry. [(3)]Measured by SEC using RI detection and PS standards. [(4)] Polydispersity index obtained from SEC.[(5)]Initiation efficiency calculated as $M_{n,th}/M_{n,exp.}$ [(6)] Initiation by AIBN with [DEPN]/[AIBN]=2.5.

Once the residual styrene was stripped off, the PS-DEPN containing medium was charged with fresh *n*-butyl acrylate and heated to 120°C. The sample obtained after 6h of reaction and 86% conversion was characterized by SEC (Figure 5).

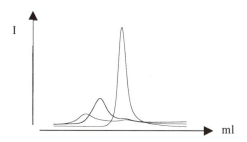

Figure 5 : SEC traces of PS-b-PBuA diblock copolymer (2.1 and 2.2, Table II) as a function of nBuA conversion.

As anticipated, this copolymer was contaminated with a non negligible amount of residual PS precursor chains (18%), and its molar mass distribution was broad owing to a slow cross-addition compared to the propagation of *n*-butyl acrylate. Even though this overall behavior was forecast by our kinetic calculations, we were intrigued by the presence of such a large amount of PS homopolymer in the resulting material since a PBuA block of large size was targeted.

Using the UV detection of the SEC line, aliquots sampled out at various times were characterized and compared. As a matter of fact, after experiencing a rapid initial decrease, the peak due to residual PS-DEPN was found to remain constant with respect to that of the copolymer, as soon as the conversion in nBuA reaches 30%. Approximately for that conversion monomer, the reaction medium was noticed to become strongly turbid, denoting the occurrence of a phase separation. If the disappearance of PS-DEPN suddenly levelled off after crossing a certain monomer conversion, this means that the medium underwent a strong partition induced by the phase separation, causing n-butyl acrylate to be unevenly distributed in the medium. Because the initiated PBuA blocks grew fast to large size, the incompatibility between PS and PBuA that has resulted might have driven the residual PS-DEPN within PS-rich particles : actually, the latter did not coalesce because of their steric stabilization by the PBuA blocks of PS-b-PBuA copolymers. Owing to the preference of nBuA to stay in the poly(BuA)-rich phase, residual PS-DEPN chains had therefore little chance to add this monomer.

Since it was previously demonstrated that the difference in the rates of cross-addition and propagation tends to narrow with smaller initial concentration of PS-DEPN, poly(BuA) blocks of molar masses much larger than 100,000 g.mol^{-1} were targeted (Table III), the polymerization being discontinued at relatively low conversion. As in the previous case, one can notice the presence of residual PS-DEPN in approximately same proportions (16%), confirming that the partition of the medium hindered the access of the monomer to PS-DEPN.

Table III : Characteristics of PS(20,000g/mol)-b-PBuA(variable size aimed) Diblock Copolymers using varying [PS-DEPN].

Exp	$M_{n,th}$[1] (g/mol)	Time (h)	ρ[2] (%)	$M_{n,exp}$[3] (g/mol)	I_p[4]	R[5]
3.1	106,000	6	86	65,900	2.5	18
3.2[6]	84,900	3	86	40,400	2.0	30
3.3	155,000	4	27	74,800	1.6	16
	470,000	12	90	324,000	2.0	16

[1]Calculated from conversion. [2]Conversion measured by gravimetry. [3]Measured by SEC using RI detection and PS standards. [4]Polydispersity index .[5] Remaining macroinitiator estimated by SEC using UV detection.[6] The PS-DEPN macroinitiator was precipitated in methanol.

In an attempt to modify the ratio of the rate of cross-addition to that of propagation, a series of experiments were carried out at temperatures lower than 120°C (Table IV). Because the various reactions involved in such block copolymer synthesis have different energies of activation, one might expect to favor one step (initiation) over the other one (propagation) by decreasing temperature.

Table IV : Characteristics of PS(20,000g.mol⁻¹)-b-PBuA(100,000g.mol⁻¹) Diblock Copolymers obtained at various temperatures

Exp	$M_{n,th}$[1] (g/mol)	Time (h)	ρ[2] (%)	$M_{n,exp}$[3] (g/mol)	I_p[4]	Temperature[5] (°C)	R[6]
4.1	106,000	6	86	65,900	2.5	120	18
4.2	93,400	133	71	67,000	1.3	95	12
4.3	42,400	253	20	32,900	1.2	80	14

[1]Calculated from conversion. [2]Conversion measured by gravimetry. [3]Measured by SEC using RI detection and PS standards. [4]Polydispersity index . [5] Temperature of polymerization of the second block. [6] Remaining macroinitiator estimated by SEC using UV detection.

The results obtained were rather disappointing since no significative change could be perceived in the amount of residual DEPN. To better demonstrate the interrelation between the incomplete consumption of PS-DEPN and the incompatibility that arises in the reaction medium because of a too slow cross-addition, PS-DEPN of shorter size shorter than 20,000 g.mol⁻¹ were used as macroinitiators.

Table V : Characteristics of PS(variable size aimed)-b-PBuA(20,000g.mol⁻¹) Diblock Copolymers

Exp.	1st block size (g.mol⁻¹)	2nd block size (g.mol⁻¹)	$M_{n,th}$[1] (g/mol)	Time (h)	ρ[2] (%)	$M_{n,exp}$[3] (g/mol)	I_p[4]	R[5]
5.1	3,900	20,000	22,500	1	93	20,800	1.5	-
5.2	4,300	20,000	22,300	5	90	23,600	1.4	-
5.3	7,000	20,000	21,400	9	70	28,200	1.3	12
5.4	20,000	100,000	104,900	6	86	65,900	2.5	18

[1]Calculated from conversion. [2]Conversion measured by gravimetry. [3]Measured by SEC using RI detection and PS standards. [4]Polydispersity index .[5]Remaining macroinitiator estimated by SEC using UV detection.

As the incompatibility between polymeric chains and thus the order-disorder transition of the corresponding block copolymers are governed by the product $\chi \cdot N$ (17), (χ being the interaction parameter and N the overall degree of polymerization), two different polymers, each of them of small size should be less prone to separate in distinct phases. When PS-DEPN of 3,000 to 4,000 g.mol⁻¹ molar mass were used as macroinitiator, no phase separation could be actually detected and as a result no unreacted macroinitiator remained in the reaction medium after growing the poly(BuA) block. The slow cross-addition step was mirrored in rather large polydispersities (1.4-1.5), but excellent control over the sample molar mass could be achieved. In contrast, the use of PS-DEPN precursor of slightly higher molar mass (7,000g.mol⁻¹) resulted in the occurrence of phase separation and the presence of unreacted precursor chains at the time the polymerization of BuA was discontinued.

344

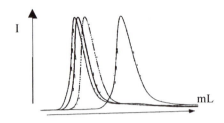

Figure 6 : SEC traces of PS-b-PBuA using PS-DEPN precursor of short size (5.1 and 5.2, Table V).

Another means to demonstrate that the incompatibility between growing poly(BuA) and remaining PS-DEPN prejudices the consumption of the latter was to grow a statistical P(S-co-BuA) block in between PS and PBuA blocks. Instead of thoroughly stripping off residual styrene after the growth of the first PS-DEPN block, *n*-butyl acrylate was directly introduced in the reaction medium that was then heated to 120°C.

Table VI : Characteristics of PS(20,000g.mol⁻¹)-b-P(S-co-BuA) (100,000g.mol⁻¹) Copolymers

Exp.	1st block size (g/mol)	$M_{n,th}^{(1)}$ (g/mol)	Time (h)	$\rho^{(2)}$ (%)	$M_{n,exp}^{(3)}$ (g/mol)	$I_p^{(4)}$	$R^{(5)}$
6.1	7,600	73,200	10.5	64	40,700	1.8	<10
6.2	10,100	79,200	10.5	66	61,900	1.4	<10
6.3	18,300		9.5	63			

[1]Calculated from conversion. [2]Conversion measured by gravimetry. [3]Measured by SEC using RI detection and PS standards. [4]Polydispersity index . [5] Remaining macroinitiator estimated by SEC using UV detection.

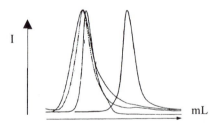

Figure 7 : SEC traces of PS-b-P(S-co-BuA) (6.1, 6.2 and 6.3, Table VI).

In contrast to the styrene-free cases, the incompatibility could be alleviated to some extent and less than 10% of PS-DEPN macroinitiator was found unreacted after the growth of the second PBuA block. Other means to alleviate the incompatibility that develops in the use of the polymerization of the 2nd monomer are currently investigated.

Conclusion

Free radical polymerizations mediated by DEPN were utilized to prepare PS/PBuA block copolymers by sequential addition of the corresponding monomers. The various factors that control the cross-addition step were thoroughly analyzed. It appears that the order of polymerization of the two monomers matters, the best copolymers formed being obtained upon polymerizing *n*BuA first. When styrene was polymerized first, the differences between the rate of cross-addition and that of propagation of *n*BuA resulted in a mesophase separation that prejudiced the thorough consumption of the first block. However, well-defined PS-b-PBuA could be prepared using PS precursors of short size.

Acknowledgement

The authors gratefully acknowledge ELF-ATOCHEM and CNRS for financial support.

Experimental Section

Materials Styrene, *n*-butyl acrylate were purified by vacuum distillation over CaH_2 ; AIBN was purified by recristallization from ether ; DEPN was used as received from Elf Atochem.

Synthesis of PBuA-b-PS diblock

a.Preparation of the PBuA macroinitiator. In a schlenk, 11 mg AIBN and 67 mg DEPN were added to 7.5 mL of *n*-butyl acrylate . The mixture was stirred in an oil bath at 120°C. The polymer obtained was dried under vacuum for several hours prior to characterization.

b.Growth of the PS block. A schlenk was charged with 4.5 g of the previously described PBuA macroinitiator and 7.5 mL of styrene. The system was degassed and heated to 120°C.

Synthesis of PS-b-PBuA diblock

a.Preparation of the PS macroinitiator. In a schlenk, 22 mg AIBN and 0.143 g DEPN were added to 6 mL of styrene . The mixture was stirred in an oil bath at 120°C. The polymer obtained was dried under vacuum for several hours prior to characterization.

b.Growth of the PBuA block. A schlenk was charged with 0.8 g of PS macroinitiator and 4.8 mL of *n*-butyl acrylate. The system was degassed and heated to 120°C.

All other attempts of diblock synthesis with PS as the first block were made using the same procedure as that described above.

References

1. Georges, M.K.; Veregin, R.P.N.; Kazmaier, P.M.; Hamer, G.K.; Saban, M. *Macromolecules* **1994**, *27*, 7228.
2. Georges, M.K.; Veregin, R.P.N.; Kazmaier, P.M.; Hamer, G.K. *Polymer Prepr. (Am. Chem. Soc., Div. Polym. Chem.)* **1994**, *35*, 870.
3. Odell, P.G.; Veregin, R.P.N.; Michalak, L.M.; Brousmiche, D.G.; Georges, M.K. *Macromolecules* **1995**, *28*, 8453.
4. Veregin, R.P.N.; Odell, P.G.; Michalak, L.M.; Georges, M.K. *Macromolecules* **1996**, *29*, 4161
5. Baldovi, M.V.; Moktat, N.; Scaiano, J.C. *Macromolecules* **1996**, *29*, 5497.
6. Benoit, D.; Grimaldi, S.; Finet, J.P.; Tordo, P.; Fontanille, M.; Gnanou, Y. in *Controlled Radical Polymerization*; Matyjaszewski, K., Ed.; ACS Symposium Series 685; American Chemical Society ; Washington, DC, **1998**; Chapter 14.
7. Benoit, D.; Chaplinski, V.; Braslau, R.; Hawker, C.J.; *J. Am. Chem. Soc.* **1999**, *16(121)*, 3904.
8. Benoit D., Harth E., Fox P., Waymouth R.M., Hawker C.J.; *Macromolecules* **2000**, *33*, 363.
9. Grimaldi S., Finet J.P., Lemoigne F., Zeghdaoui A., Tordo P., Benoit D., Fontamille M., Gnanou Y.; *Macromolecules* **2000**, *to be published*.
10. Robin, S; Gnanou, Y.; Polym. Prepr. 1999, 40(2), 387.
11. Buback M.; Gilbert R.G.; Hutchinson R.A.; Klumperman B.; Kuchta F.D.; Manders B.G.; O'Driscoll K.F.; Russel G.T.; Schwer J.; *Macromol. Chem. Phys.* **1995**, *196*, 3267.
12. Lyons, R.A.; Hutovic, J.; Piton, M.C.; Christie, D.I.; Clay, P.A.; Manders B.G.; Kable, S.H.; Gilbert, R.G. *Macromolecules* **1996**, 29,1918.
13. Shipp, D.A.; Matyjaszewski, K. *Macromolecules* **1999**, *32*, 2948.
14. Benoit, D. Ph.D. Thesis, Bordeaux-I University, France, **1997**.
15. Fisher, H. *J. Polym. Sci., Part A : Polym. Chem.* **1999**, *37*, 1885.
16. Benoit, D.; Grimaldi, S.; Robin, S.; Finet, J.P.; Tordo, P.; Gnanou, Y. *J. Am. Chem. Soc., submitted.*
17. Leibler, L. *Macromolecules* **1980**, *13*, 1602.

Chapter 24

Functionalized Polymers by Atom Transfer Radical Polymerization

Scott G. Gaynor and Krzysztof Matyjaszewski

Center for Macromolecular Engineering, Department of Chemistry,
Carnegie Mellon University, 4400 Fifth Avenue, Pittsburgh, PA 15213

The use of controlled radical polymerization, specifically, atom transfer radical polymerization (ATRP) has provided a means for the synthesis of a variety of well-defined polymers. Among these, includes a variety of functionalized monomers and polymers with well-defined end groups. Functional end groups can be introduced by use of a functionalized initiator, or through the transformation of the halogen end group by simple organic chemistry techniques.

Introduction

The development of a wide variety of controlled radical polymerization systems, has allowed for the preparation of polymers with predefined molecular weights, low polydispersities, novel compositions (block copolymers), and with easily definable end groups. The benefit of using a radical polymerization as compared to living ionic polymerization systems is that monomers containing a variety of functional groups can be polymerized directly, without requiring the use of protecting groups. Our group has focused on the development of atom transfer radical polymerization and this chapter will focus on the preparation of polymers with a variety of functional groups, incorporated either in the monomers themselves or at the chain ends.

Atom transfer radical polymerization (ATRP) involves the reversible activation and deactivation of an alkyl halide initiator by a transition metal catalyst.*(1-4)* By correct selection of the catalyst, it has been demonstrated that well-defined polymers of styrenes, (meth)acrylates, acrylonitrile, acrylamides and dienes can be prepared.*(5)*

Methacrylates

2-Hydroxyethyl Methacrylate

The controlled polymerization of 2-hydroxyethyl methacrylate (HEMA)*(6)* was at first difficult, when compared to the polymerization of MMA. Due to the polar nature of the monomer, it was found that polymerization of HEMA in bulk using a Cu(I)Br/2,2'-bipyridine (bpy) catalyst was extremely fast and uncontrolled. To slow the polymerization and allow for better control of the polymerization, solvents were employed to dilute the system, but were not overly successful. For example the use of DMF gave polymer whose molecular weight increased with conversion but significant termination was observed in the first order kinetic plots and the molecular weight distributions broadened as the polymerization progressed. After trying other solvents, it was found that the use of a mixed solvent system (70% methyl ethyl ketone /30% *n*-propanol) resulted in homogeneous polymerization mixtures and well-defined polymers. For example in a 50:50 monomer/solvent mixture, HEMA was polymerized using ethyl 2-bromoisobutyrate (EBiB) as the initiator and Cu(I)Br/2bpy as the catalyst system; molecular weights up to $M_n = 60,000$ were obtained with polydispersities less than $M_w/M_n = 1.5$.

2-(Dimethylamino)ethyl Methacrylate

2-(Dimethylamino)ethyl methacrylate is similar in structure to HEMA but the hydroxyl group is replaced with a tertiary amine. This monomer is less polar than HEMA, which allows for it to be soluble in organic solvents such as dichlorobenzene or anisole. This simplified the polymerization conditions. By using EBiB or 2-bromopropionitrile (BPN) with a Cu(I)Br/HMTETA catalyst (HMTETA = Hexamethyltriethylenetetramine, Figure 1), well-defined polymers were obtained in a 50:50 mixture of solvent and monomer.*(7)*

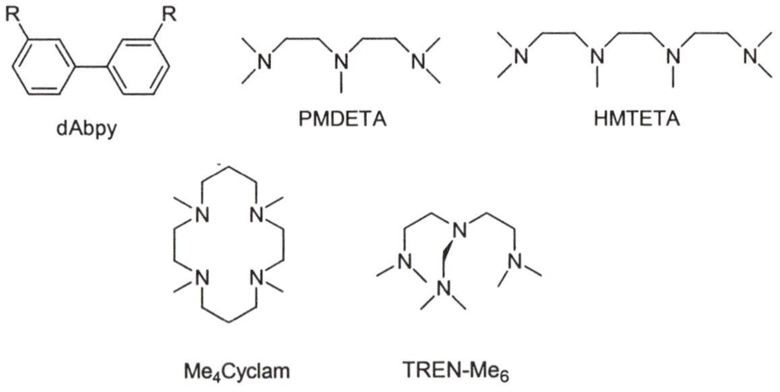

Figure 1. Ligands used in copper mediated ATRP.

Although molecular weights increased with conversion and polydispersities were low, there was some increase of M_w/M_n with conversion and better control was observed at lower temperatures. These results suggest that there may be destruction of the alkyl halide end group by quaternization with the amine in the repeat units/monomer. This quaternization may be avoided by the use of a Cu(I)Cl catalyst. Employing chloride catalyst results in a "halogen switch" where the bromine end group from the initiator is replaced by chlorine. This has two benefits: the rate of initiation compared to propagation is faster resulting in polymers with lower polydispersities and quaternization with the alkyl halide end group should be slower with chlorine (*vide infra*).

Acrylates

2-Hydroxyethyl Acrylate

The acrylate analog of HEMA, 2-hydroxyethyl acrylate (HEA) was simpler to polymerize than HEMA, as it could be directly polymerized in bulk or in aqueous solution.*(8)* One problem that was observed, however, was that the monomer required relatively careful purification, so as to remove any residual acrylic acid or diacrylate from the monomer. Polymerization was relatively straight forward using methyl 2-bromopropionate as the initiator and Cu(I)Br/2bpy as the catalyst; the polar nature of the monomer rendered the catalyst completely soluble. The polymerization was well controlled with obtained molecular weights up to M_n ~80,000 and narrow molecular weight distributions (M_w/M_n < 1.3). Additionally, it has been possible to prepare amphiphilic block copolymers containing HEA, either by direct polymerization of HEA or by protecting the –OH group with trimethylsilyl chloride; the TMS protected monomer allows for easier handling and characterization of the polymers in organic solvents.

Glycidyl Acrylate

Other functional groups can be readily incorporated in acrylic monomers, with the epoxy functional glycidyl acrylate as an example.*(9)* This monomer was polymerized under conditions similar to those used for other acrylate monomers; this monomer was polymerized in bulk at 90 °C, obtaining molecular weights up to M_n ~ 50,000 and M_w/M_n < 1.25.

Acrylic Acid (Protected)

To date, the direct polymerization of acrylic acid has proven difficult, although the polymerization of the sodium salt of methacrylic acid has been reported.*(10)* The reasons for this failure are not definitively known, but may be one, or a combination of, the following: protonation of the (amine) ligands, reaction of the acid with the Cu(II) deactivator generating a complex incapable of deactivating the growing radical, and formation of a lactone through nucleophilic displacement of the halogen by a penultimate carboxylic acid group.

To avoid these difficulties, and to allow for easier characterization, we have used the *t*-butyl protected form of the acrylic acid, i.e., *t*-butyl acrylate (tBA).*(11,12)* The polymerization of tBA is straight forward and can be used to prepare a wide variety of block and statistical copolymers. Deprotection of the *t*-butyl group is simple; treatment with acid eliminates isobutene and generates the acid. The formation of the

350

acid was confirmed by ^{1}H NMR (disappearance of the resonances assigned to the *t*-butyl group) and IR spectroscopy (the presence of the characteristic absorption of the carboxylic acid from 2400 –3900 cm^{-1}), Figure 2. In addition, other protecting groups, such as benzyl, trimethylsilyl and tetrahydropyranyl, have been used for the protection of methacrylic acid.*(13)*

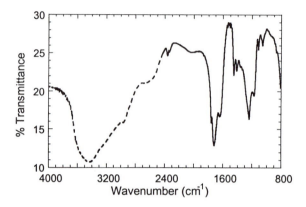

Figure 2. IR spectrum of poly(acrylic acid) formed by deprotection of tBA.

Acrylamide

Like the polymerization of acrylic acids, the polymerization of acrylamides was not overly successful; polymerization using bpy or PMDETA (PMDETA = pentamethyldiethylenetriamine) did not yield high polymer.*(14,15)* The proposed difficulties are low equilibrium constant for ATRP, competing complexation of the catalyst by the polymer/monomer with the ligand rendering the catalyst inactive and formation of a lactam by displacement of the halogen end group.

When the cyclic ligand, 1,4,8,11-tetramethyl-1,4,8,11-tetraazacyclotetradecane (Me$_4$Cyclam) was used, high conversions of the acrylamide monomer was obtained, but the polymerization was uncontrolled.*(15)* However, this still allowed for the preparation of block copolymers by the use of macroinitiators prepared by ATRP. The catalyst would abstract the halogen from the macroinitiator forming a radical which would add acrylamide monomer by propagation. As the failure of the Me$_4$Cyclam ligand was the result of slow deactivation of the growing radical by the Cu(II) halide/Me$_4$Cyclam complex, long segments of the polymer would grow before being deactivated. The result is a polymer with a well-defined segment (from the macroinitiator) and an ill-defined polyacrylamide segment.

Subsequently, the catalyst Cu(I)/TREN-Me$_6$ in conjunction with methyl 2-chloropropionate was used for the polymerization of dimethyl acrylamide (DMAm) and was found to give polymers of controlled molecular weight and narrow polydispersity.*(16)* However, the polymerization reached limited conversions, although the final conversion increased with increasing catalyst:initiator ratio. In all cases, the molecular weights were in good agreement with theoretically predicted molecular weights, Figure 3.

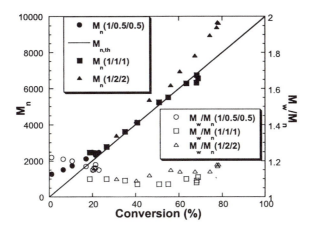

Figure 3. Molecular weight behavior as a function of conversion for the polymerization of dimethyl acrylamide by ATRP.

Other Functional Monomers

Acrylonitrile

The polymerization of acrylonitrile (AN) poses difficulties due to the fact that the polymer is not soluble in its own monomer, but only in very polar solvents. The use of DMF as a solvent was not feasible as it apparently coordinates with the copper catalyst and results in much slower polymerizations. To avoid coordination, the polymerization was conducted in ethylene carbonate. The resulting polymer had molecular weights which corresponded well with theory, and had very low polydispersities.*(17)* It was later discovered that single electron transfer from the Cu(I) catalyst to the acrylonitrile radical was occurring, slowly, and irreversibly, deactivating the growing chains and generating additional deactivator, Cu(II).*(18)* This suggests that not all of the polyacrylonitrile chains may have 2-halopropionitrile end groups. Its use as a macroinitiator for block copolymers may thus be inefficient, but other polymers prepared by ATRP can be employed as initiators for acrylonitrile, thus yielding block copolymers.

4-Vinylpyridine

The polymerization of 4-vinylpyridine (4VP)*(19)* faced obstacles similar to those incurred with acrylamides: coordination of the catalyst, reaction with the halogen end group. The polymerization of 4VP was found to be very slow when using either bpy or PMDETA as the ligands for the Cu(I) catalyst. However, the polymerization

352

proceeded at a reasonable rate when TREN-Me$_6$ was used. What was observed however, was a significant influence on the polymerization by the type of halogen used, bromine vs. chlorine.

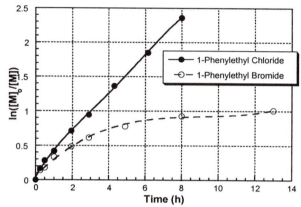

Figure 4. Comparison of the kinetics for the polymerization of 4VP using 1-phenylethyl halide initiators. 50% 4VP in 2-propanol, [4VP]$_o$:[1-PEX]$_o$:[CuX/TREN-Me6] = 95:1:1, 40 °C.

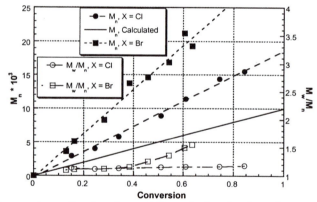

Figure 5. Comparison of molecular weight evolution for the polymerization of 4VP using 1-phenylethyl halides.

In Figure 4, one can see that the use of 1-phenylethyl chloride as the initiator resulted in the polymerization proceeding up to nearly 90% conversion, while for the bromide analog, the reaction stopped at about 70%. This was ascribed to the destruction of the halide end group either by the formation of the salt or through elimination of HBr. The effects of this side reaction are observable in the plot of molecular weight versus conversion, Figure 5. For 1-PEBr, the molecular weights increase linearly, but at a faster rate than for 1-PECl indicating that some of the starting initiator was destroyed, and that the polydispersities increased for the bromine system, but remained low for chlorine.

Monomers Containing Biologically Active Groups
ATRP has also been used to prepare well-defined polymers whose monomers contain functional groups that have biological activity. These include the polymerization of a sugar containing methacrylate to yield a glycopolymer.*(20)* Additionally, monomers containing either uridine or adenosine groups have been successfully polymerized.*(21)*

Functional Initiators

One of the simplest ways to introduce functionality at the end of a polymer chain prepared by ATRP is to use an initiator that contains a functional group. A variety of functional groups have been used including hydroxy, cyano, epoxy, allyl, vinyl acetate, lactone, and amide., Table I.*(5,22,23)*

Table I. Functional Initiators for ATRP

Initiator	Initiator Name	Monomer[a]	Conv. (%)	$M_{n, SEC}$	M_w/M_n
∕∕⌒Br	Allyl Bromide	MA	89	6 220	1.34
HO⌒O—C(=O)—CH(Br)	Hydroxyethyl 2-bromopropionate	MA	97	4 560	1.30
		Sty	48	7 500	1.10
(epoxide)—O—C(=O)—CH(Br)	Glycidyl 2-bromopropionate	MA	93	4 020	1.23
		Sty	62	6 800	1.12
(lactone)—Br	α-Bromo-butyrolactone	MA	83	4 120	1.13
		Sty	41	4 000	1.17
∕∕⌒O—C(=O)—CH2—Cl	Vinyl Chloroacetate	MA	70	3 260	1.34
		Sty	94	5 800	1.12

a) Sty = Styrene, MA = Methyl Acrylate

Initiators which contain a carboxylic acid group have been used but with reduced initiator efficiencies.*(24)* For example, the use of 2-bromopropionic acid yielded polymers that had relatively low polydispersities, but whose molecular weights were ten times higher than that predicted by theory; the molecular weight did increase linearly with conversion, however. Silyl protected derivatives of 2-bromopropionic acid were more effective, but hydrolysis by moisture present in the reaction mixture (from monomer, ligand, etc.) formed the free acid and lowered the initiator efficiency; the trimethylsilyl protected acid had an initiator efficiency of f=0.6, M_w/M_n = 1.2; *t*-butyldimethylsilyl protected acid, f=0.8, M_w/M_n = 1.1. Use of the

354

more stable t-butyl ester, *t*-butyl 2-bromopropionate, gave the best results with *f*=1 and M_w/M_n = 1.15. Deprotection of the *t*-butyl group was accomplished by treatment with HCl. Other functional initiators have been used by Haddleton, et al. which are based on phenols.*(25)* Hedrick, et al. have used protected thiols as initiators, which after appropriate treatment yielded polymer chains with a thiol end group,*(26)* and Fukuda et al. have prepared anthracene functionalized polymers.*(27,28)*

Halogen End Group Transformation

Nucleophilic Substitution

Nucleophilic substitution of the halogen end group has been performed with a variety of nucleophiles.*(29-32)* The most successful has been the replacement of the halogen with an azide.*(29,30)* The azide can be introduced either by reaction with sodium azide in a polar solvent, such as DMF, or by reaction with trimethylsilyl azide and tetrabutylammonium fluoride in THF. Substitution is fast and quantitative, resulting in all chains having an azide end group; Figure 6 shows a MALDI-TOF MS spectrum for a poly(methyl acrylate) with one azide end group.

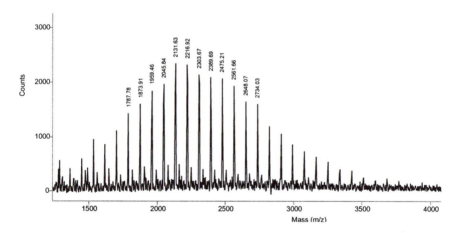

Figure 6. MALDI-TOF MS of Poly(methyl acrylate)-N_3.

The azide groups can be transformed into other, more useful functional groups, or they can be employed directly. For example, a hyperbranched polyacrylate, which contained one bromine for each repeat unit, was treated with TMS-N_3/TBAF in THF to obtain a highly branched polymer that contained only azide groups. This polymer was then heated separately in a DSC and in a DMTA. At about 180 °C, a large exotherm was observed in the DSC which corresponded to a significant increase in the shear modulus of the sample ($4*10^2$ to 10^6 Pa). This experiment was done to demonstrate that the highly branched acrylate could be used as a low molecular weight precursor and upon heating (or UV exposure), form a rigid, highly crosslinked materials.

The azide groups were also reduced to amines. For styrenes, this was readily accomplished by treatment with $LiAlH_4$.(33) This methodology was employed for the preparation of telechelic α,ω-diaminopolystyrenes (M_n = 5,100; M_w/M_n = 1.2) which were reacted with 1eq. of terephthaloyl chloride to make the polyamide (M_n = 23,000; M_w/M_n = 2.5). For acrylates, the use of hydrides is precluded by the presence of the ester groups, therefore, more mild reduction conditions were used. Reduction was accomplished by reacting the azide with triphenylphosphine to form the phosphoranimine, followed by reduction with water to yield the amine, Scheme 1. Both the phosphoranimine and the amine were isolated and analyzed.

Scheme 1

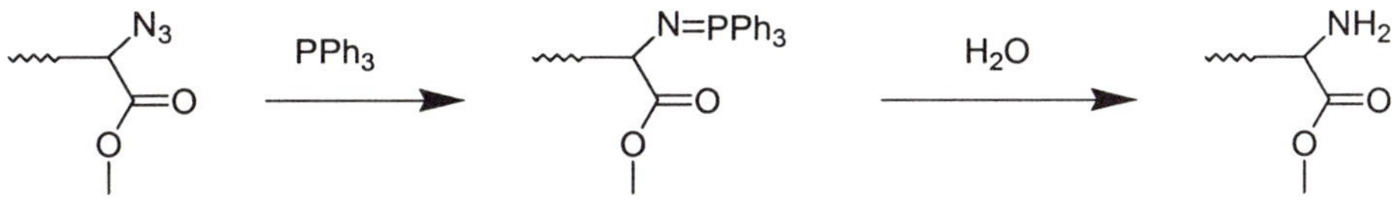

Other groups have reported a variety of techniques to transform the halogen to different, more useful, functional groups.(34,35)

Radical Reactions

Dehalogenation
As ATRP is based on a radical mechanism, it is possible to take advantage of many known radical reactions to functionalize the active end group of the polymer chain. For example, the active radical chains can react with transfer agents, abstracting a hydrogen and thus rendering the chain inactive by replacing the halogen with a hydrogen.(36) This was demonstrated by the addition of tributyltin hydride at the end of a polymerization of methyl acrylate. Figures 7 and 8 show the results of electrospray ionization mass spectroscopy (ESI-MS) of poly(methyl acrylate) before and after removal of the halogen. In Figure 7, the series of peaks corresponds to the expected masses as predicted by MW = (167 + 23 + n*86), where 167 is the mass of the initiator (methyl 2-bromopropionate), 23 is from sodium (dopant), and 86 is the mass of the MA repeat unit; for n = 15, MW = 1480. In Figure 8, the series corresponds to MW = (88 + 23 + n*86), the mass of the polymer with bromine replaced by a hydrogen; n =15, MW = 1401.

The above dehalogenation was accomplished in a single pot reaction by addition of the tin hydride transfer agent. The same material was also obtained by purification of the polymer and dehalogenation as a second step. The dehalogenation was done in a similar manner as above, i.e., using Cu(I)/PMDETA to generate the radicals by reaction with the alkyl halide, Scheme 2. It was also performed in the absence of

Cu(I), but by using AIBN as the radical source. The use of AIBN also provided quantitative removal of the halogen. It is proposed that upon formation of the radical (either by reaction of the catalyst/halogen end group or decomposition of AIBN) the radical reacts with the tin hydride to form the dead polymer chain and the tin radical. This tin radical can then abstract a halogen to reform the alkyl radical and the cycle repeats itself. Thus, only a catalytic amount of the AIBN or Cu(I) catalyst is required for complete removal of the halogen end groups.

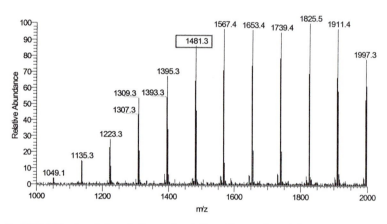

Figure 7. ESI-MS of poly(methyl acrylate)-Br. (Reproduced from Ref. 36. Copyright 1999 Wiley-VCH, STM)

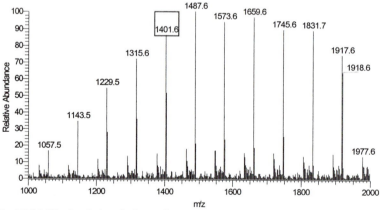

Figure 8. ESI-MS of poly(methyl acrylate)-H. (Reproduced from Ref. 36. Copyright 1999 Wiley-VCH, STM)

Allylation

In the same manner as the dehalogenation reaction, the halogen can also be replaced by an allyl group.*(37)* Addition of allyl stannane at the end of a polymerization will introduce an allyl group by the radical cycle similar to that described for the dehalogenation.

Scheme 2

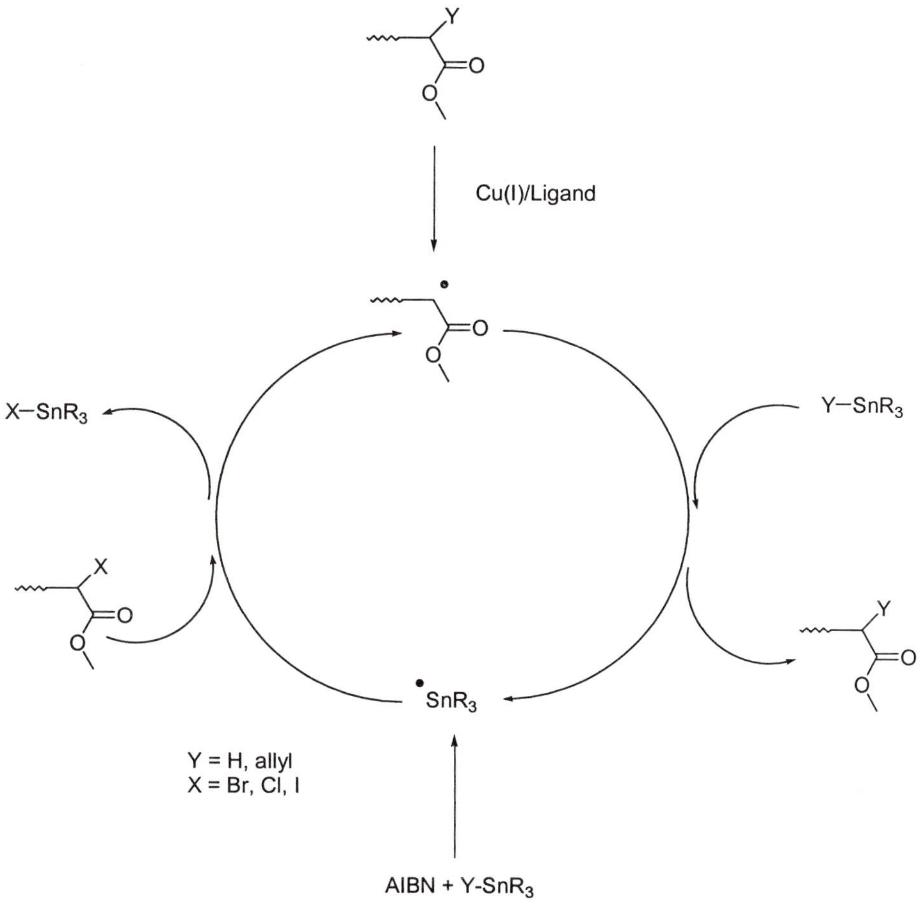

$$Y = H, \text{allyl}$$
$$X = Br, Cl, I$$

AIBN + Y-SnR$_3$

In a similar manner, but one which is not catalytic, Sawamoto et al. have shown that silyl enol ethers can react with polymeric radicals to form ketone end groups on poly(methyl methacrylate).(38) The silyl enol ethers can contain a variety of functional groups on the R' group, Scheme 3.

Scheme 3

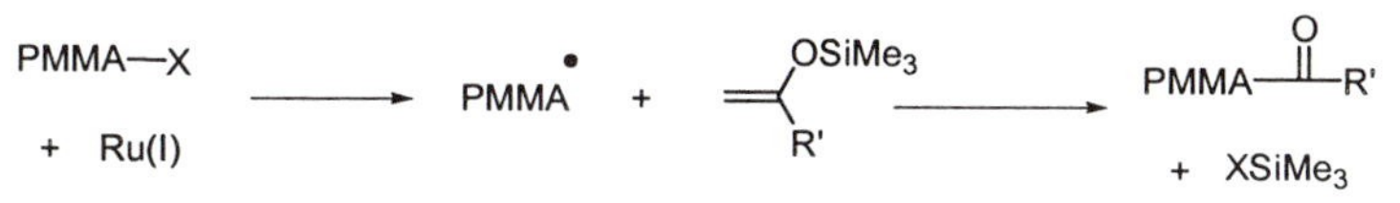

Addition of Less Reactive Monomers

By using functionalized monomers that do not homopolymerize by ATRP due to very slow activation of the halogen by reaction with the transition metal catalyst, it is possible to introduce a functional group at the chain end. This has been demonstrated using 1,2-epoxyhexene or allyl alcohol.*(37)* As the radicals are generated by the ATRP process, they add to the functionalized monomer (allyl alcohol is shown in Scheme 4). Since these monomers have low rate constants of propagation (k_p), they are rapidly deactivated after only addition of one unit of the monomer. The monomer is chosen so that the resulting C-Br (or Cl) bond is not susceptible to homolysis by the catalyst that is being used for the ATRP reaction. The result is that the chain now has a functional group incorporated at the chain end as well as an "inactive" halogen end group.

Scheme 4

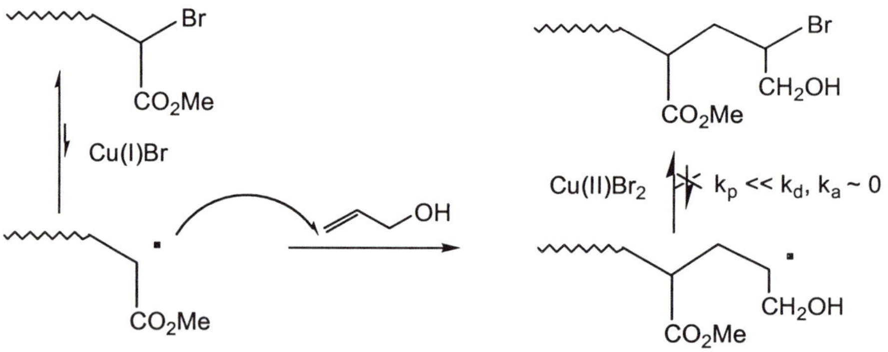

In a similar manner to the addition of allyl alcohol and 1,2-epoxyhexene to the polymer chain end, Kallitsis et al. have reported the end functionalization of polystyrene with maleic anhydride and this functionalized polymer's subsequent use in blends with Nylon 6.*(39)*

Conclusion

ATRP is a powerful method for preparing a wide variety of well-defined polymers. Among these is the synthesis of polymers of precise functionality, where the functional groups are incorporated as the monomer units, in the initiator, or by transformation of the halogen end groups.

Acknowledgements

This research has been partially supported by the Industrial Members of the ATRP Consortium at Carnegie Mellon University.

References

(1) Kato, M.; Kamigaito, M.; Sawamoto, M.; Higashimura, T. *Macromolecules* **1995**, *28*, 1721.
(2) Wang, J.-S.; Matyjaszewski, K. *J. Amer. Chem. Soc.* **1995**, *117*, 5614.
(3) Matyjaszewski, K.; Wang, J.-S. *Macromolecules* **1995**, *28*, 7901.
(4) Matyjaszewski, K.; Patten, T.; Xia, J.; Abernathy, T. *Science* **1996**, *272*, 866.
(5) Gaynor, S. G.; Matyjaszewski, K. *ACS Symposium Series* **1998**, *685*, 396.
(6) Beers, K. L.; Boo, S.; Gaynor, S. G.; Matyjaszewski, K. *Macromolecules* **1999**, *32*, 5772.
(7) Zhang, X.; Xia, J.; Matyjaszewski, K. *Macromolecules* **1998**, *31*, 5167.
(8) Coca, S.; Jasieczek, C.; Beers, K. L.; Matyjaszewski, K. *J. Polym. Sci., Polym. Chem. Ed.* **1998**, *36*, 1417.
(9) Coca, S.; Jasieczek, C.; Matyjaszewski, K. *Macromol. Chem. Phys.* **1997**, *198*, 4011.
(10) Ashford, E. J.; Naldi, V.; O'Dell, R.; Billingham, N. C.; Armes, S. P. *Chem. Commun.* **1999**, 1285.
(11) Coca, S.; Davis, K.; Miller, P.; Matyjaszewski, K. *Polym. Prepr. (Am. Chem. Soc., Div. Polym. Chem.)* **1997**, *38(1)*, 689.
(12) Davis, K.; Matyjaszewski, K. *Macromolecules* **2000**, *in press*,
(13) Zhang, X.; Xia, J.; Matyjaszewski, K. *Polym. Prepr. (Am. Chem. Soc., Div. Polym. Chem.)* **1999**, *40(2)*, 440.
(14) Li, D.; Brittain, W. J. *Macromolecules* **1998**, *31*, 3852.
(15) Teodorescu, M.; Matyjaszewski, K. *Macromolecules* **1999**, *32*, 4826.
(16) Teodorescu, M.; Matyjaszewski, K. *Macromol. Rapid Commun.* **2000**, *21*, 190.
(17) Matyjaszewski, K.; M., J. S.; Paik, H.-J.; Gaynor, S. G. *Macromolecules* **1997**, *30*, 6398.
(18) Matyjaszewski, K.; Jo, S. M.; Paik, H.-j.; Shipp, D. A. *Macromolecules* **1999**, *32*, 6431.
(19) Xia, J.; Zhang, X.; Matyjaszewski, K. *Macromolecules* **1999**, *32*, 3531.
(20) Ohno, K.; Tsujii, Y.; Fukuda, T. *J. Polym. Sci, Polym. Chem.* **1998**, *36*, 2473.
(21) Marsh, A.; Khan, A.; Haddleton, D. M.; Hannon, M. J. *Macromolecules* **1999**, *32*, 8725.
(22) Matyjaszewski, K.; Coessens, V.; Nakagawa, Y.; Xia, J.; Qiu, J.; Gaynor, S. G.; Coca, S.; Jasieczek, C. *ACS Symposium Series* **1998**, *704*, 16.
(23) Haddleton, D. M.; Waterson, C.; Derrick, P. J.; Jasieczek, C. B.; Shooter, A. J. *Chem. Commun.* **1997**, 683.
(24) Zhang, X.; Matyjaszewski, K. *Macromolecules* **1999**, *32*, 7349.
(25) Haddleton, D. M.; Derrick, T. J.; Eason, M. D.; Waterson, C. *Polym. Prepr. (Am. Chem. Soc., Div. Polym. Chem.)* **1999**, *40(2)*, 327.
(26) Carrot, G.; Hillborn, J.; Hedrick, J. L.; Trollsas, M. *Macromolecules* **1999**, *32*, 5171.
(27) Wang, J. S.; Greszta, D.; Matyjaszewski, K. *Amer. Chem. Soc., PMSE* **1995**, *73*, 416.
(28) Ohno, K.; Fujimoto, K.; Tsujii, Y.; Fukuda, T. *Polymer* **1999**, *40*, 759.
(29) Coessens, V.; Nakagawa, Y.; Matyjaszewski, K. *Polym. Bull.* **1998**, *40*, 135.
(30) Coessens, V.; Matyjaszewski, K. *J. Macromol. Sci.* **1999**, *A36*, 667.

(31) Coessens, V.; Matyjaszewski, K. *J. Macromol. Sci.* **1999**, *A36*, 653.

(32) Coessens, V.; Matyjaszewski, K. *Macromol. Rapid Commun.* **1999**, *20*, 127.

(33) Nakagawa, Y.; Gaynor, S. G.; Matyjaszewski, K. *Polym. Prepr. (Am. Chem. Soc., Div. Polym. Chem.)* **1996**, *37(1)*, 577.

(34) Weimer, M. W.; Frechet, J. M. J.; Gitsov, I. *J. Polym. Sci., Polym. Chem.* **1998**, *36*, 955.

(35) Malz, H.; Komber, H.; Voigt, D.; Hopfe, I.; Pionteck, J. *Macromol. Chem. Phys.* **1999**, *200*, 642.

(36) Coessens, V.; Matyjaszewski, K. *Macromol. Rapid Commun.* **1999**, *20*, 66.

(37) Coessens, V.; Matyjaszewski, K.; Pyun, J.; Miller, P.; Gaynor, S. G. *Macromol. Rapid Commun.* **2000**, *21*, 103.

(38) Ando, T.; Kamigaito, M.; Sawamoto, M. *Macromolecules* **1998**, *31*, 6708.

(39) Koulouri, E. G.; Kallitsis, J. K.; Hadziioannou, G. *Macromolecules* **1999**, *32*, 6242.

Chapter 25

Copolymerization of *n*-Butyl Acrylate with Methyl Methacrylate and PMMA Macromonomers by Conventional and Atom Transfer Radical Copolymerization

Sebastian G. Roos[1], Axel H. E. Müller[1,2,4], and Krzysztof Matyjaszewski[3]

[1]Institut für Physikalische Chemie, Universität Mainz, D–55099 Mainz, Germany
[2]Makromolekulare Chemie II, Universität Bayreuth,
D–95440 Bayreuth, Germany
[3]Department of Chemistry, Carnegie Mellon University,
4400 Fifth Avenue, Pittsburgh, PA 15213

ABSTRACT: The reactivity ratios of *n*-butyl acrylate (nBuA) with methyl methacrylate (MMA) and ω-methacryloyl-PMMA macromonomers (MM) in conventional and atom transfer radical copolymerization (ATRP) have been determined. For the copolymerization of nBuA with MMA, good agreement of the ratios is observed between conventional and controlled radical copolymerization, indicating that chemoselectivities in both processes are similar. The relative reactivity of the MM $(1/r_{nBuA})$ in conventional copolymerization is significantly lower than of MMA. It depends on the concentration of the comonomers but is not significantly influenced by the length of the MM. At high concentrations the relative reactivity decreases due to diffusion control of the MM addition. In ATRP the relative reactivity of the MM is much closer to the value of MMA. This is explained by the different time scales of monomer addition in both processes: whereas the frequency for monomer addition is in the range of milliseconds for conventional polymerizations, it is in the range of seconds or minutes in ATRP, thus diffusion control is less important here. This gives the opportunity to copolymerize at much higher concentrations than in conventional radical copolymerization. In addition, two-dimensional chromatography shows that the graft copolymers obtained by ATRP are much more homogeneous in terms of MWD and number of side-chains.

Graft copolymers offer all properties of block copolymers but are usually easier to synthesize. Moreover, the branched structure leads to decreased melt

---

[4]Corresponding author.

viscosities which is an important advantage for processing. Depending on the nature of their backbone and side chains, they can be used for a wide variety of applications, such as impact-resistant plastics, thermoplastic elastomers, adhesives, compatibilizers, and polymeric emulsifiers. The state-of-the-art technique to synthesize graft copolymers is the copolymerization of macro-monomers (MM) with the low molecular weight monomers.[1] It allows for the control of the polymer structure which is given by (i) the chain length of side-chains which is controlled by the synthesis of the macromonomer by living polymerization, (ii) by the chain length of backbone which can be controlled in a living copolymerization, and (iii) by the average spacing of the side chains which is determined by the molar ratio of the comonomers and the reactivity ratio of the low-molecular-weight monomer, $r_1 = k_{11}/k_{12}$. However, the distribution of spacings may not be very easy to control due to the incompatibility of the polymer backbone and the macromonomers.

In the past, we have used conventional radical copolymerization for the synthesis of poly(*n*-butyl acrylate)-*graft*-PMMA and other acrylic graft copolymers.[2,3] Obviously, a control of backbone chain length is not possible for this mechanism. Thus, we have used ATRP to synthesize the graft copolymers. It should be noted that conventional and controlled polymerizations lead to different copolymer structures. In conventional radical copolymerization the polymers show a chemical heterogeneity of first order. The chemical composition of different polymer molecules is different due to the short period of time needed to form a polymer and the shift of the comonomer ratio during polymerization. In a living polymerization all chains grow simultaneously with the same chemical composition but this changes during the polymerization leading to a heterogeneity of second order, i.e. a compositional shift within all of the chains.

In order to control the structure of graft copolymers the reactivity ratios of the copolymerization of n-butyl acrylate (nBuA) and PMMA-MM in the ATRP were investigated and compared to those obtained in conventional radical copolymerization. For comparison, the same was done for the copolymerization of nBuA and MMA.

Experimental Section

Reagents: The purification of reagents has been published elsewhere.[4] Meth-acryloyl-terminated PMMA macromonomers (PMMA-MM) of low polydispersity carrying a UV label were prepared by group transfer polymerization using a functionalized initiator.[5]

Copolymerizations: All copolymerizations were conducted at a weight ratio $m_{nBuA}:m_{MM} = 2$. Conventional radical copolymerizations were performed in butyl acetate at 60 °C using AIBN as initiator; initial total monomer concentrations ranged from 4-33 wt-%. ATRP was performed in ethyl acetate at 90 °C using methyl α-bromopropionate as initiator, CuBr as catalyst, and 4,4'-di(5-nonyl)-2,2'-bipyridine (dNbpy) as ligand, initial total monomer concentration was 50%. Details have been reported elsewhere.[4]

Analysis: Comonomer conversions were determined from the reaction solution by GC using a poly(methylsiloxane) capillary column. Decane was used as an internal standard. The conversion of the macromonomer was determined by

GPC analysis of the copolymer.[4] PMMA and PnBuA standards were used for GPC calibration. The graft copolymers were characterized by 2-dimensional chromatography.[6] An HPLC and a SEC apparatus are connected by a dual-loop automatic injector. The HPLC was used under critical conditions of adsorption (LACCC)[7] for PnBuA: eluent, THF:acetonitrile (53:47 by weight); flow, 0.01 ml/min, 35 °C; column set: 25 cm × 4 mm, RP 18 (YMC S 5 μm), 120 Å and 300 Å (reverse phase). The elute of the HPLC was collected in two sample loops of exactly the same size (100 μl) and was immediately injected onto the SEC. Conditions for SEC: THF as eluent at a flow rate of 2 ml/min, RT, column set: 2 x 30 cm, 5 μm PSS SDVgel, linear and 100 Å. Detectors: TSP UV3000 diode array detector and PL EMD-960 evaporative light scattering detector (ELSD) at 40 °C with a gas flow of 3.5-5 l/min. The PSS 2D SEC software (Polymer Standards Service GmbH, Mainz) was used for collecting and evaluating the raw data.

Results and Discussion

Determination of reactivity ratios of MMA and nBuA in ATRP: Using methyl 2-bromopropionate (MBP) as initiator and CuBr/dNbipy as catalyst, the monomer reactivity ratios were obtained by two different methods: the Jaacks method[8] for high monomer feed ratios ([nBuA]$_0$/[MMA]$_0$ = 8 and 0.1, respectively) and the Kelen-Tüdös[9] method and non-linear (NLLS) optimization [10] with varying feed monomer composition. No differences were found to the Kelen-Tüdös values. In Table 1 the reactivity ratios obtained are compared with an average of the known literature data of conventional radical copolymerization of the mentioned monomers determined by the Kelen-Tüdös method and nonlinear optimization. The nonlinear optimization produces more accurate values.

Table 1: Reactivity ratios of the copolymerization of the conventional and controlled radical polymerization of nBuA and MMA.[4]

system	ATRP		conventional radical polymerization	
method	Kelen-Tüdös; NLLS	Jaacks	Kelen-Tüdös[a]	NLLS
Ref.	4	4	11-13	9
r_{nBuA}	0.36 ± 0.12	0.39 ± 0.01	0.26 ± 0.14	0.30 ± 0.03
r_{MMA}	2.07 ± 0.09	2.19 ± 0.03	2.15 ± 0.37	1.79 ± 0.18

[a] mean and standard deviation of six values

It is important to realize that the reactivity ratios for the conventional polymerizations were obtained at different temperatures and solvents. However, for this pair of comonomers the ratios are not expected to significantly depend on these parameters within the limits of experimental error. The ratios show a good agreement for both polymerization techniques. Using NiBr$_2$(PPh$_3$)$_2$ as catalyst, Moineau et al.[14] recently reported comparable NLLS reactivity ratios (r_{MMA} = 1.7; r_{nBuA} = 0.34). Thus, we conclude that the different mechanisms do not affect the propagation reaction. This means that the equilibrium between dormant and active species does not affect the selectivity of the growing radicals. Similar observations have been made with other comonomer pairs.[15-18]

Conventional radical copolymerization of nBuA with PMMA-MM:
Figure 1 shows a typical molecular weight distribution of a PnBuA-g-PMMA graft copolymer and that of the PMMA-MM used.

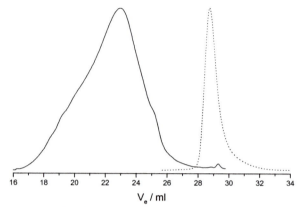

Figure 1: Apparent molecular weight distribution of PnBuA-g-PMMA graft copolymer ((———)$M_{n,app}$ = 158,000, PDI = 3.1) and PMMA macromonomer used ((• • •)M_n = 10,900, PDI = 1.19).

Figure 2 shows a typical time-conversion plot. The macromonomer is converted faster than nBuA due to the methacryloyl end group of the MM. Figure 3 shows the accompanying Jaacks plot for determination of the relative reactivity, $1/r_{nBuA}$.

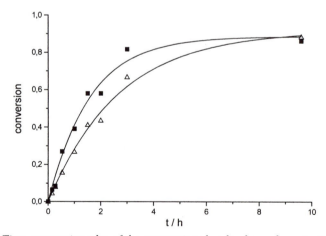

Figure 2: Time-conversion plot of the conventional radical copolymerization of nBuA (△) and PMMA-MM (■) in butyl acetate at 60 °C. $[nBuA]_0$ = 0.77 mol/l, $[nBuA]_0/[MM]_0$ = 170, $[AIBN]_0$ = $5 \cdot 10^{-3}$ mol/l; $M_n(MM)$ = 10,900.

The dependence of the relative reactivity of the PMMA-MM on the total weight concentration of monomers, $w_{nBuA} + w_{MM}$, was studied. The reciprocal value of r_{nBuA} (as determined by the Jaacks method) is equivalent to the relative reactivity of the MM ($1/r_{nBuA} = k_{nBuA-MM}/k_{nBuA-nBuA}$). In all copolymerizations with nBuA the reactivity of the MM ($1/r_{nBuA} < 1.6$) is much lower than that of MMA ($1/r_{nBuA} \approx 3.3$[10]).

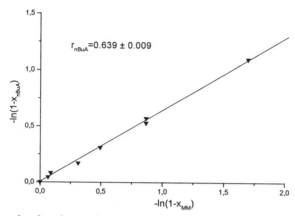

Figure 3: Jaacks plot obtained from the time-conversion plot in Figure 2.

Figure 4 shows that the apparent reactivity of the MM initially increases with increasing total concentration of monomers (and thus total concentration of polymer). After reaching a maximum the reactivity decreases at very high concentration of monomers. Let us first discuss the apparent reactivity decrease at high concentration. A similar effect was observed by Radke and Müller[5] in the copolymerization of PMMA-MM with MMA. It was attributed to the increased viscosity of the reaction solution leading to diffusion control of the mobility of the MM lowering its apparent reactivity. At very high concentrations (38 wt-% of MM in bulk nBuA) the resulting product was reported to be a polymer blend of nBuA and unreacted PMMA-MM, which was not incorporated into the backbone.[3] The initial increase of the reactivity is in excellent agreement to earlier results on the conventional copolymerization of nBuA and PMMA-MM in toluene.[19] Since the MM and the resulting graft copolymers are incompatible, the polymer coils will try to avoid contacts at low concentration. Thus, the actual concentration of macromonomers near the growing radical will be lower than its stoichiometric concentration, similar to the "bootstrap" model which Harwood[20] used in order to explain deviations from the terminal model in the copolymerization of monomers with different polarity. However, in contrast to reports on the copolymerization of the ω-styryl macromonomer of poly(2,6-dimethyl-1,4-phenylene ether) and MMA,[21] no evidence for micelle formation could be found by viscosity measurements. With increasing polymer concentration the polymer coils start to overlap and the "bootstrap effect" will become less important than the viscosity effect. Using two macromonomers of different length it was shown that $1/r$ does not strongly depend on the MM chain length (see Figure 4). Further experiments showed that the molar ratio of comonomers also does not strongly effect the MM reactivity.

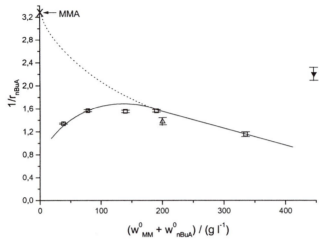

Figure 4: Dependence of PMMA macromonomer reactivity on the total initial concentration of monomers. Conventional radical copolymerization with nBuA in butyl acetate at 60 °C with $[nBuA]_0/[MM]_0 = 170$ and $M_n(MM) = 10900$ (□) and $M_n(MM) = 5600$ (△). ATRP with nBuA in ethyl acetate at 90 °C with $[nBuA]_0/[MM]_0 = 83$ and $M_n(MM) = 5600$ (▼); (×) copolymerization with MMA.

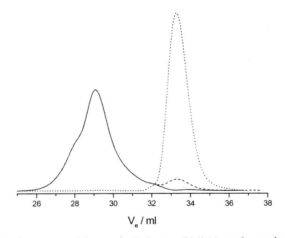

Figure 5: GPC eluograms of the crude PnBuA-g-PMMA graft copolymer (- - -), the pure graft copolymer ($M_{n,app} = 47200$, PDI = 1.66) (——), and macromonomer used ($M_n = 5600$, PDI = 1.16) (• • •).

Copolymerization by ATRP: nBuA and PMMA-MM were copolymerized by ATRP in ethyl acetate at 90 °C $[nBuA]_0:[MM]_0:[MBP]_0:[CuBr]_0:[dNbpy]_0:[Cu]_0 = 500:5.8:1:10:20:10)$. Metallic copper was added in order to reduce $CuBr_2$ formed via termination reactions.[22] The MM used had $M_n = 5600$. Figure 5 shows the GPC eluogram of the graft copolymer obtained. For the calculation of M_n and PDI of the pure graft copolymer the unreacted UV-labeled macromonomer was numerically subtracted from the crude copolymer.

Figure 6 show the conversions of the comonomers versus time. The macromonomer is first converted faster than nBuA due to its more reactive methacryloyl end group. The conversion of the MM stopped at 90 % due to the very high viscosity of the reaction solution (ca. 45% solid content) which reduces the mobility of the MM.

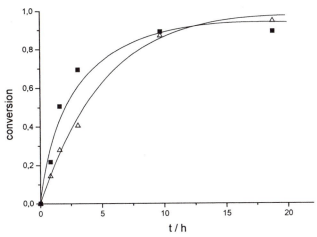

Figure 6: Time-conversion plot of the atom transfer radical copolymerization of nBuA ($\triangle$) and PMMA-MM ($\blacksquare$), synthesized with $[nBuA]_0 = 2.33$ mol/l, $[nBuA]_0/[MM]_0 = 83$ in ethyl acetate at 90 °C.

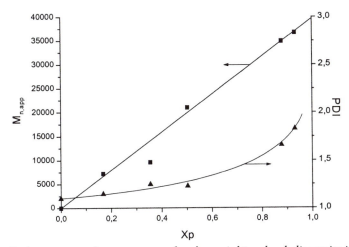

Figure 7: Apparent number-average molecular weight and polydispersity index versus average conversion for the time conversion plot in Figure 6.

Figure 7 shows the apparent molecular weights and polydispersities of the pure copolymers (after subtracting residual MM) versus the average conversion, $(w^0_{nBuA} \cdot x_{p,nBuA} + w^0_{MM} \cdot x_{p,MM})/ (w^0_{nBuA} + w^0_{MM})$. With increasing conversion the molecular weight raises linearly indicating a controlled copolymerization. However, the PDI increases slowly with conversion indicating some side reactions. Both termination and transfer to polymer[23,24] could contribute to this effect.

Figure 8 shows the Jaacks plot obtained from the experiment of Figure 6. The relative reactivity of the MM ($1/r_{nBuA} = 2.2$) is significantly higher than that found in conventional radical polymerization at comparable concentrations ($1/r_{nBuA} \approx 1.2$, Figure 4). It is much closer to that of MMA ($1/r_{nBuA} = 3.2$, Table 1).

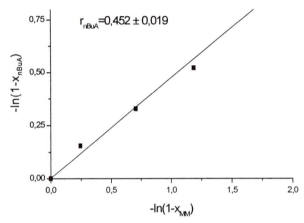

Figure 8: Jaacks plot obtained from the time conversion plot in Figure 6.

In conventional radical polymerization the lower reactivity of macromonomers compared to MMA was explained by the diffusion control of the reaction. One polymer molecule is built in less than a few seconds and the time interval between two consecutive monomer additions is in the range of milliseconds. Thus, a MM has only a short time to move to the reactive chain end. Especially at the higher monomer concentrations (up to 33%) leading to high viscosities at higher conversions this effect might become important. In ATRP it takes hours to build the same degree of polymerization due to the reversible deactivation. Thus, the time interval for monomer addition is in the range of seconds to minutes, leaving enough time for the monomer and macromonomer to move to the dormant chain end. Once the dormant chain end is converted to a radical the MM has the same chance to react than the low molecular weight monomer. Consequently, the relative reactivity is closer to the value for the low molecular weight model, MMA.

GPC and 2D-Chromatography. A comparison of Figures 1 and 5 reveals that the MWDs of the copolymers obtained in ATRP are significantly narrower (although not extremely narrow) than those obtained in conventional radical copolymerization.

Recently, we were able to show by two-dimensional chromatography of the products that the polymers obtained in ATRP are much more homogeneous in structure than those obtained by conventional polymerization.[6] The principle of this method is based on separation according to chemical composition by Liquid Adsorption chromatography (LACCC)[7] followed on-line by separation according to total size by SEC. It has been shown that LACCC of block copolymers allows for the independent determination of the size of block B under conditions where block A is not separated according to molecular weight because size exclusion and adsorption cancel each other.[25] In our case column and solvent conditions were such that separation according to the size of the PnBuA backbone does not occur. However, polymers with increasing number of PMMA arms elute at decreasing elution volume (SEC mode). Subjecting the eluate to on-line SEC separates graft copolymers with exactly one arm from residual macromonomers.

Fig. 9 shows a two-dimensional chromatogram of the graft copolymer obtained by conventional polymerization, indicating that four different species are present in the product. Integration of the peaks allows for quantitative determination of the composition and shows that we only find 63% of the desired product under these conditions.

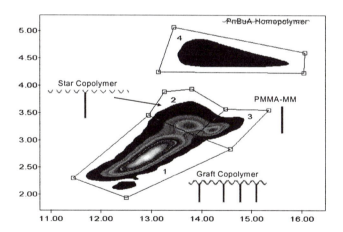

Figure 9: 2D chromatogram of PnBuA-g-PMMA obtained by conventional radical polymerization. The ordinate corresponds to LACCC elution volume, the abscissa to SEC elution volume. Graft copolymer (1; 63%), star copolymer (2; 17%), PMMA MM (3; 8%), PnBuA homopolymer (4; 9%).

In contrast, ATRP leads to a >90% yield of graft copolymer (Fig. 10). Less than 1% PMMA MM remain in the reaction product, which promises better mechanical properties (as thermoplastic elastomer) of the product. As the chromatogram is scaled linear, the peak of PMMA MM and in the same way star copolymer and PnBuA homopolymer peaks vanish in the signal noise.

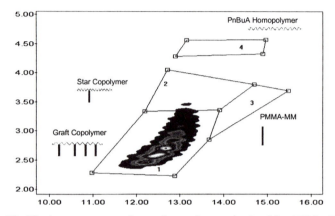

Figure 10: 2D chromatogram of a graft copolymer obtained by ATRP. Graft copolymer (1; 91%, $M_{n,app}$ = 72600, PDI = 1.8), star copolymer (2; 6%), PMMA MM (3; 1%, M_n = 5900, PDI = 2.1), PnBuA homopolymer (4; 1%).

Conclusions

Both conventional and controlled radical polymerization result in comparable reactivity ratios for low molecular weight monomers where diffusion control is absent. This indicates that the selectivity of the corresponding radicals is independent of the mechanism. The relative reactivity of macromonomers, however, strongly decreases at high concentrations in the conventional radical copolymerization of nBuA and PMMA-MM, due to the diffusion control. In controlled radical copolymerization a much higher relative reactivity of the MM is observed at these concentrations. This is explained by different time scales for monomer addition leading to the absence or a much later onset of diffusion control in ATRP. Accordingly, copolymerization of macromonomers is facilitated in ATRP. As a result, lower polydispersities of the graft copolymers are observed.

Acknowledgement: This work was supported by the *Deutsche Forschungsgemeinschaft* within the *Sonderforschungsbereich 262*.

References:

1)   Schulz, G. O.; Milkovich, R. *J. Appl. Polym. Sci.* **1982**, *27*, 4773.
2)   Radke, W.; Roos, S.; Stein, H. M.; Müller, A. H. E. *Macromol. Symp.* **1996**, *101*, 19.

3) Roos, S.; Müller, A. H. E.; Kaufmann, M.; Siol, W.; Auschra, C. in: "*Applications of Anionic Polymerization Research*" ; Quirk, R. P., Ed.; Am. Chem. Soc.: Washington, DC, 1998; Vol. 696, pp 208.

4) Roos, S. G.; Müller, A. H. E.; Matyjaszewski, K. *Macromolecules* **1999**, *32*, 8331.

5) Radke, W.; Müller, A. H. E. *Makromol. Chem., Macromol. Symp.* **1992**, *54/55*, 583.

6) Roos, S. G.; Schmitt, B.; Müller, A. H. E. *Polym. Prepr. (Am. Chem. Soc., Div. Polym. Chem.)* **1999**, *40 (2)*, 140.

7) Pasch, H.; Much, H.; Schulz, G.; Gorshkov, A. V. *LC GC international* **1992**, *5*, 38.

8) Jaacks, V. *Makromol. Chem.* **1972**, *161*, 161.

9) Kelen, T.; Tüdös, F.; Turcsányi, B.; Kennedy, J. P. *J. Polym. Sci., Polym. Chem. Ed.* **1977**, *15*, 3047.

10) Dube, M. A.; Penlidis, A. *Polymer* **1995**, *36*, 587.

11) Bevington, J. C.; Harris, D. O. *J. Polym. Sci., Part B* **1967**, *5*, 799.

12) Brosse, J.-C.; Gauthier, J.-M.; Lenain, J.-C. *Macromolecular Chem.* **1983**, *184*, 505.

13) Emelie, B.; Pichot, C.; Guillot, J. *Makromolecular Chem.* **1991**, *192*, 1629.

14) Moineau, G.; Minet, M.; Dubois, P.; Teyssié, P.; Senninger, T.; Jérôme, R. *Macromolecules* **1999**, *32*, 27.

15) Greszta, D.; Matyjaszewski, K. *Polym. Prepr. (Am. Chem. Soc., Div. Polym. Chem.)* **1996**, *37(1)*, 569-70.

16) Arehart, S. V.; Greszta, D.; Matyjaszewski, K. *Polym. Prepr. (Am. Chem. Soc., Div. Polym. Chem.)* **1997**, *38(1)*, 705.

17) Matyjaszewski, K. *Macromolecules* **1998**, *31*, 4710.

18) Haddleton, D.; Crossman, M. C.; Hunt, K. H.; Topping, C.; Waterson, C.; Suddaby, K. S. *Macromolecules* **1997**, *30*, 3992.

19) Stein, H. M., Diplomarbeit, Universität Mainz Mainz 1992.

20) Harwood, J. *Makromol. Chem., Macromol. Symp.* **1987**, *10/11*, 331.

21) Percec, V.; Wang, J. H. *Makromol. Chem., Macromol. Symp.* **1992**, *54/54*, 561.

22) Matyjaszewski, K.; Coca, S.; Gaynor, S. G.; Wei, M.; Woodworth, B. E. *Macromolecules* **1997**, *30*, 7348.

23) Ahmad, N. M.; Heatley, F.; Lovell, P. A. *Macromolecules* **1998**, *31*, 2822.

24) Roos, S. G.; Müller, A. H. E. *Macromol. Rapid Commun.* , submitted.

25) Falkenhagen, J.; Much, H.; Stauf, W.; Müller, A. H. E. *Polym. Prepr. (Am. Chem. Soc., Div. Polym. Chem.)* **1999**, *40(2)*, 984.

Chapter 26

Quantitative Derivatizations of 1-Chloro-1-phenylethyl Chain End of Polystyrene Obtained by Quasiliving Atom Transfer Radical Polymerization

Béla Iván and Tamás Fónagy

Department of Polymer Chemistry and Material Science,
Chemical Research Center, Hungarian Academy of Sciences,
Pusztaszeri u. 59–67, P.O. Box 17, H–1525 Budapest, Hungary

Low molecular weight polystyrene with 1-chloro-1-phenylethyl terminal functionality was prepared in high yields by quasiliving atom transfer radical polymerization. The resulting endgroup was converted via carbocationic intermediates to unreactive methyl chain ends with trimethylaluminum and to allyl termini with allytrimethylsilane in the presence of $TiCl_4$. ^1H NMR analysis indicated that both methylation and allylation were quantitative, and chain scission and coupling can be excluded on the basis of GPC measurements. It was also demonstrated that the allyl endgroup can be transformed to α-hydroxyl or epoxy chain ends, i. e. new endfunctional polystyrenes and macromonomers were obtained.

Introduction

There has been significant interest in endfunctional polymers in both academia and industry for a long time. Polymers with terminal functional groups have already been used in several applications, such as macroinitiators, macromonomers, curing agents and chain extenders etc. (see e. g. Refs. 1 and 2 and references therein). In the field of polymer science, the major interest in endfunctional polymers is focused on the preparation of structurally well-defined macromolecular architectures.

Polymers with exact chain end functionality can be prepared by quasiliving polymerizations. However, functional groups obtained by direct endquenching (functionalization) processes are often not suitable for desired subsequent processes and transformations. Therefore, quantitative post-polymerization reactions are of great importance for obtaining functional polymers with designed microstructure.

Chlorine-terminated polystyrene (PSt-Cl) prepared either by quasiliving carbo-cationic [3-7] or atom transfer free radical [8-10] polymerizations belongs to this class of polymers since direct utilization of this endfunctional polymer is mainly limited to the synthesis of certain block copolymers [5-10]. Recently, successful transformation of PSt-Cl and bromine-terminated polystyrene to PSt with reactive terminal azide and amino groups were reported [11-13].

Since the 1-chloro-1-phenylethyl terminus can be easily transformed to the corresponding polystyryl carbocation, quantitative carbocationic derivatizations are expected to yield polystyrenes with exact chain end functionalities. This study deals with quantitative methylation and allylation of polystyrene obtained by quasiliving atom transfer radical polymerization. Subsequent conversion of the allyl end group to primary hydroxyl and epoxy functional groups is also reported.

Experimental Section

Materials

Styrene (Aldrich, 99%) was distilled under reduced pressure at 30 °C. Solvents (hexane, dichloromethane and tetrahydrofuran, all from Chemolab, Hungary) were purified by conventional methods. 3-Chloroperbenzoic acid (Aldrich, 57-86%) was purified by washing with phosphate buffer (pH=7.37), then with distilled water and was dried overnight under vacuum at room temperature. 1-Chloro-1-phenylethane (Acros, 99%), 2,2'-bipyridine (Aldrich, 99%), copper(I) chloride (Aldrich, 99%), trimethylaluminum (Aldrich, 2M solution in heptane), titanium(IV) chloride (Aldrich, 99.9%), allyltrimethylsilane (ATMS) (Fluka, 97%), 9-borabicyclo[3.3.1]nonane (Aldrich, 0.5 M solution in tetrahydrofuran), hydrogen peroxide (Reanal, Hungary, 30%), potassium hydroxide (Reanal, 86%) and methanol (Chemolab, 99.8%) were used as received.

Procedures

Synthesis details are described in the Result and Discussion section.

Characterization

Molecular weight distributions were determined by gel permeation chro-matography (GPC) with a Waters/Millipore liquid chromatograph equipped with a Waters 510 pump, Ultrastyragel columns of pore sizes 1×10^5, 1×10^4, 1×10^3 and 500 Å, and a Viscotek parallel differential refractometer/viscometer detector. Tetrahydrofuran was used as the mobile phase with a flow rate of 1.5 ml/min.

Calibration was made with narrow MWD polystyrene standards. ^{1}H NMR analyses were performed by a Varian 400 MHz equipment.

Results and Discussion

Synthesis of chlorine terminated polystyrene (PSt-Cl) oligomer

As it has been widely demonstrated quasiliving atom transfer radical polymerizations (ATRP) yield terminal carbon-halogen bonds:

$$P\text{-}X + Mt^nL_m \rightleftharpoons P\cdot + Mt^{n+1}L_mX$$
$$(+M)$$

where P-X is the nonpropagating (inactive) polymer (X = Cl, Br), Mt^nL_m is a transition metal complex, P· is the propagating (living) radical, M is the monomer and $Mt^{n+1}L_mX$ is a complex formed by abstracting X from P-X [14-16]. The quasiliving term means that only a fraction of the chains are active (living, propagating) and these are in equilibrium with inactive (nonliving, nonpropagating) chains [17,18]. The quasiliving equilibrium is shifted to the left in quasiliving ATRP and lowering the polymerization temperature exclusively yields halogen-terminated chains.

During the first step of our investigations, low molecular weight polystyrene containing chlorine terminus (PSt-Cl) was synthesized by quasiliving ATRP of styrene with the 1-chloro-1-phenylethane/CuCl/2,2'-bipyridine (1:1:3) initiating system in bulk at 130 °C for four hours. The monomer/initiator ratio was 20:1. After workup, purification and drying the resulting polymer was characterized by GPC (M_n=2400 M_w/M_n=1.39) and ^{1}H NMR.

The ^{1}H NMR spectrum in Figure 1A indicates that the only detectable chain end is the expected 1-chloro-1-phenylethyl group. The functionality of PSt-Cl was calculated from this spectrum. The integrals of methyl protons at 0.9-1.1 ppm and methine protons at 4.2-4.6 ppm were compared and the ratio was found 3:1 within experimental error. Similar to GPC and MWD curves reported in several cases in the literature for polystyrenes obtained by ATRP loss of terminal functionality due to termination by recombination cannot be detected by GPC measurements either. As shown in Figure 2A there is no bimodality or high MW shoulder in the MWD of the PSt-Cl sample prepared with nearly 100% styrene conversion, which should be observed in the case of significant recombination reactions. These results indicate that polystyrene oligomers with 1-chloro-1-phenylethyl termini can be synthesized in high yield and close to the expected functionality by quasiliving ATRP of styrene. These endfunctional polymers are suitable starting materials for further quantitative derivatizations in order to obtain a variety of new functional polystyrenes.

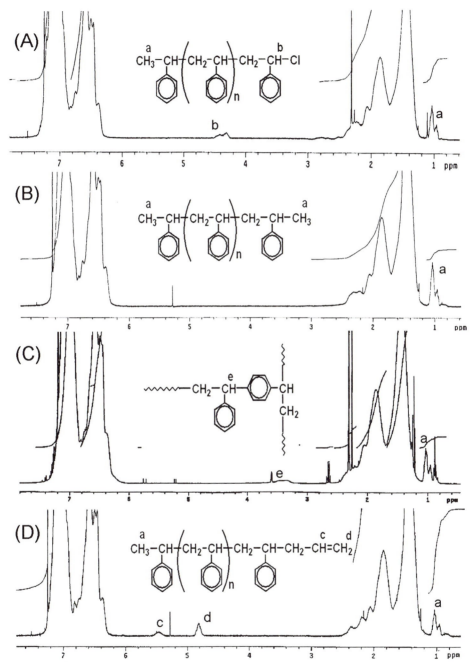

Figure 1. ¹H NMR spectra of chlorine- (A), methyl- (B, for synthesis details see Table 1 line 2) and allyl-terminated (D, Table 1 line 5) and branched (C, Table 1 line 4) polystyrenes

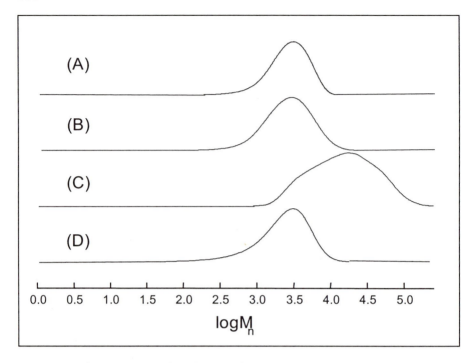

*Figure 2. Molecular weight distribution of chlorine- (A), methyl- (B) and allyl (D)
terminated and branched (C) polystyrenes*

Quantitative methylation of 1-chloro-1-phenylethyl terminated polystyrene with trimethylaluminum

Chain end modifications are generally performed in order to enhance or modify the reactivity of the endgroup but sometimes the relatively high reactivity of functional groups can result in problems during applications. For example, hydrogen chloride may eliminate from the chlorine-terminated polystyrene by heat or light. In this study we aimed at transferring the chlorine termini quantitatively to an unreactive group in order to solve these problems. Trimethylaluminum (Me$_3$Al) was reported as methylating agent for *tert*-chlorine containing compounds [19,20]. It occurred to us that Me$_3$Al might also lead to quantitative methylation of PSt-Cl possessing terminal benzyl chloride.

In the course of our experiments PSt-Cl was dissolved in dichloromethane (0.1 g/ml), Me$_3$Al was added in 2 M heptane solution, and the reaction mixture was allowed to react for 30 mins under dry argon atmosphere. The reaction temperature and the excess of Me$_3$Al in different experiments are listed in Table 1. After the reaction the polymers were precipitated into methanol filtered and dried under vacuum at room temperature.

Table 1. Excess of reagents, reaction temperature (T) and reaction time (t) during methylation and allylation of PSt-Cl, number average molecular weight (M_n) and polydispersity (M_w/M_n) of the resulting polymers

	Reagent	Excess to PSt-Cl	Solvent	T [°C]	t [min]	$M_n/10^3$	M_w/M_n	End group
1	$(CH_3)_3Al$	3X	CH_2Cl_2/hex (60/40)	-78	20	2.6	1.49	-Cl
2	$(CH_3)_3Al$	4X	CH_2Cl_2	-78	30	2.2	1.56	$-CH_3$
3	$(CH_3)_3Al$	4X	CH_2Cl_2	-20	30	2.3	1.55	$-CH_3$
4	$TiCl_4$ ATMS[a,b]	8X/5X	CH_2Cl_2/hex (60/40)	-78	20	8.2	2.56	(alkylation)
5	ATMS $TiCl_4$[a,b]	5X/3X	CH_2Cl_2/hex (60/40)	-78	20	2.6	1.46	$-CH_2-CH=CH_2$
6	ATMS $TiCl_4$[a]	5X/3X	CH_2Cl_2/hex (60/40)	r. t.	20	1.8	1.72	$-CH_2-CH=CH_2$
7	ATMS $TiCl_4$[a]	5X/3X	CH_2Cl_2	r. t.	20	2.2	1.41	$-CH_2-CH=CH_2$

[a]The order of reagents indicate the sequence of addition
[b]Nucleophilic additive was added

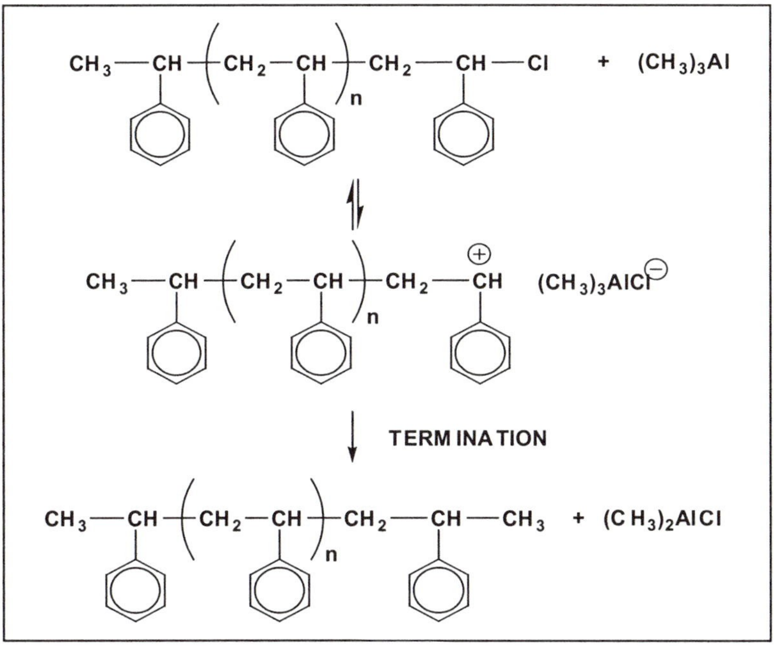

Scheme 1. Transformation of chlorine termini of polystyrene to methyl group by reacting PSt-Cl with Me₃Al

As Figure 1B shows the ^{1}H NMR signal of methine protons next to the chlorine atom at 4.2-4.6 ppm disappeared after the reaction with Me$_3$Al (in fourfold excess in dichloromethane at –78 °C), and the signal of methyl protons at 0.9-1.1 ppm increased. Using the integral of aromatic protons at 6.4-7.2 ppm as reference, the integral of the methyl protons at 0.9-1.1 ppm doubled within experimental error upon treatment with Me$_3$Al. This indicates that the chlorine termini of PSt-Cl was quantitatively converted to methyl groups. The mechanism of this reaction is shown in Scheme 1. It is worth to mention that the transformation was not successful in dichloromethane/hexane (40:60 v/v) mixture probably because of the lower polarity of this solvent mixture. As it is exhibited in Scheme 1 the intermediate of this reaction is a carbocation which could attack the nucleophilic aromatic rings of the polymer backbone leading to branched polymers. However, Figure 2B indicates that the molecular weight distribution did not change during this reaction, and this excludes the coupling of PSt-Cl macromolecules and confirms the quantitative methylation of PSt-Cl.

Quantitative allylation of 1-chloro-1-phenylethyl terminated polystyrene with allyltrimethylsilane

Preparation of allyl-ended polymers are often targeted due to their wide application possibilities as curing agents, intermediates of other functional polymers etc. Polystyrenes with allyl chain ends have already been synthesized with high yield by ATRP using allyl halogenides as initiators [21].

As it was shown in previous studies [22-24], quasiliving carbocationic polymerization (QLCCP) of isobutylene and styrene followed by endquenching with allyltrimethylsilane (ATMS) yields quantitatively allyl-ended polymers. Due to the fact that the nonpropagating (nonliving) chain end in QLCCP of olefins is also a chlorine atom, the 1-chloro-1-phenylethyl group obtained by quasiliving ATRP can presumably be modified under QLCCP conditions [25-28]. This possibility was also mentioned by Matyjaszewski and coworkers [29]. In the course of our investigations systematic experiments were carried out in order to reveal the effect of reaction conditions on the outcome of the reaction between PSt-Cl and ATMS in the presence of TiCl$_4$. The allylation is expected to occur as shown in Scheme 2. Thus PSt-Cl and a nucleophilic additive was dissolved in dichloromethane/hexane (40:60 v/v) mixture and it was cooled to –78 °C under dry nitrogen atmosphere. (For the role and effect of nucleophilic additives in QLCCPs see Ref. 23). Then TiCl$_4$ was added in the first experiment, and the reaction mixture was aged for 10 mins before charging prechilled ATMS. As shown in Figure 2C this process led to MWD broadening, and the ^{1}H NMR spectrum (Figure 1C) indicated intermolecular alkylation.

When TiCl$_4$ was added to a mixture of PSt-Cl/nucleophile/ATMS the MWD did not change as exhibited in Figure 2D. The ^{1}H NMR spectrum of the product is shown in Figure 1D. Comparing Figures 1A, 1C and 1D reveals that the 1-chloro-1-phenylethyl endgroups (4.3-4.5 ppm) are quantitatively transformed to allyl chain ends (4.8 and 5.45 ppm), and there is no sign (3.3-3.5 ppm) of intermolecular

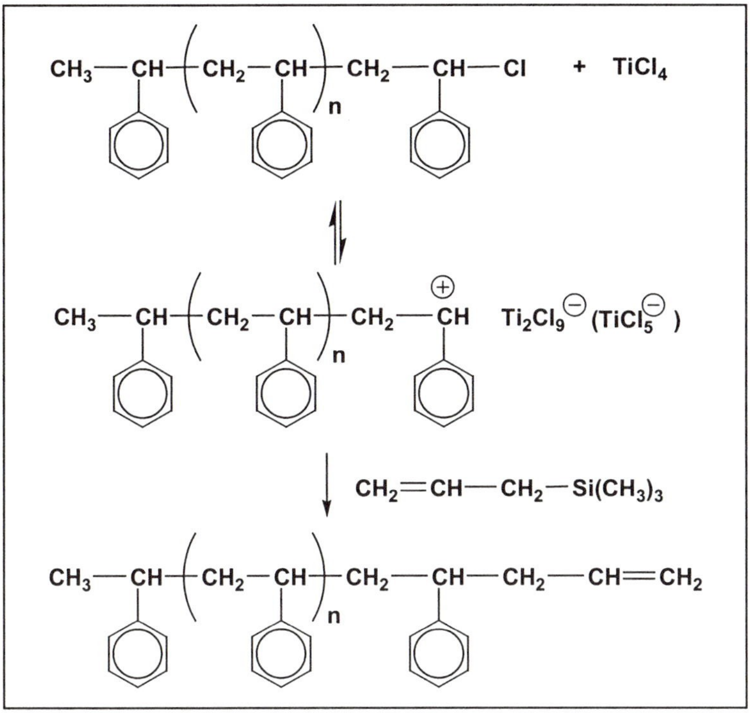

Scheme 2. Synthesis of allyl-terminated polystyrene

alkylation. The functionality of allyl-ended polystyrene (PSt-allyl) was calculated from its ^{1}H NMR spectrum (Figure 1D), and it was found that the integral ratio of methyl protons (0.9-1.1 ppm) and terminal (4.8 ppm) and internal (5.45 ppm) olefinic protons was 3:2:1 within experimental error. These experiments clearly indicate the importance of the order of addition of the reacting materials. Charging TiCl$_4$ last leads to quantitative allylation and the absence of undesired side reactions.

After establishing the role of the order of the reacting materials, the role of nucleophile additives, solvent and temperature were also investigated. It was found that the post polymerization transformation is not so sensitive to these conditions. In all cases listed in Table 1 (in the absence nucleophile, dichloromethane and room temperature) quantitative allylation without intermolecular alkylation occurred.

Further quantitative modifications of allyl-ended polystyrene

It was reported earlier that the allyl chain end of polyisobutylene prepared by endquenching of living polyisobutylene chains with ATMS can be converted to other functional groups by a variety of chemical reactions. The PSt-allyl was expected to be transformed to α-hydroxyl- or epoxy-ended polymer similar to allyl-ended polyisobutylene by hydroboration/oxidation [23] and epoxidation [23], respectively.

The PSt-allyl (2.5 g) was dissolved in 25 ml tetrahydrofuran which contained the 9-borabicyclo[3.1.1]nonane (9-BBN) in tenfold excess (0.5 M) under argon atmosphere. After the reaction for 5 hours at room temperature, twofold dilution was made with tetrahydrofuran then potassium hydroxide solution (2.3 g KOH in 10 ml methanol) and 6 ml hydrogen peroxide (30 v/v%) were added. The expected reaction is shown in Scheme 3. The appearance of α-hydroxyl chain ends was confirmed by ^{1}H NMR spectroscopy. Comparing Figures 3A and 3B reveals that the allyl endgroups (4.8 and 5.45 ppm) are quantitatively transformed to α-hydroxyl chain ends (3.4 ppm). The GPC curves of the starting PSt-allyl and the resulting hydroxyl-ended polystyrene indicate that cleavage or coupling of macromolecules did not occur under the conditions of hydroboration of PSt-allyl. It has to be mentioned that the hydroxyl endgroup can be converted to other functional groups by subsequent quantitative derivatization reactions. For example, reacting the hydroxyl-ended polymer with methacryloyl chloride methacrlylate-ended polystyrene, i. e. a useful macromomer, can be obtained which can be copolymerized with other monomers to yield graft copolymers.

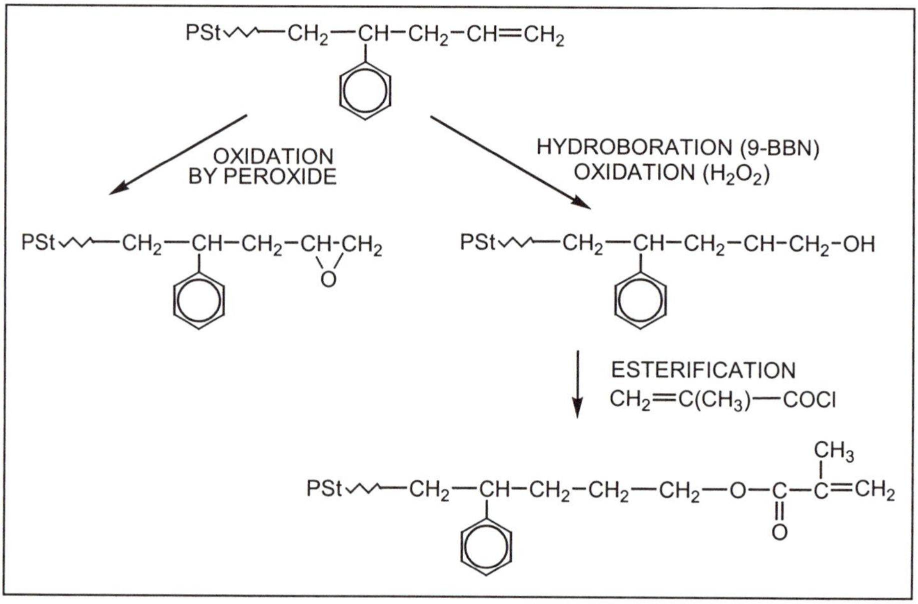

Scheme 3. Synthesis of α-hydroxyl-, epoxy- and methacrylate-terminated polystyrenes

The epoxy-ended polystyrene was prepared by reacting the PSt-allyl (0.075 g/ml) with threefold m-chloroperbenzoic acid in dichloromethane for 5 hours. As shown

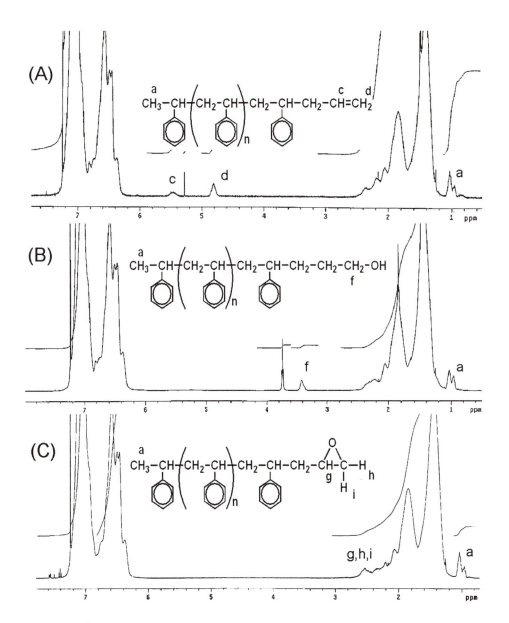

Figure 3. ^1H NMR spectra of allyl- (A), hydroxyl- (B) and epoxy-terminated (C) polystyrenes

382

by the ^{1}H NMR spectrum of the resulting polymer (Figure 3C) the conversion reached 100%, i. e. the signals of allyl protons at 4.8 and 5.45 ppm have disappeared completely, and the signals belong to the protons next to the epoxy group appeared at 2.4-2.7 ppm.

Conclusions

Polystyrene oligomers with 1-chloro-1-phenylethyl terminal functionality were prepared in high yields by quasiliving ATRP without detectable side reactions, such as recombination. The resulting endgroup was quantitatively converted to unreactive methyl chain ends with trimethylaluminum. This is an easy way to eliminate problems arising from the relatively high reactivity of the chlorine termini. The terminal 1-chloro-1-phenylethyl functionality was also transformed to allyl chain ends with allytrimethylsilane in the presence of TiCl$_4$ under quasiliving carbocationic polymerization conditions. It was found that the order of addition of the reacting compounds is the determining factor for obtaining the desired derivatization. It was demonstrated that the allyl end-group can be transformed to α-hydroxyl and epoxy endgroups by hydroboration/oxidation and oxidation with m-chloroperbenzoic acid, respectively.

Acknowledgements

Partial support by the National Research Fund (OTKA T014910 and T029711) is gratefully acknowledged. Technical assistance by Mrs. E. Tyroler is also acknowledged.

References

1. Goethals, E. J. (Ed.) *Telechelic Polymers*, CRC Press, Boca Raton, 1989
2. Rempp. P.; Franta, E.; Herz J.-E. *Adv. Polym. Sci.* **1984**, *58*, 1
3. Ishihama, Y.; Sawamoto, M.; Higashimura, T. *Polym. Bull.* **1990**, *24*, 201
4. Kennedy J. P.; Iván, B. *Designed Polymers by Carbocationic Macromolecular Engineering Theory and Practice*, Hanser, New York, 1992
5. Iván, B.; Chen, X.; Kops, J.; Batsberg, W. *Macromol. Rapid Commun.* **1998**, *19*, 15
6. Chen, X.; Iván, B.; Kops, J.; Batsberg, W. *Macromol. Rapid Commun.* **1998**, *19*, 585
7. Coca, S.; Matyjaszewski, K. *J. Polym. Sci., Part A: Polym. Sci.* **1997**, *35*, 3595
8. Gao, B.; Chen, X.; Iván, B.; Kops, J.; Batsberg, W. *Polym. Bull.* **1997**, *39*, 559
9. Coca, S.; Matyjaszewski, K. *Macromolecules* **1997**, *30*, 2808
10. Coca, S.; Paik, H-J.; Matyjaszewski, K. *Macromolecules* **1997**, *30*, 6513

11. Matyjaszewski, K.; Nakagawa, Y.; Gaynor, S. G. *Macromol. Rapid Commun.* **1997**, *18*, 1057

12. Coessens, V.; Nakagawa, Y.; Matyjaszewski, K. *Polym. Bull.* **1998**, *40*, 135

13. Matyjaszewski, K.; Coessens, V.; Nakagava, Y.; Xia, J.; Qiu, J.; Gaynor, S.; Coca, S.; Jasieczek, C. *ACS Symp. Ser.* **1998**, *704*, 16

14. Wang, J.-S.; Matyjaszewski, K. *J. Am. Chem. Soc.* **1995**, *117*, 5614

15. Wang, J.-S.; Matyjaszewski, K. *Macromolecules* **1995**, *28* , 7572

16. Wang, J.-S.; Matyjaszewski, K. *Macromolecules* **1995**, *28* , 7901

17. Iván, B. *Macromol. Symp.* **1994**, *88*, 201

18. Iván, B. *Macromol. Chem., Macromol. Symp.* **1993**, *67*, 311

19. Kennedy, J. P.; Desai, N. V.; Sivaram, S. *J. Am. Chem. Soc.* **1973**, *95*, 6386

20. Kennedy, J. P.; Melby, E. G. *J. Org. Chem.* **1975**, *40*, 1099

21. Nakagawa, Y.; Matyjaszewski, K. *Polymer J.* **1998**, *30*, 138

22. Wilczek, L.; Kennedy, J. P. *J. Polym. Sci., Polym. Chem. Ed.* **1987**, *25*, 3255

23. Iván, B.; Kennedy, J. P. *J. Polym. Chem., Part A: Polym. Chem.* **1990**, *28*, 89

24. Miyashita, K.; Kamigaito, M.; Sawamoto, M.; Higashimura, T. *J. Polym. Sci., Part A: Polym. Chem.* **1994**, *32*, 2531

25. Iván, B. *Macromol. Symp.* **1998**, *132*, 65

26. Everland, H.; Kops, J.; Nielsen, A.; Iván, B. *Polym. Bull.* **1993**, *31*, 159

27. Feldthusen, J.; Iván, B.; Müller, A. H. E.; Kops, J. *Macromol. Rapid Commun.* **1997**, *18*, 417

28. Gao, B.; Chen, X.; Iván, B.; Kops, J.; Batsberg, W. *Polym. Bull.* **1997**, *39*, 559

29. Nakagawa, Y.; Gaynor, S. G.; Matyjaszewski, K. *Polym. Prepr.* **1996**, *37(1)*, 577

Chapter 27

Synthesis of Alternating Copolymers of N-Substituted Maleimides with Styrene via Atom Transfer Radical Polymerization

Fu-Mian Li, Guang-Qiang Chen, Ming-Qiang Zhu, Peng Zhou, Fu-Sheng Du, and Zi-Chen Li

Department of Polymer Science and Engineering, College of Chemistry and Molecular Engineering, Peking University, Beijing 100871, China

It has been demonstrated that atom transfer radical polymerization (ATRP) can be used successfully to obtain well-defined copolymers between donor monomers such as styrene, and acceptor monomers such as N-(2-acetoxyethyl)maleimide(AEMI) or N-phenylmaleimide(PhMI). This polymerization system affords copolymers with designed molecular weights with narrow molecular weight distributions in the range of 1.16~1.36. The copolymers obtained possess a predominately alternating structure. The copolymerization of N-substituted maleimides with isobutyl vinyl ether (iBVE) or vinyl benzoate (VBz) via ATRP has also been attempted.

As a novel precision polymerization, atom transfer radical polymerization (ATRP) has received rapidly increasing interest recently, since it provides polymers with predetermined molecular weights and low polydispersities[1, 2]. The merit of the ATRP system is that it can be performed by an ordinary radical polymerization procedure. A general ATRP catalyst is composed of a simple alkyl halide as an initiator and copper(I)-bipyridine complex as a catalyst. The mechanism of ATRP is presumed to be a reversible homolytic cleavage of a carbon-halogen bond catalyzed by a transition metal complex. Many vinyl monomers have been polymerized by ATRP for the preparation of traditional homopolymers, random copolymers, block/graft copolymers and gradient copolymers. In addition, ATRP also shows its potential in the synthesis of a variety of new well-defined macromolecular architectures such as comb[3], star and dendritic macromolecules[4]. However, the reports on the synthesis of well-defined alternating copolymers via ATRP are scarce[5].

It is well known that an alternating copolymer can be easily obtained by the copolymerization of an electron-rich monomer and an electron-deficient monomer through the formation of charge transfer complexes (CTC)[6]. The interaction between acceptor and donor monomers relies more or less on the temperature and polarity of solvent as well as the steric and charge effects of monomeric substituted groups. There are many studies in the literature which report the CTC formation prior to the alternating copolymerization of the CTC forming monomer pair[7]. One of the extensively studied systems is maleic anhydride (MAn) and styrene (St). The polymerization can be initiated by radical initiators such as AIBN or by UV light irradiation, but the polymerization process is uncontrolled[8,9]. Therefore, whether a controllable alternating copolymerization of CTC forming monomer pair takes place via ATRP is of concern. Matyjaszewski et al. reported on the copolymerization of donor monomers such as isobutene and isobutyl vinyl ether with acceptor monomers such as butyl acrylate via ATRP, showing that the polymerization was controlled[10]. In this article, the copolymerization of styrene (St) , iBVE and VBz with N-substituted maleimides, i.e. N-(2-acetoxyethyl)maleimide (AEMI) and N-phenylmaleimide (PhMI), via ATRP have been reported aiming to evaluate whether the CTCs-forming monomer pair perform the controllable/'living' copolymerization and whether the resulted copolymers possess the predominantly alternating structure.

Experimental Section

Materials N-(2-acetoxyethyl)maleimide (AEMI), m.p. 77-78° (lit. 79 °)[11], N-phenylmaleimide (PhMI), m.p. 88.5-89.5 ° (lit. 89-89.8 °)[12] , and 1-phenylethyl bromide (1-PEBr), n_D^{20} =1.5614 (lit. 1.5600) [13] were prepared according to the literature. Styrene (St , Beijing Chemicals Co.), isobutyl vinyl ether(iBVE, Aldrich) and vinyl benzoate (VBz, a gift from Shinetsu Vinyl Acetate Corp., Japan), were distilled under reduced pressure. Maleic anhydride (MAn, Lushun Chemicals Co.) was recrystallized from benzene. CuBr (Beijing Chemicals Co.) was purified by subsequently washing with acetic acid, ethanol, and then dried in vacuo. 2,2'-Bipyridine (bipy, Beijing Chemicals Co.), methyl 2-bromopropionate (MBP, Aldrich) and other reagents were used as received.

Atom transfer radical polymerization The bulk polymerization was conducted in a sealed glass tube. A representative feed was AEMI (or PhMI): St: 1-PEBr: CuBr: bipy = 50: 50: 1: 1: 2 (in molar ratio). After the mixture was degassed three times, the tube was sealed under vacuum, then was kept in an oil bath of 80□ to conduct the polymerization. After an interval of time, CHCl$_3$ was added to cease the polymerization. The polymer was obtained by precipitation from methanol. The crude product was dissolved in CHCl$_3$ again, and the CHCl$_3$ solution was passed through a silica column to remove the catalyst. The polymer was recovered by precipitating from a large excess of methanol, and then dried under vacuum at 60°. The polymer obtained was white fine powder.

Thermal polymerization The bulk polymerization was conducted at 80 ° in a sealed glass tube. The monomer feed was AEMI(or PhMI): St=1:1(in molar ratio). The entire procedure is similar to that of ATRP.

Characterization The charge transfer complex between donor and acceptor monomers was recorded by Shimadzu UV2101PC spectrophotometer. The polymerization kinetics were measured gravimetrically. The molecular weights and molecular weight distributions were measured on a Waters 150-C gel permeation chromatography equipped with three Ultrastyragel columns (1×10^4Å, 1×10^3 Å, and 500 Å pore size) at room temperature. THF was used as an eluent and polystyrene standards were used for calibration. The copolymer composition was determined by analyzing the nitrogen content. The residue of copolymerization product was detected by GC-MS.

Results and Discussion

Formation of CTC between N-substituted maleimides and donor monomers

The CTC formation of AEMI with St was confirmed by absorption spectroscopic method. Shown in Figure 1 are UV-VIS spectra for AEMI in 1,2-dichloroethane by adding various amounts of St. As it is seen that new absorption bands are observed in the wavelength 340nm after St has been added to AEMI in 1,2-dichloroethane. iBVE and VBz are known as electron-donating monomers, however, the CTCs formation of them with N-substituted maleimides have not been apparently detected by UV spectroscopic method due to their relatively poor electron-donating capability.

Copolymerization of AEMI and PhMI with St

We tried the copolymerization of MAn and St by the general procedure of ATRP, but the polymerization did not take place. This may be ascribed to the reaction of MAn with one of the components of the ATRP catalyst such as Cu(I) or/and 2,2'-pyridine. Instead of MAn, N-substituted maleimides, e.g. AEMI and PhMI, are less prone to react with the ligand of ATRP catalytic system under the polymerization condition. The structures are illustrated in Scheme 1.

Scheme 1

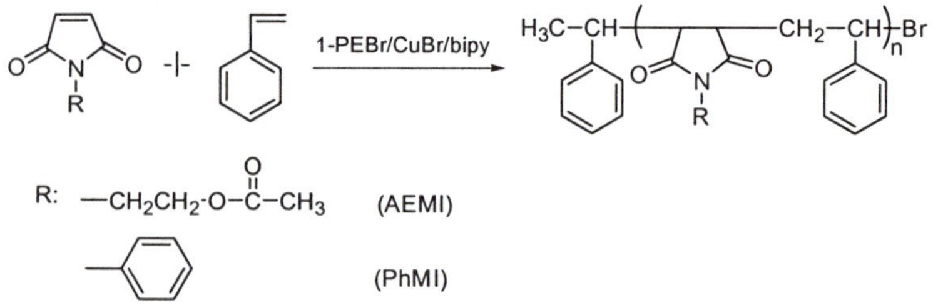

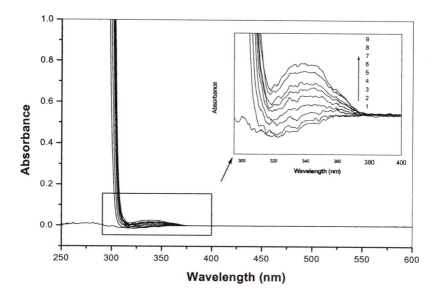

Figure 1 The UV spectra of AEMI in 1,2-dichloroethane after the addition of St [AEMI]$_0$=3.3x10^{-4}mol/L, From bottom to top, the concentrations of St were 0, 0.15, 0.30, 0.45, 0.60, 0.75, 0.90, 1.20, 1.50 mol/L

Both AEMI and PhMI are miscible in St at 80 ° in equimolar amounts. The complex of Cu(I) and bipy was also dissolved in the polymerization system for AEMI and PhMI at 80 °, giving a dark brown and red brown color, respectively, but a trace of CuBr always remains as residue in the polymerization system.

The copolymerization of AEMI with St by ATRP occurres smoothly. The plot shown in Figure 2 is the relationship between the polymerization conversion and time for the copolymerization of St with AEMI and PhMI initiated by PEBr-Cu(I)Br-bipy catalytic system. A linear relationship between ln([M]$_0$/[M]) and polymerization time was observed until higher polymerization conversions, 60-70%, which indicated that the polymerization kinetics were of a first-order nature. This means that the number of active species remains constant during the copolymerization of St with AEMI or PhMI via ATRP. From Figure 2, it can also be seen that the copolymerization rate of AEMI with St is faster than that of PhMI with St. This may be attributed to the fact that the solubility of Cu(I)-bipy complex in the AEMI-St system was higher than that in the PhMI-St system.

Figure 3 shows the plots of number average molecular weights (M$_n$) and molecular weight distribution (M$_w$/M$_n$) versus conversion for the copolymerization of St with AEMI. For the ATRP of PhMI and St, similar plots

388

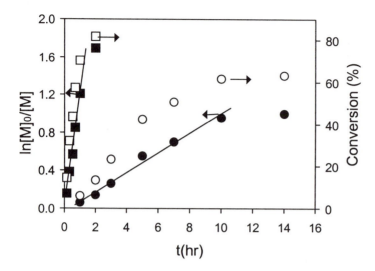

Figure 2 Kinetics of the copolymerization of AEMI and PhMI with St by ATRP
[AEMI]$_0$ or [PhMI]$_0$:[St]$_0$:[1-PEBr]:[CuBr]:[bipy]= 50:50:1:1:2 , 80°
(■ □: AEMI; ● O : PhMI)

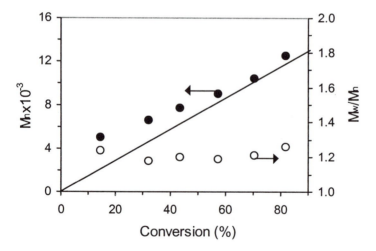

Figure 3 M$_n$ and M$_w$/M$_n$ versus polymerization conversion plot for the
copolymerization of AEMI with St by ATRP
[AEMI]$_0$:[St]$_0$:[1-PEBr]:[CuBr]:[bipy]= 50:50:1:1:2 , 80°

have been obtained. It is seen that the plot of M$_n$ determined by GPC against the
polymerization conversion is nearly a linear relationship and matches the

theoretical calculated M_n. This indicates that the anticipated average molecular weights of P(AEMI-*co*-St) or P(PhMI-*co*-St) can be controlled by terminating the polymerization at an appropriate conversion. It is seen that the M_n versus conversion plot deviates from the linearity, especially at lower polymerization conversion, which may be due to low initiator efficiency [14]. It is also seen that P(AEMI-*co*-St) possesses narrow molecular weight distributions (M_w/M_n) in the range of 1.16 to 1.36. The observation of the copolymerization kinetics together with the linear evolution of molecular weights versus conversion suggests that the contribution of chain breaking, transfer as well as termination reactions during the copolymerization can be neglected until higher conversions. This result supports the conclusion that the copolymerization process of AEMI and PhMI with St mediated by Cu(I)-bipy catalyst is controllable and of a 'living' polymerization nature.

In comparison, the thermal copolymerization of AEMI and PhMI with St at 80° in the absence of PEBr-Cu(I)-bipy catalytic system was also demonstrated. Table I summarized the related results. It can be seen that the copolymer PhMI-St-3 obtained by ordinary thermal radical initiation possesses a higher molecular weight, up to 1.0×10^7, with a broader M_w/M_n of 2.1, whereas the M_w/M_n of PhMI-St-1 and PhMI-St-2 copolymers obtained by ATRP was as narrow as 1.3. With regard to the sample of AEMI-St-3 obtained by thermal polymerization, it was not soluble in THF or chloroform at room temperature, which implies that the copolymer is crosslinked. This may be ascribed to chain transfer reactions occurring on the methyl group of

Table 1 The molecular weights, molecular weight distribution and composition of AEMI-St and PhMI-St copolymers obtained by ATRP [a]

Copolymer	Time (hr)	Yield (%)	M_n (theory)	M_n [b] (found)	M_w/M_n [b]	nitrogen content (%)	AEMI or PhMI in copolymer (mol%) [d]
AEMI-St-1	0.33	32.1	4,600	6,600	1.18	4.61	47.2
AEMI-St-2	1.0	70.2	10,000	12,500	1.25	4.73	48.5
AEMI-St-3 [c]	0.5	93.5	-	Gel	Gel	4.82	49.4
PhMI-St-1	3.0	23.1	3,200	3,500	1.26	4.75	47.0
PhMI-St-2	10	61.4	8,500	7,400	1.16	4.95	49.0
PhMI-St-3 [c]	2.5	96.0	-	1.0×10^7	2.10	5.01	49.6

[a] [monomers]$_0$: [1-PEBr] : [CuBr] : [bipy] =100 : 1 : 1 : 2, 80°.
[b] via GPC,THF as an eluent, at room temperature.
[c] Thermal polymerization, the monomer feed was 1:1(molar ratio), 80°.
[d] Calculated based on the nitrogen content by element analysis.

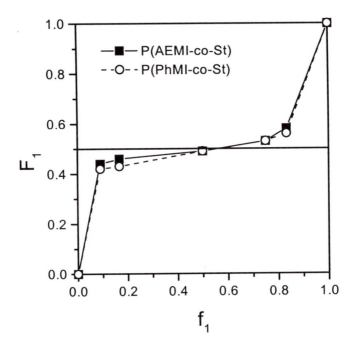

Figure 4 Relationship between the comonomer feed and the composition of AEMI-St and PhMI-St copolymers for via ATRP
[Monomers]$_0$:[1-PEBr]:[CuBr]:[bipy]= 100:1:1:2 , 80°

acetyl moiety of AEMI monomer[15]. The fact that no obvious gelation was observed for ATRP of AEMI and St may be attributed to the low conversion of the monomers and low molecular weights of the resulting copolymers. This result might also imply that the ATRP polymerization of AEMI and St could be controlled at least better than the thermal polymerization.

It can be seen from Table 1 that the compositions of the copolymers obtained via ATRP are always close to 1:1 in equimolar comonomer feed at different polymerization conversions. In order to further clarify whether the AEMI-St and PhMI-St copolymers possess an alternating structure, the copolymerization of AEMI and PhMI with St at different comonomer feeds was undertaken. Figure 4 shows the relationship between the monomer feed and the composition of the copolymers obtained at lower conversion. The results show that the copolymers possess a predominantly alternating structure over a large range of the monomer feeds. Thus, the copolymers obtained by ATRP at different molar ratios of comonomer feeds possess an alternating structure.

Copolymerization of N-substituted maleimides with iBVE and VBz
iBVE and VBz are known as electron-rich monomers, which would coplymerize with electron-deficient monomers to give predominantly alternating copolymers via CTCs formation. This was demonstrated by an ordinary radical polymerization of iBVE with AEMI initiated by AIBN which gave high conversions, up to 80%, within 30 minutes with high molecular weights up to 2×10^4 and broader molecular weight distribution of 2.96 as shown in Table 2. The structure of the obtained copolymer is predominantly alternating nature. However, the MBP-initiated copolymerization of N-substituted maleimides with iBVE and VBz catalyzed by Cu(I)-bipy complex did not work as well as the polymerizations of AEMI or PhMI with St. Table 2 summarizes the results for the copolymerizations of iBVE and VBz with AEMI via ATRP. It is seen that, for all runs, only around 10% conversions with lower molecular weights was achieved, even though the copolymerization was conducted for a long time. This might be attributed to some impurities produced during the polymerization which acts as a retarder of the radical polymerization.

In order to explore the reason of the low conversion of iBVE with AEMI in ATRP system, GC and GC-MS analyses of the copolymerization system were carried out. It was detected by GC-MS that acetaldehyde and isobutylene were formed during the copolymerization. It is well-known that acetaldehyde acts as an active chain transfer agent hindering the radical polymerization[16]. The retarding effect of acetaldehyde on ATRP was also proved by the fact that the ATRP of BA or St was inhibited by the addition of a trace of acetaldehyde.

Table 2 The molecular weights, molecular weight distributions and composition of AEMI-iBVE and AEMI-VBz copolymers obtained by ATRP [a]

Copolymer	Time (h)	Yield (%)	M_n (theory)	M_n^b (found)	M_w/M_n^b	nitrogen content (%)	AEMI or PhMI in copolymer $(mol\%)^e$
AEMI-iBVE-1	24	11.8	1670	5143	1.53	4.83	48.3
AEMI-iBVE-2[c]	24	9.1	1099	3812	1.40	4.97	50.3
AEMI-iBVE-3[d]	0.5	80	-	23097	2.96	4.81	48.1
AEMI-VBz-1	24	10.4	1721	4624	1.44	4.08	48.0
PhMI-iBVE-1	12	8.3	1141	3088	1.20	4.93	47.9

[a] [monomers] :[MBP]:[CuBr]:[bipy] =100:1:1:2; 80°C in anisole (total monomers wt% was 50%).
[b] via GPC, THF as an eluent, at room temperature.
[c] AEMI:iBVE=3:1(in molar ratio).
[d] Thermal polymerization, 60°C.
[e] Calculated based on the nitrogen content by element analysis.

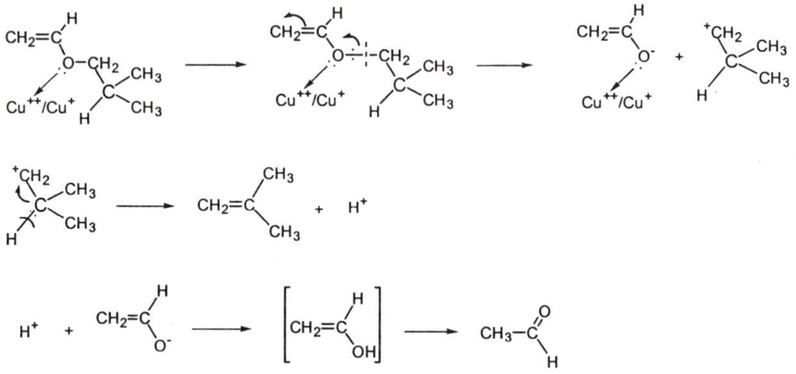

Scheme 2

The occurrence of acetaldehyde would be ascribed to the cleavage of iBVE in the presence of Cu(I)-bipy complex. This was firmly clarified by a control experiment, that is, when iBVE was heated at 80^0C in the presence of Cu(I)-bipy complex, an apparent amount of acetaldehyde was detected by GC. It is illustrated as shown in Scheme 2.

In conclusion, controlled/'living' ATRP of AEMI and PhMI with St has been successfully carried out using PEBr/Cu(I)Br/bipy catalytic system. The copolymers obtained possess a designed molecular weight with narrow molecular weight distribution. The copolymers are predominantly alternating in structure clarified by classical method. The transition metal-mediated copolymerization of other CTC-forming monomer pairs has also been attempted. It is demonstrated by GC-MS that under the present polymerization conditions, the Cu(I)-bipy complex can catalyze the cracking of iBVE resulting in acetaldehyde which acts as an active chain transfer agent to retard the ATRP of iBVE with AEMI.

Acknowledgements. Financial support for this research by the National Natural Science Foundation of China (No. 59773018) is appreciated.

References and Notes

1. Wang, J. S.; Matyjaszewski, K. *J. Am. Chem. Soc.* **1995**, *117*, 5614.
2. Matyjaszewski, K; Wang, J. S. *Macromolecules*, **1995**, *28*, 7901.
3. Husseman, M.; Malmström; Namara, M. Mc.; Mate, M.; Mecerreyes, D.; Benoit, D. G.; Hedrich, J. L.; Mansky, P.; Huang, E.; Russell, T. And Hawker, C. J. *Macromolecules*, **1999**, *32*, 1424.
4. Hedrich, J. L.; Trollsås, M.; Hawker, C. J.; Atthoff, B.; Claesson, H.; Heise, A.; Miller, R. D.; Mecerreges, D.; Jérôme,R.; Dubois, Ph. *Macromolecules*, **1998**, *31*, 8691.
5. (a) Gaynor, S. G.; Matyjaszewski, K. in *Controlled Radical Polymerization*, ACS Symp. Series 685; American Chemical Society, Washington D.C., 1998, p 258,

p396. (b) Patten, T. E.; Matyjaszewski, K. *Adv. Materials.* **1998**, 10, 901. (c) Matyjaszewski, K.; Coca, S.; Gaynor, S. G.; Jo, S. M.; Nakagawa, Y. US patent No. 5789487, 1998. (d) Matyjaszewski, K.; Gaynor, S.G.; Coca, S., PCT Int. Appl. WO 98 40415, 1998.

6. (a) Tsuchida, E.; Tomono, T., *Makromol. Chem.*, **1971**,141, 265. (b) Bamford, G. H.; Tipper, C. F., Eds., *Comprehensive Chemical Kinetics* Vol. 14A, Elsevier, New York, 1975, pp. 333-481.

7. (a) Schmidt-Naake, G.; Drache, M.; Leonhardt, K., *Macromol. Chem. Phys.*, **1998**, 199, 353-361. (b) Seiner, J. A., Litt, M., *Macromolecules*, **1977**, 7, 4. (c) Hallensleben, M. L., *Makromol. Chem.*, **1970**, 144, 267.

8. Hirai, H.; Gotoh, Y. "Alternating Radical Copolymerization" in: *Polymeric Materials Encyclopedia*, CRC Press, Inc., 1996.

9. Cowie, J. M. G. *"Alternating Copolymers"*, Plenum Press, New York, 1985.

10. Coca, S.; Matyjaszewski, K. *Polym. Prepr.* **1996**, *37(1)*, 573.

11. Yamada, M.; Takase, I.; Tatsuo, T.; Hayashi, K. et al., *Kobunshi Kagaku,* **1969** *22(2)*, 166 ; *Chem. Abstr 63*: 490a.

12. Cava, M. P.; Deana, A. A.; Muth, K.; Michell, M. J., *Org. Syn.*, *Coll. Vol. 5*, 944.

13. Ladini, D.; Rolla, F. *J. Org. Chem.*, **1980**, *45*, 3527.

14. Matyjaszewski K. *Polym. Prepr.*, **1997**, *38(2)*, 383.

15. Glass, J. E., Zutty, N. L., *J. Polym. Sci.*, **1966**, A-1, 4, 1223.

16. Clarke, J. T; Howard, R. O.; Stockmayer, W. H., *Makromol. Chem.*, **1961**, 44/46, 427.

Chapter 28

Controlled Radical Copolymerization of Styrene and Acrylonitrile

S. Brinkmann-Rengel and N. Niessner

BASF AG, ZKT/C, 67056 Ludwigshafen, Germany

The copolymerization of styrene and acrylonitrile using 2,2,6,6-tetramethyl-piperidine-N-oxyl (TEMPO) as a stable free radical is discussed. Although the homopolymerization of acrylonitrile is inhibited by TEMPO, a linear increase in the molecular weight with conversion is observed in styrene acrylonitrile copolymerizations. The temperature window for controlled styrene acrylonitrile copolymerization is small since TEMPO acts as an inhibitor at temperatures below 100 °C and control is lost above 140 °C. The polymerization rate strongly depends on temperature. Batch copolymerizations using azeotropic monomer feed compositions performed at 110 °C proceed initially with a rate of approximately 4 %/h, whereas at 120 °C rates of approximately 9%/h are observed. The composition of the copolymers prepared by controlled radical and conventional free-radical polymerization are comparable over the complete composition range. Using the terminal model similar copolymerization parameters are obtained from conventional and controlled radical copolymerization in the temperature range between 100 and 130 °C. No major difference in sequence distribution of the styrene centered triads of polymers prepared by both processes are observed. The comparison of experimentally observed triad fractions and predictions from theory show that the terminal model fails.

Introduction

Due to the easy access to a variety of monomers that can be polymerized and the simplicity of the technical process, radical polymerization is the most thoroughly studied polymerization process known. Nevertheless, it is not possible to obtain polymers with defined structures or low polydispersities because transfer and termination reactions always impede control over molecular weight and architecture. Up to now methods like anionic, group transfer, ring-opening or cationic polymerizations were used for the synthesis of "living" polymers with defined molecular weights and small polydispersities. Recently, several approaches were developed that enable controlled radical polymerization. These methods include stable free radical polymerization (SFRP), atom transfer radical polymerization (ATRP) and reversible addition fragmentation chain transfer (RAFT). Using these polymerization methods, the molecular weight of the polymers increases linearly as a function of conversion and polydispersities of well below 1.5 can be obtained (1-5). The as-prepared polymers contain chain ends that can be used to prepare block or graft copolymers, or to introduce functionality into the polymer. Using these techniques polymers with complicated topologies such as star or dendritic structures are accessible. The mechanism of TEMPO-mediated controlled radical polymerization is shown in Scheme 1. A stable free radical such as TEMPO is added to the polymerization mixture and an equilibrium between the free propagating chain and free stable radical versus propagating chain ends capped with TEMPO is established. By the reversible capping of the polymer radical, the overall radical concentration in the reaction medium is reduced. Thus termination processes are limited. Nevertheless, termination or transfer can occur just as in conventional radical polymerization.

Scheme 1. *Mechansim of TEMPO-mediated polymerization*

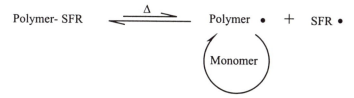

Up to now a few papers dealt with the TEMPO mediated copolymerization in detail (6-14). Fukuda investigated poly(styrene)-b-poly(styrene-co-acrylonitrile) block copolymers in 1996 (15). He observed that although the homopolymerization of acrylonitrile was inhibited by TEMPO, the copolymerization of styrene and acrylonitrile could be performed in the presence of TEMPO. He suggested that penultimate unit effects could influence the strength of the acrylonitrile-TEMPO bond and thus enhance the dissociation of the TEMPO-growing polymer chain adduct. The microstructure of a polymer produced at the azeotropic composition was similar to that obtained from free-radical polymerization. As the focus of the paper was the preparation of block copolymers, the copolymerization of styrene and acrylonitrile was not discussed in more detail. Additional work was performed in the working groups of G. Schmidt-Naake and R. Mühlhaupt (16, 17). The objective of this work is to investigate the copolymerization behavior of styrene and acrylonitrile in the presence of TEMPO over the whole composition range and a range of temperatures.

Controlled Radical Copolymerization of Styrene and Acrylonitrile

Molecular Weight. First the "livingness" of styrene acrylonitrile copolymerization in the presence of TEMPO was investigated. For this purpose polymerizations starting from the azeotropic monomer feed composition of free-radical copolymerization containing 25 wt.% of acrylonitrile were performed. Several control experiments in absence of TEMPO were performed at different temperatures using thermal, azo or peroxide initiation. Samples were taken at small and large conversion. The results are reported in Table 1. All polymerizations carried out according to the conventional bulk polymerization result in molecular weights above 300,000 g/mol and polydispersities around 2. However, in the presence of TEMPO, considerably smaller molecular weights are observed which correspond well with the molecular weight predicted from the monomer to TEMPO ratio.

Table 1. Molecular Weight and Polydispersities of SAN Copolymers

T [°C]	Initiator	Controller	t [h]	Conversion [%]	M_n [g/mol]	PDI
95	BPO	-	4	16	367,000	1.99
95	AIBN	-	4	12	408,000	2.05
110	-	-	1	7	330,000	1.99
110	BPO	TEMPO	2	3	7,300	1.26
110	BPO	TEMPO	8	50	44,600	1.40
130	BPO	TEMPO	2	4	8,600	1.33
130	BPO	TEMPO	8	65	51,200	1.43

The difference in molecular weight and molecular weight distribution of samples prepared in the presence or absence of TEMPO can easily be visualized when the SEC traces are superimposed. Figure 1 shows that the molecular weight of samples prepared by the conventional process is significantly larger and molecular weight distribution is broader compared to the samples obtained from TEMPO-mediated polymerization. It is also shown, that the peak maximum obtained from controlled radical polymerization completely shifts to higher molecular weights with conversion. There is neither a shoulder on the high molecular weight side indicating uncontrolled growth nor a shoulder on the low molecular weight side indicating large amounts of irreversibly terminated chains. Thus it is believed that the copolymerization of styrene and acrylonitrile proceeds in a quasi-living fashion.

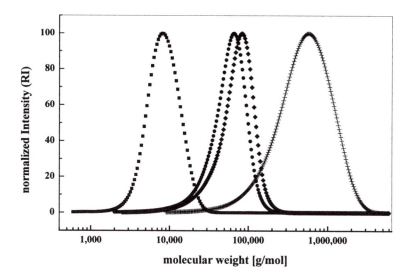

Figure 1. *SEC traces of polymers prepared by conventional and controlled radical copolymerization (25 wt.% acrylonitrile). Controlled radical copolymerization: sample at 5 % conversion (■), 50 % conversion (●), 65 % conversion (♦); conventional radical copolymerization: thermal initiation (+).*

The molecular weight dependence on conversion for the controlled radical copolymerization of styrene and acrylonitrile (75 : 25) is shown in Figure 2. The increase in molecular weight is almost linear up to conversions of about 60 %. At higher conversions the molecular weight levels off to smaller values due to inevitable side reactions at long reaction times. As soon as the TEMPO-polymer bond is dissociated and the growing chain radical is free, transfer and termination reactions can occur just as in conventional free-radical polymerization. Additionally the thermal decomposition of the stable free radical has to be taken into account, especially at high reaction temperatures and long reaction times (*18*). For both temperatures investigated, the dependence of molecular weight on conversion of the linear part is similar. Thus, molecular weight can be predetermined by the composition of the reaction mixture, independent of the polymerization temperature.

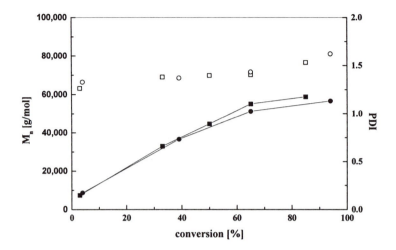

Figure 2. *Number average molecular weight as a function of conversion for polymerizations carried out in the presence of TEMPO at temperatures of 110 °C (-■-) and 120 °C (-●-); polydispersity index at 110 °C (□) and at 120 °C (○). All polymerizations were performed using an acrylonitrile content of 25 wt.%.*

Polydispersity. From the evolution of the polydispersity index with conversion it is also obvious that side reactions occur, especially at high conversions / long reaction times. Although the polydispersity index increases at high conversion, it is still well below 2, even when the copolymerization is carried out to conversions as high as 95 %. For conversions up to approximately 70 %, the polydispersity index does not rise above the critical value of 1.5.

Polymerization Rate. In the kinetic plot an induction period of up two hours is observed. In the following polymerization stage $\ln([M]_0/[M]_t)$ increases linearly with reaction time before it levels off. At longer reaction times the deviation from the linear behavior is even larger. Figure 3 shows that the polymerization carried out at 120 °C is already significantly faster than that performed at 110 °C. All polymerizations were performed using an acrylonitrile content of 25 wt.% in the monomer feed.

The apparent rate of polymerization can be estimated when plotting conversion versus the reaction time. For this evaluation only data obtained at reaction times of up to 10 hours are taken into account, as those correspond to the fairly linear increase observed in the kinetic plot. For the polymerization performed at 110 °C a value of 4.2 %/h is obtained, for the polymerization performed at 120 °C 9.6 %/h.

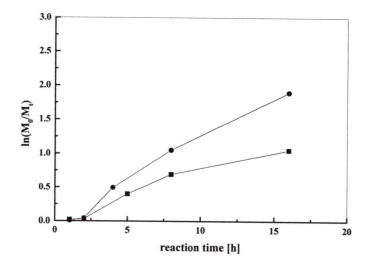

Figure 3. *Kinetic plot for the controlled radical copolymerization of styrene and acrylonitrile (75 : 25 by wt.) performed at 110 °C (-■-) and 120 °C (-●-).*

Copolymerization Diagram

The results shown so far were obtained at acrylonitrile contents of 25 wt.%, corresponding to the azeotropic mixture of free-radical polymerization. Here, the composition of the monomer feed should equal the composition of the resulting copolymer even when the polymerization is carried out in a batch process without the addition of monomers. In the following section the copolymerization behavior shall be investigated as a function of monomer feed composition.

In order to study this, a series of experiments was carried out over the whole composition range, varying the acrylonitrile content between 17 and 95 wt%. Conversions were kept below 5 %, assuming that monomer feed composition can be treated as being constant in this range.

For comparison reference experiments were carried out without TEMPO to ensure that only data that was obtained using the same experimental conditions is compared. Figure 4 shows the results obtained from bulk polymerizations carried out at 120 °C. It is obvious that the composition of the copolymers is not significantly altered by the presence of TEMPO. Polymers with comparable acrylonitrile contents are obtained regardless of the kind of process - controlled or conventional free-radical polymerization - applied.

400

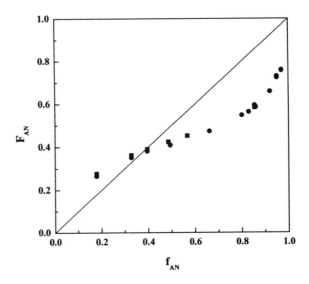

Figure 4. *Copolymerization diagram for controlled radical (●) and conventional free-radical copolymerization (■) of styrene and acrylonitrile at 120 °C.*

Copolymerization Parameters

Various methods are reported in literature to describe the free-radical copolymerization of styrene and acrylonitrile. The terminal model accounts for the different reactivity of polymer chains containing styrene and acrylonitrile terminal units towards the monomer molecules, respectivly. For the description of copolymer composition the terminal model is useful for many systems of varying structure and reactivity (*19*). On the other hand this model is often not successful in the prediction of the sequence distribution and kinetic parameters such as the overall copolymerization rate.

The penultimate model takes the influence of the second-last monomer into account and has proven to coincide better with experimentally determined sequence distribution in the styrene-acrylonitrile system (*20, 21*).

A third model discussed is the complex participation model which assumes that a complex of styrene and acrylonitrile monomer is added to the growing polymer chain. However, it does not agree with experimental sequence distribution data of the system (*22*).

Data reported in literature for the copolymerization parameters of styrene and acrylonitrile is rather different. From the polymer handbook, various sets of copolymerization parameters obtained by different methods are reported. Depending on the polymerization method applied (emulsion, bulk, solution polymerization, thermal or peroxide initiation), on the polymerization temperature investigated, and on the determination method, reactivity ratios ranging from 0.29 to 0.55 are reported for r_S and from 0.02 to 0.17 for r_{AN}. It is also discussed in literature, that the copolymerization of styrene and acrylonitrile is temperature dependent (23). Although no difference in copolymer composition is observed, the sequence distribution of polymers obtained at different temperatures shows temperature dependence. As most of the data reported so far was obtained at polymerization temperatures well below those applied for controlled radical copolymerization in the presence of TEMPO, the apparent reactivity ratios were also investigated for a set of reference samples obtained under comparable reaction conditions.

Using the methods of Fineman and Ross (24) and Kelen and Tüdös (25), the apparent polymerization parameters were estimated (Table 2 and Table 3).

Table 2. Apparent Copolymerization Parameters Obtained for Conventional Free-Radical Polymerization Performed in Bulk

T [°C]	Fineman and Ross		Kelen and Tüdös	
	r_{AN}	r_S	r_{AN}	r_S
72	0.08	0.43	0.06	0.38
95	0.09	0.44	0.07	0.37
110	0.09	0.45	0.07	0.41
123	0.09	0.44	0.07	0.40
128	0.08	0.48	0.07	0.41

The apparent copolymerization parameters obtained by the method of Fineman and Ross are slightly higher compared to those obtained by the method of Kelen and Tüdös. However, as data points for linear regression are more evenly partitioned using the method of Kelen and Tüdös, the latter method is regarded as more reliable.

Comparing the results obtained for conventional and controlled radical polymerization, the apparent acrylonitrile reactivity ratios obtained are within the same range. The apparent styrene copolymerization parameters are slightly higher for controlled radical polymerization. However, this difference is within the experimental error. Nevertheless, in the case of TEMPO-mediated polymerization a slight increase of the apparent styrene copolymerization parameter with temperature is observed (Table 3), whereas the acrylonitrile copolymerization parameter remains virtually constant over the temperature range investigated in this work (Figure 5).

Table 3. Apparent Copolymerization Parameters Obtained for TEMPO-mediated Copolymerization of Styrene and Acrylonitrile

T [°C]	Fineman and Ross		Kelen and Tüdos	
	r_{AN}	r_S	r_{AN}	r_S
99	0.08	0.48	0.07	0.43
111	0.07	0.52	0.08	0.45
122	0.09	0.53	0.08	0.45
133	0.10	0.77	0.09	0.53

The temperature window in which TEMPO-mediated copolymerization can be performed is limited. At low temperatures the N-oxyl radical will serve as an inhibitor. As discussed above, the polymerization is already significantly slower when going from 120 to 110 °C. At 100 °C the polymerization rate is decreased even further. In the bulk polymerizations investigated here, control was already lost at temperatures around 140 °C. At this temperature complete conversion is observed in reaction times of less than one hour. Molecular weights of more than 100,000 g/mol as well as broad polydispersities are obtained. For comparison, the ratio of stable free radical to initiator and monomer was kept constant in this work. Using higher levels of N-oxyl radical, higher polymerization temperatures might be accessible.

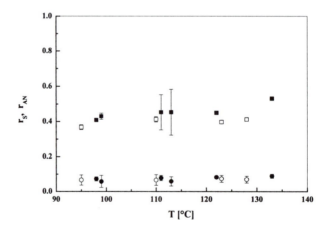

Figure 5. *Copolymerization parameters obtained by the method of Kelen and Tüdös. Controlled radical copolymerization: r_S (■), r_{AN} (●); conventional free-radical polymerization: r_S (□), r_{AN} (○).*

Sequence Distribution

In the following section the influence of TEMPO on the sequence distribution in the copolymers shall be investigated. For this purpose the samples obtained from polymerizations carried out at 120 °C were analyzed by ^{13}C NMR spectroscopy. The theoretical sequence distribution was calculated using the penultimate model (20). Over the whole temperature range investigated, the data for conventional radical copolymerization corresponds well with the data reported in reference 23. Using the equations given there, the penultimate reactivity ratios calculated for 120 °C are used for the comparison of theoretically predicted and experimentally observed sequence distribution. For the theoretical predictions using the terminal model, it was assumed that r_{SS} equals r_{AS} and r_{AA} equals r_{SA}. For the terminal model r_S was set to 0.45 and r_{AN} to 0.08 as obtained from controlled radical copolymerization at 122 °C.

The copolymerization of styrene and acrylonitrile was also simulated using a Monte Carlo algorithm and the PREDICI software package. The simulation results and comparison with the experimental data shall be reported elsewhere (26).

Figure 6 shows the overlay of the theoretically predicted monomer sequence distribution of the styrene centered triads calculated using the terminal and penultimate models and the experimental results. In order to provide additional data points, the experimental values determined by Ferrando and Longo are also included. The results obtained from conventional and controlled radical copolymerization in our lab are very close to those obtained by Ferrando and Longo. There is no significant difference in the triad distribution for conventional and controlled radical copolymerization. The styrene sequences data $f_{SSA+ASS}$ from controlled radical copolymerization corresponds better with the theoretical predictions calculated from the penultimate model. This can also been seen for styrene rich monomer feed compositions in the results for styrene sequences of length three (f_{SSS}). A penultimate effect was also supported in model reactions performed by Tirrell and co-workers. They determined the rates of addition of styrene and acrylonitrile to substituted propyl radicals (27, 28) and found a good agreement with the results obtained by Hill and co-workers (22).

The comparison of theoretical predictions for acrylonitrile centered triads are shown in Figure 7. Again, the results obtained in our lab for conventional and controlled radical copolymerization are comparable. The acrylonitrile sequences of length one (f_{SAS}) and two ($f_{AAS+SAA}$) show again that the copolymerization behavior of styrene and acrylonitrile can be better described by the penultimate model than by the terminal model. The results obtained in our lab show slight deviations from the theoretical behavior at acrylonitrile contents smaller than the azeotropic mixture. However, this deviation is observed in samples prepared by both processes, conventional free-radical polymerization and controlled radical copolymerization. The cause of the deviation is yet unknown but is ascribed to the specific experimental conditions applied.

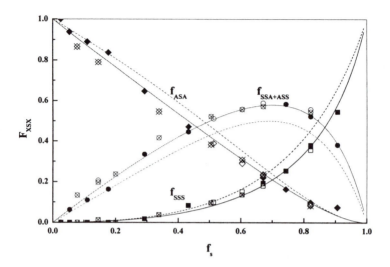

Figure 6. *Styrene centered triads as a function of monomer feed composition (T = 120 °C). Conventional free-radical copolymerization: data reported in ref. 23 for T = 115 °C (solid symbols), this work (open symbols); controlled radical copolymerization (cross-centered symbols); theoretical distribution terminal model r_S =0.45, r_{AN} = 0.08 (---), and penultimate model (r_{SS}, r_{SA}, r_{AA}, r_{AS} calculated from ref. 23 for T = 120 °C)(—).*

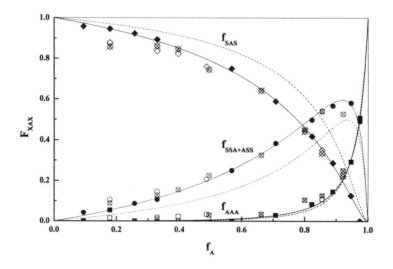

Figure 7. *Acrylonitrile centered triads as a function of monomer feed composition (T = 120 °C). Symbols and parameters as specified in Figure 6.*

Conclusions

The TEMPO-mediated copolymerization of styrene and acrylonitrile was investigated over the whole composition range. Although the acrylonitrile homopolymerization is inhibited, the copolymerization can be performed in a quasi-living manner. However, the temperature window for the polymerization is rather small. In our experiments, inhibition predominates at temperatures below 100 °C and control is lost at temperatures above 140 °C. The composition of polymers prepared by conventional and controlled radical polymerization are comparable. The copolymerization parameters determined by the methods of Fineman and Ross and Kelen and Tüdös do not show significant differences for the two processes, although for r_S consistently slightly higher values are obtained from controlled radical copolymerization. Although the copolymerization parameters are practically constant over the temperature range investigated, r_S obtained from controlled radical copolymerization reflects the loss of control above 130 °C. The triad distributions of styrene and acrylonitrile centered triads do not largely differ when TEMPO is present in the reaction mixture. Results for sequence distribution correspond to predictions obtained from the penultimate model.

Experimental

Materials. Styrene and acrylonitrile monomers were used as supplied. The initiators dibenzoylperoxide (70 %) and 1,1-di-tert-butyl-3,3,5-trimethylcyclohexane (50 %) and the stable free radical 2,2,6,6-tetramethylpiperidine-1-oxyl were used without further purification.

Polymerizations. All polymerizations were performed in bulk. For conventional radical polymerization 1,1-di-tert-butyl-3,3,5-trimethylcyclohexane was used as initiator for reactions carried out at 70 °C. When not stated otherwise, thermal initiation was applied at all other temperatures when the conventional process was applied. Polymerizations using TEMPO as a mediator were carried out in the presence of dibenzoylperoxide. The ratio of monomer to mediator to initiator were 1400 : 1.5 : 1. All polymerizations were performed in sealed glass ampoules. The reaction mixtures were placed into the glass tubes, degassed and sealed under nitrogen. Before placing the sealed ampoules in an oil bath at the polymerization temperature, they were placed into protective aluminum containers. To stop polymerization, the aluminum containers were withdrawn from the oil bath and quenched to room temperature. All polymers were precipitated into methanol, filtered and dried in vacuo.

Characterization. The conversion of the polymerizations was determined gravimetrically using a HR73 Mettler Moisture Analyzer. Molecular weights and molecular weight distributions were determined by size exclusion chromatography using THF as eluent and polystyrene calibration. The acrylonitrile content of the polymers was determined by elemental analysis and from NMR spectroscopy. The polymer sequence distribution was determined by ^{13}C NMR spectroscopy.

Literature Cited

1   M. K. Georges, R. P. N. Veregin, P. M. Kazmaier, G. K. Hamer *TRIP* **1994**, 2, 66-72

2   C. J. Hawker *Acc. Chem. Res.* **1997**, 30, 373-382

3   T. E. Patten, K. Matyjaszewski *Adv. Mat.* **1998**, 10, 901-915

4   J. Chiefari, Y. K. Chong, F. Ercole, J. Krstina, J. Jeffery, T. P. T. Le, R. T. A. Mayadunne, G. F. Meijs, C. L. Moad, G. Moad, E. Rizzardo, S. Thang *Macromolecules* **1998**, 31, 5559-5562

5   Y. K. Chong, T. P. T. Le, G. Moad, E. Rizzardo, S. H. Thang *Macromolecules* **1999**, 32, 2071-2074

6   M. K. Georges, R. P. N. Veregin, P. M. Kazmaier, G. K. Hamer *Macromolecules* **1993**, 26, 2987-2988

7   C. J. Hawker, E. Elce, J. Dao, W. Volksen, T. P. Russell, G. G. Barclay *Macromolecules* **1996**, 29, 2686-2688

8   G. Schmidt-Naake, S. Butz *Macromol. Rapid Commun.* **1996**, 17, 661-665

9   N. Ide, T. Fukuda *Macromolecules* **1997**, 30, 4268-4271

10 H. Baethge, S. Butz, G. Schmidt-Naake *Macromol. Rapid Commun.* **1997**, 18, 911-916

11 S. Butz, H. Baethge, G. Schmidt-Naake *Macromol. Rapid Commun.* **1997**, 18, 1049-1055

12 N. Ide, T. Fukuda *Macromolecules* *1999*, 32, 95-99

13 H. Baethge, S. Butz, C. Han, G. Schmidt-Naake *Angew. Makromol. Chem.* **1999**, 267, 52-56

14 E. Yoshida, Y. Takiguchi *Polymer J.* **1999**, 31, 429-434

15 T. Fukuda, T. Terauchi, A. Goto, Y. Tsujii, T. Miyamoto *Macromolecules* **1996**, 29, 3050-3052

16 M. Baumann, Diploma Thesis, Technische Universität Clausthal, Clausthal-Zellerfeld, Germany, 1997

17 M. Baumert, Diploma Thesis, Albert-Ludwig-Universität, Freiburg, Germany, 1996

18 T. Fukuda, A. Goto, K. Ohno, Y. Tsujii in Controlled Radical Polymerization; K. Matyjaszewski, Ed.; ACS Symp. Ser. 685; American Chemical Society: Washington, DC, 1998; pp. 180-199

19 D. A. Tirrell in Comprehensive Polymer Science; A. Geoffrey, J. C. Bevington, Ed.; Pergamon Press; Oxford, UK, 1989; Vol. 2, pp. 195-207

20 D. J. T. Hill, A. P. Lang, J. H. O'Donnell, P. W. O'Sullivan *Eur. Polym. J.* **1989**, 25, 911-915

21 A. Kaim *J. Macromol. Sci. – Pure Appl. Chem.* **1998**, A35, 577-588

22 D. J. T. Hill, J. H. O'Donnell, P. W. O'Sullivan *Macromolecules* **1982**, 15, 960-966

23 A. Ferrando, A. Longo *Polymer Preprints* **1997**, 38 (1), 798-799

24 M. Fineman, S. D. Ross *J. Polym. Sci.* **1950**, 5, 259-265

25 T. Kelen, F. Tüdös *J. Macromol. Sci. – Chem.* **1975**, A9, 1-27

26 J. Cao, S. Brinkmann-Rengel, N. Nießner, Y. Yang *to be published*

27 G. S. Prementine, D. A. Tirrell *Macromolecules* **1987**, 20, 3034-3038

28 D. A. Cywar, D. A. Tirrell *Macromolecules* **1986**, 19, 2908-2911

Chapter 29

Synthesis of Oligomers by Stable Free Radical Polymerization of Acrylates, Methacrylates, and Styrene with Alkoxyamine Initiators

Helmut Keul, Dirk Achten, Birte Reining, and Hartwig Höcker[1]

Lehrstuhl für Textilchemie und Makromolekulare Chemie
der Rheinisch-Westfälischen Technischen Hochschule Aachen,
Worringerweg 1, 52056 Aachen, Germany

A controlled polymerization of styrene (St) and methyl acrylate (MA) yielding oligomers of $M_n \leq 2000$ was achieved using an alkoxyamine as the initiator. For the polymerization of methyl methacrylate (MMA) side reactions leading to unsaturated end groups prohibit the control of the polymerization. The extent of the side reactions is explained by a sterically hindered combination of TEMPO with the active PMMA chain end. Copolymerization of MA with St reveals an increase of the polymerization rate compared to those of the homopolymerizations and a good control of the MA/St copolymer composition within a wide range. The copolymerization parameters were found to be in good agreement with those observed in the free radical polymerization. For the copolymerization of MMA with St a controlled polymerization was achieved at molar fractions of styrene higher than 50 %.

Introduction

Nitroxide mediated living radical polymerization is a process suitable for the preparation of well defined polymers[1-5], especially on the basis of styrene. Two initiation procedures have been employed: (i) a bimolecular initiating system

[1]Corresponding author.

409

comprising a free radical initiator, styrene, and a stable nitroxide, e. g., TEMPO (2,2,6,6-tetramethylpiperidin-1-oxyl) and (ii) a unimolecular initiating system comprising an alkoxyamine prepared in advance[1-8].

With unimolecular initiators long induction periods are usually avoided. The molecular weight of the polymers is predetermined by the monomer/alkoxyamine ratio and the conversion. Recently, the syntheses of alkoxyamines with structures resembling the dormant species of styrene, acrylate, methacrylate and acrylonitrile have been performed[6].

The main topics addressed in the publications dealing with alkoxyamine initiators are: (i) the synthesis of alkoxyamines[6-8], (ii) the synthesis of polymers with complex architecture using these alkoxyamines[9,10], (iii) the initiation efficiency of these alkoxyamines (R=NO-R′) as a function of R and R′[1c,6,7,11,12], and (iv) the thermal stability resp. the thermal decomposition of the dormant and active species[1c,13,14,15].

In this contribution the results concerning the homopolymerization of methyl acrylate and methyl methacrylate and the copolymerization of these monomers with styrene with alkoxyamine initiators will be presented in comparison to the homopolymerization of styrene with the goal to prepare oligomers. Special emphasis will be given to side reactions.

TEMPO mediated living radical polymerization in the ideal case exerts fast initiation followed by propagation with nearly no irreversible termination and

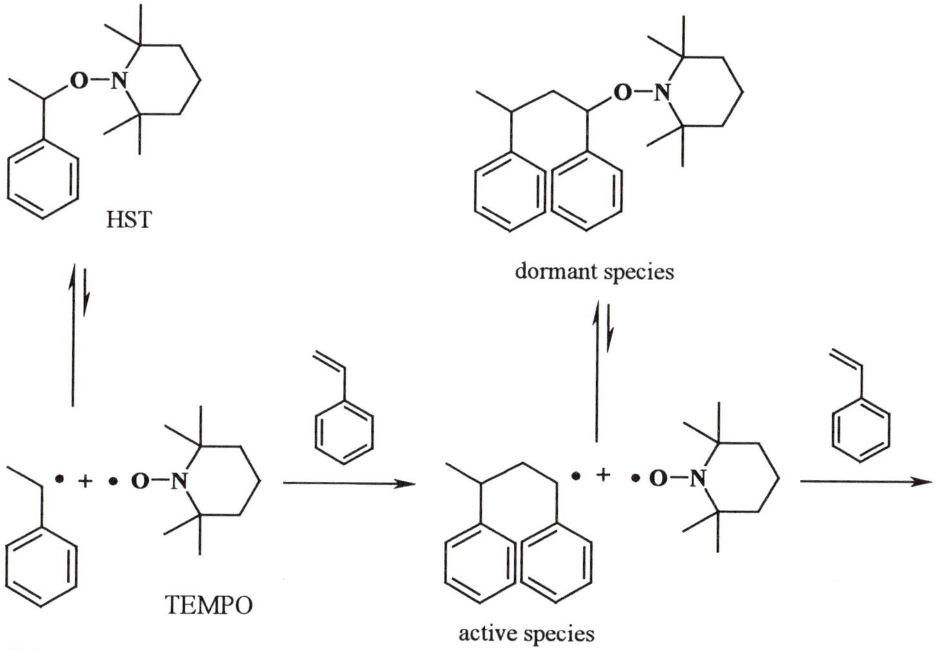

Scheme 1: Reversible alkoxyamine C-O bond homolysis and propagation in nitroxide mediated living radical polymerization.

transfer reactions. At elevated temperatures alkoxyamines, the initiators for this polymerization procedure, suffer a homolytical cleavage of the most labile bond - the C-O bond - producing two radicals.

1′-Phenyl-1′-(2,2,6,6,-tetramethylpiperidinyloxy)-ethane (HST) upon thermal cleavage results in the nitroxide radical (TEMPO) a stable free radical which does not initiate the polymerization and a phenylethyl radical (Scheme 1).

The phenylethyl radical is expected to recombine with the nitroxide radical or react with styrene present in the system in the sense of an initiation reaction. The newly formed radical (active species) is quenched in a nearly diffusion controlled process by the nitroxide radical with formation of the dormant species. Active chains and dormant chains are in equilibrium which is strongly shifted to the side of the dormant chains with the consequence of a low stationary radical concentration. The low concentration of radicals and a high rate of exchange between dormant and active species in comparison to the rate of monomer addition results in a living radical polymerization[16]. Because of the persistent radical effect[17] and the occurence of termination a continuous increase in nitroxide concentration is to be expected shifting the equilibrium towards the side of the dormant species, thus decreasing the overall rate of polymerization. The only way to reach high conversion is the removal of TEMPO from the system either by adding an additional radical source[18] or by decomposition of the nitroxide, i.e., by a sulfonic acid[19]. In the case of the polymerization of styrene radicals formed by thermal self initiation compensate the amount of TEMPO produced by termination reactions. The radicals obtained by thermal self initiation control the rate of polymerization since their concentration is higher than the concentration of radicals produced by the homolysis of the alkoxyamine by one order of magnitude[20].

Results and Discussion

In order to study the influence of high initiator concentrations necessary for the synthesis of oligomers and the nature of the alkoxyamine on the reaction course we have polymerized styrene with alkoxyamine initiators based on styrene (HST), methyl acrylate (HMAT), and methyl methacrylate (HMMAT).

These initiators were prepared according to a modified literature procedure[6a].

HST was obtained analytically pure by fractional destillation of the crude product. A waxy solid with a melting point of 41-43 °C was obtained. ^{1}H NMR / CDCl$_3$: δ = 0.66, 1.03, 1.17, 1.37 (br. s, CH$_3$, 12H), 1.29-1.48 (m, CH$_2$, 6H), 1.47 (d, ^3J = 6 Hz, 3H, CH$_3$), 4.77 (q, ^3J = 6 Hz, CH, 1H), 7.20-7.37 (5H, CH$_{arom}$).

HMAT was obtained analytically pure by the same procedure as a waxy solid with a melting point of 28°C. ^{1}H NMR /CDCl$_3$: δ = 1.04, 1.13, 1.20, 1.47 (br. s, CH$_3$, 12H), 1.33-1.47 (m, CH$_2$, 6H), 1.40 (d, ^3J = 9 Hz, CH$_3$, 3H), 3.7 (s, CH$_3$, 3H), 4.39 (m, CH, 1H).

In analogy HMMAT was obtained as a waxy solid with a melting point of 32-34 °C. ^{1}H NMR /CDCl$_3$: δ = 0.99, 1.15, (s, CH$_3$, 12H), 1.27-1.57 (m, CH$_2$, 6H), 1.47 (s, CH$_3$, 6H), 3.71 (s, CH$_3$, 3H) (Figure 1).[21]

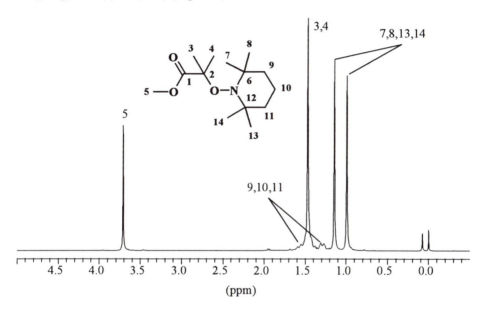

Figure 1. ^{1}H NMR spectrum of HMMAT in CDCl$_3$.

The polymerization of styrene was performed in Schlenk tubes at 130 °C with the various indicated alkoxyamine initiators. The conversion was determined gravimetrically after evaporation of residual monomer in high vacuum. These samples were used for the molecular weight determination by means of GPC using polystyrene standards.

The conversion index vs. time plots obtained for a monomer to initiator ratio of 40 at 130 °C are shown in Figure 2a. Compared with HST as initiator, the polymerization with HMAT as initiator shows an induction period probably caused by slow initiation. This results in a higher polydispersity index obtained with HMAT as initiator than with HST (Figure 2b).

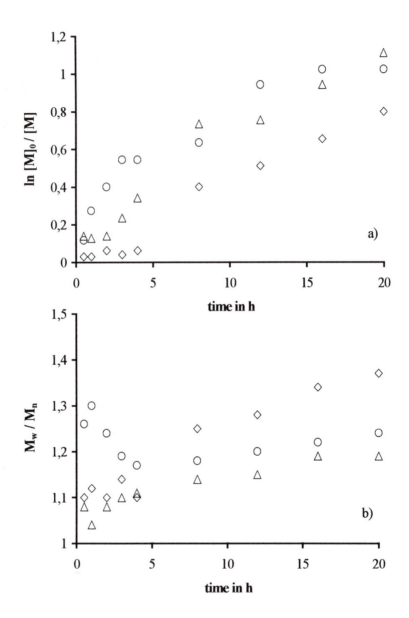

Figure 2. Polymerization of styrene with the alkoxyamine initiators HST (O), HMAT (◇), and HMMAT (△). Polymerization conditions: $[St]_0/[I]_0 = 40$, $T = 130°C$. (a) conversion index vs. time; (b) polydispersity index (M_w/M_n) vs. time.

The alkoxyamine initiator HMMAT shows a high rate for the homolytic cleavage leading to a high initial radical concentration (the reason will be discussed later). This has two consequences: (i) a high initial monomer conversion and (ii) afterwards a reduced polymerization rate due to excess TEMPO produced in side reactions. The polymerization is in effect delayed until additional radicals formed by thermal self initiation of styrene have consumed the excess TEMPO. The fast initiation reaction leads to polymers with polydispersity indices comparable to those produced with HST as initiator. The delay of the polymerization by excess TEMPO does not influence the polydispersity indices since all chains are equally affected.

The plot of M_n vs. conversion shows for all the initiators applied a linear dependence with values close to the theoretical ones (Figure 3).

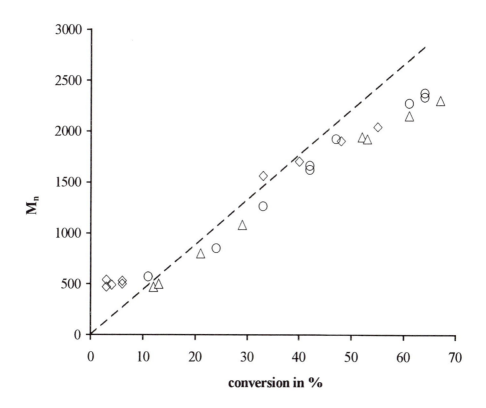

Figure 3. Plot of number average molecular weight M_n vs. conversion for the polymerization of styrene with the alkoxyamine initiators HST (O), HMAT (◊), and HMMAT (Δ).Polymerization conditions: $[St]_0 / [I]_0 = 40$, $T = 130\ °C$.

From these results we concluded that for the further investigations HST is the best suited alkoxyamine initiator.

Polymerization of Methyl Acrylate and Methyl Methacrylate with HST as the Initiator

The polymerization of methyl acrylate was performed in glass ampules in bulk at 130 °C with HST as the alkoxyamine initiator. The initial monomer/initiator ratio applied was 40. For comparison reasons, the results obtained with styrene under the same conditions are presented. After 20 h for styrene as the monomer a conversion of 65 % and for methyl acrylate a conversion of 45 % was reached. For both monomers a linear increase of the molecular weight with conversion and a narrow molecular weight distribution is observed (Figure 4). The polydispersity index of the poly(methyl acrylate), however, is higher than that of poly(styrene) (Figure 4b). This higher value of the polydispersity index can be explained by the addition of more than one monomer unit during the active period of a polymer chain[16]. In contrast to styrene MA and MMA do not significantly enhance the rate of polymerization by radicals produced by thermal self initiation. We believe that in the case of MA an acceptable rate of polymerization is only due to the high rate constant of monomer addition which is about one order of magnitude higher than that of St and MMA[22].

The polymerization of methyl methacrylate reveals completely different results. A monomer conversion of 20 % is reached after less than 1 h. Then conversion and

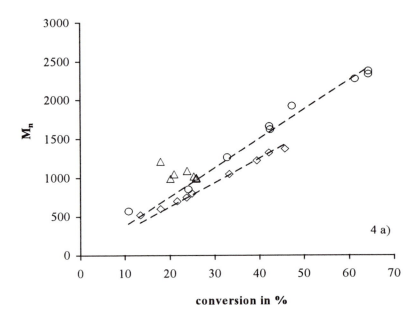

Figure 4. Polymerization of styrene (O), methyl acrylate (◇), and methyl methacryalte (△) with HST as initiator. Polymerization conditions: $[M]_0/[I]_0 = 40$, $T = 130°C$. (a) number average molecular weight vs. time.

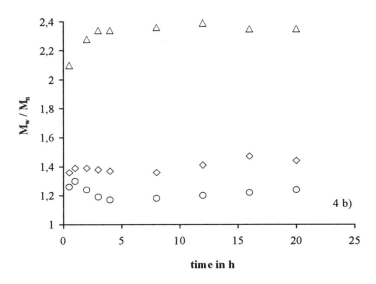

Figure 4. (b) polydispersity index vs. time.

molecular weight do not change any more significantly with time (Figure 4a). The polydispersity index for all samples is higher than 2 (Figure 4b). The ^{1}H NMR analysis of the PMMA oligomers shows a high concentration of end groups (Figure 5): (i) the resonances for the unsaturated PMMA endgroups [$H_2C=C(COOMe)-CH_2$] are centered at $\delta = 6.2$ (1H), 5.5 (1H), 3.7 (3H) and 2.5 (2H) ppm; (ii) those of the HS endgroup[$CH_3-CH(C_6H_5)-$] are centered at $\delta = 7.2$ (5H), 3.0(1H) and 1.25 (3H). The assigned δ-values are in accordance with calculated values according to incremental tables.

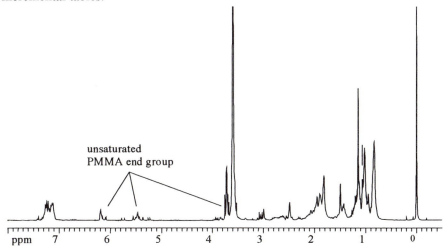

unsaturated
PMMA end group

Figure 5. ^{1}H NMR spectrum of PMMA in CDCl$_3$; initiator: HST.

Side reactions observed in nitroxide mediated polymerization of methyl methacrylate, methyl acrylate, and styrene.

What is our explanation of the lack of control observed in the polymerization of MMA? We assume in agreement with the literature[1] that the high concentration of unsaturated end groups as indicated in the ^{1}H NMR spectrum can be explained by a formal elimination of TEMPOH – which may occur in a concerted reaction or by abstraction of a hydrogen atom from α-position to the radical center by the stable TEMPO radical. Furthermore, the hydroxylamine may react with the radical species transferring a hydrogen atom to the radical center generating TEMPO and a saturated chain end. TEMPO in excess acts as an inhibitor by shifting the equilibrium toward the dormand chains. As a result, the rate of polymerization decreases to zero. The formation of excess TEMPO is observed by the typical red colour of this stable radical.

To a substantially lower extent unsaturated endgroups are observed in polystyrene and poly(methyl acrylate)[21]. The strong tendency towards side reactions leading to unsaturated endgroups and TEMPO in excess found in the polymerization of MMA cannot only be ascribed to the statistical factor - five hydrogen atoms available to be abstracted from an active PMMA chain end in comparison to two hydrogen atoms for an active PSt or PMA chain end. We believe that the combination reaction of the sterically demanding TEMPO with the active PMMA chain end - a tertiary radical - is sterically hindered in analogy to the recombination of two PMMA chain ends. Thus the addition of monomer is sterically strongly favored over the combination of an active chain with TEMPO. In analogy to the polymerization of MA more than one monomer unit may be added per activation/deactivation cycle[16]. We believe that both elimination of TEMPOH and sterically hindered endcapping of PMMA chains by TEMPO are responsible for the lack of control in the polymerization of MMA.

In order to obtain information on the stability and reactivity of chain ends in the polymerization of styrene, methyl acrylate, and methyl methacrylate we studied the conversion of the alkoxyamine initiators HST, HMAT, and HMMAT as models for the active chain ends in ethyl benzene (all polymer chains contain activated C-H bonds either by phenyl groups or by α-carbonyl groups). The conversion of the alkoxyamines, the formation of TEMPOH and TEMPO, the formation of HST and the formation of St, MA, and MMA were determined quantitatively by means of gas chromatography (GC) and / or ^{1}H NMR spectroscopy. TEMPO and TEMPOH, however, did not give well resolved peaks by GC and therefore the sum of both was determined [TEMPO(H)]. The monomers St, MA, and MMA could not be determined quantitatively by GC since their retention time is close to that of ethyl benzene; however, the concentration of the monomers could be estimated from ^{1}H NMR analysis. Additional peaks in GC with high retention time and with low intensity were assigned to dimers of the primary radicals $CH_3(R^1)(R^2)C\bullet$ or to oligomers of $(R^1)(R^2)C=CH_2$ (Scheme 2).

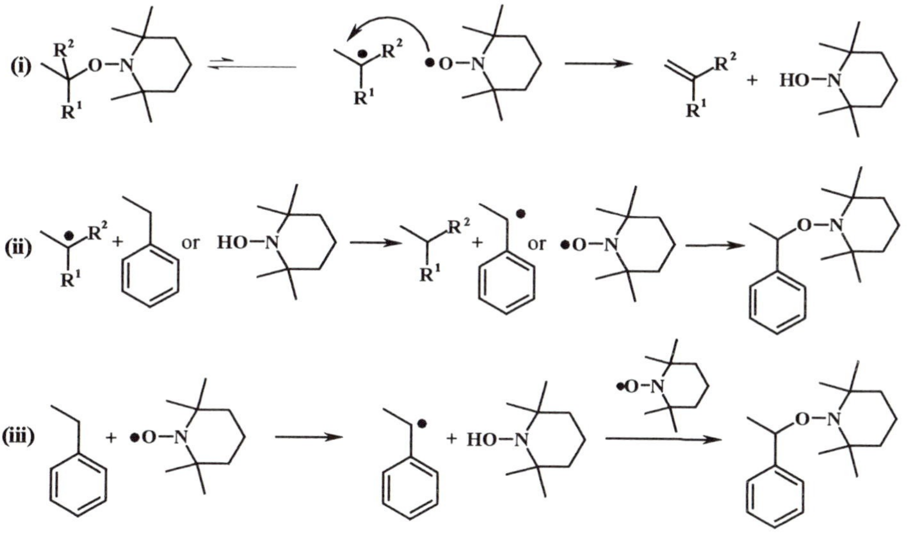

Scheme 2: Decomposition of the alkoxyamine initiators HST (R^1 = H, R^2 = C_6H_5), HMAT (R^1 = H, R^2 = $COOCH_3$), and HMMAT (R^1 = CH_3, R^2 = $COOCH_3$) in ethylbenzene.

Compared with HMAT and HMMAT for HST the lowest decrease in concentration is observed (Figure 6a). After 18 h only 40 % are consumed and consequently 40 % of TEMPO(H) is formed. A good agreement between GC- and NMR data is observed. In the interpretation of these values the formation of HST by reaction of the phenylethyl radical – produced by hydrogen atom transfer from ethylbenzene – with TEMPO has to be taken into consideration. The concentration of styrene increases to a value of about 60 % after 15 h and decreases then slightly.

For HMAT after 18 h 83 % are consumed (Figure 6b). In the same time the concentration of TEMPO(H) and HST increases to final values of 66 % and 13 %, respectively. The concentration of MA reaches a maximum value of 30 % after 10 h and decreases slightly afterwards.

For HMMAT a fast conversion is observed due to the fact that recombination of the radicals formed by homolysis is a sterically hindered process (Figure 6c). After 6 h 90 % of the alkoxyamine is consumed. The formation of TEMPO(H) and the consumption of HMMAT occurred with the same rate. The concentration of HST increases slowly to a final value of 25 % and the concentration of MMA decreases slightly after a maximum value of nearly 40 % was reached after 4 h. The decrease of the monomer concentration after a certain reaction time can be explained by oligomerization.

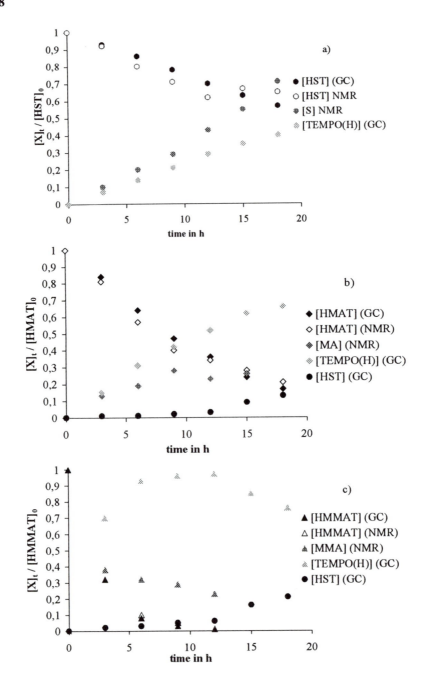

Figure 6. Conversion of alkoxyamine-initiators HST (●), HMAT (◆), and HMMAT (▲) in ethyl benzene. Reaction conditions: $[EtB]_0 / [I]_0 = 40$, $T = 130°C$.

Three reactions were taken into consideration for the explanation of the alkoxyamine consumption and the formation of HST (Scheme 2):

(i) elimination of TEMPOH and formation of styrene ($R^1 = H$, $R^2 = C_6H_5$), methyl acrylate ($R^1 = H$, $R^2 = COOCH_3$), or methyl methacrylate ($R^1 = CH_3$, $R^2 = COOCH_3$), which depends on the equilibrium constant between active and dormant species and the probability of H-abstraction, (ii) H-transfer from TEMPOH or ethyl benzene to the active species followed by combination of the resulting radicals with formation of HST, and (iii) H-transfer from ethyl benzene to TEMPO and consecutive combination of the resulting phenylethyl radical with TEMPO. Recombination or disproportionation of the carbon centered radicals produces an excess of TEMPO radicals.

The formation of TEMPOH was confirmed by the reaction of this hydroxylamine with diphenylchlorophosphate and subsequent ^{31}P NMR analysis[21]. The chemical shift of the resulting ester was found to be $\delta = -6.5$ ppm and identical with an authentic sample.

In order to verify the influence of TEMPOH on the polymerization course we performed two parallel experiments in which styrene was polymerized thermally in the presence of 0.27 mol/L TEMPOH and in the absence of any additive (Table I.). In the presence of TEMPOH the monomer conversion is 31 % and the number average molecular weight as determined by means of GPC is 2450. The radicals formed by self initiation of styrene at first act as an acceptor for the hydrogen atoms generating TEMPO. Later such radicals are trapped by TEMPO generating alkoxyamines which further initiate the polymerization. ^1H NMR analysis of the obtained oligomers confirms the existence of alkoxyamine endgroups[21].

Table I. Thermal polymerization of styrene in bulk at 130 °C for 20 h

1-Hydroxy-2,2,6,6-tetra-methylpiperidine in mol/L	Conversion in %	$\overline{M}_n$	$\overline{M}_w / \overline{M}_n$
0.27	31	2450	2.02
0.0	86	90600	2.08

Copolymerization of methyl acrylate and methyl methacrylate with styrene

As mentioned before, we believe that the reason for the lack of control in the polymerization of methyl methacrylate is due to the sterically hindered combination reaction of TEMPO and active chains that leads to an excess of TEMPO. Based on these considerations we have studied the effect of styrene as a comonomer in the polymerization of methyl methacrylate and methyl acrylate. The copolymerization of MMA with St serves two purposes: (i) a pathway to an easier combination of TEMPO with the active chain is given, if the active chain end is styrenic and (ii) the additional radicals produced by thermal self initiation of styrene consume excess

TEMPO eventually produced in side reactions, thus keeping the polymerization running.

The copolymerizations were performed in bulk in sealed glass ampules at 130 °C (some experiments at 110 °C) for 16 h. The molar fraction of styrene in the feed was varied between 0.1 and 0.9 and the molar ratio of monomers to the alkoxyamine intitator applied was 40. The conversion of the obtained samples was determined gravimetrically after evaporation of residual monomer in high vacuum and the molecular weights were determined by means of GPC using polystyrene standards.

Table II. Copolymerization of styrene and methyl acrylate

No.	$\dfrac{X_P}{\text{wt. \%}}$	f_{St}	F_{St}	$\overline{M}_{n,th.}$	$\overline{M}_{n,exp.}$	$\dfrac{\overline{M}_w}{\overline{M}_n}$
1	40	0	0	1480	1300	1,52
2	57	0,1	0,17	2180	1930	1,35
3	67	0,2	0,28	2620	2340	1,32
4	71	0,3	0,39	2830	2670	1,29
5	75	0,4	0,47	3030	2800	1,29
6	77	0,5	0,53	3150	2980	1,30
7	85	0,6	0,62	3530	3010	1,24
8	84	0,7	0,71	3540	3040	1,21
9	82	0,8	0,80	3510	3040	1,20
10	79	0,9	0,89	3430	2960	1,18
11	58	1	1	2570	2290	1,27

For the copolymerization of MA with St (Table II) with increasing molar fraction of styrene in the feed (f_{St}) from 0.1 to 0.9 the conversion obtained after 16 h increases from 57 % to a maximum of 85 % at f_{St} = 0.6 and decreases again to 79 % at f_{St} = 0.9. However, all conversions are higher than the conversions obtained in the homopolymerization of each MA and St. The polymerizaion rate is dependent on the radical concentration which is controlled by the equilibrium constant of the alkoxyamine homolysis, the rate of formation of additional radicals by thermal self initiation and further by the rate constants for the monomer addition (k_{12}, k_{11}, k_{22} k_{21}). As the rate of the formation of additional radicals is controlled by $[St]^3$ [23] and should increase with increasing concentration of styrene, we believe that the observed polymerization rates for the copolymerization of St and MA are best explained by the values of the homo- and cross propagation rate constants. For the system MA/St at 130 °C with HST as initiator we determined the copolymerization parameters by the method of Fineman and Ross[24] to be $r_{St} = 0.9$ and $r_{MA} = 0,2$. The copolymer composition was determined by means of ^{1}H NMR spectroscopy for conversions < 10 %. Although the accuracy of the applied method is limited, these values are close to the values found in the literature for the free radical polymerization at 60 °C: $r_{St} = 0.7 \pm 0.1$ and $r_{MA} = 0.2 \pm 0.1$[25]. This was expected by several authors before[4c,5a,26]. Recently, for another living radical system, ATRP, a good agreement of the copolymerization parameters of MMA and nBuA with those of the free radical copolymerization was found[27].

The molecular weight shows the same dependence on the molar fraction of styrene as the conversion whereas the polydispersity index decreases from 1.35 to 1.18 with increasing molar fraction of styrene from 0.1 to 0.9.

Table III. Copolymerization of styrene and methyl methacrylate

No.	X_P wt. %	f_{St}	F_{St}	$\overline{M}_{n,th.}$	$\overline{M}_{n,exp.}$	$\dfrac{\overline{M}_w}{\overline{M}_n}$
1	22	0	0	940	920	2,42
2	31	0,1	0,14	1330	1230	1,87
3	38	0,2	0,25	1640	1510	1,72
4	43	0,3	0,35	1860	1610	1,66
5	49	0,4	0,46	2130	1780	1,64
6	52	0,5	0,52	2260	1960	1,58
7	48	0,6	0,63	2100	1860	1,57
8	58	0,7	0,74	2540	2160	1,45
9	61	0,8	0,86	2690	2270	1,41
10	63	0,9	0,89	2780	2320	1,36
11	66	1	1	2920	2460	1,30

For the system MMA/St (Table III) with increasing molar fraction of styrene in the feed an increase of the monomer conversion within 16 h from 31 % to 63 % is observed. In the same time the number average molecular weight of the copolymers increases from 1230 to 2320 and the polydispersity index decreases from 1.87 to 1.36. The conversion stays below the conversion obtained in the homopolymerization of styrene. Here again the rate of polymerization is determined by the homo- and cross propagation rate constants and by the concentration of styrene. Furthermore, with increasing styrene concentration side reactions that lead to TEMPOH and subsequently to TEMPO decrease. (This is observed in the concentration of unsaturated end groups as determined by means of ^{1}H NMR spectroscopy.)

In order to examine if these copolymerizations are controlled polymerizations we have performed a kinetical study with a molar fraction of styrene in the feed of f_{St} = 0.5. In Figure 7 we observe for the polymerizations performed at 130 °C a rapid increase of the conversion within 4 h for St/MA to 60 % and St/MMA to 40 % and thereafter a slower increase. At lower temperatures (110 °C) the rate of polymerization is slower as expected. The plot of the polydispersity index vs. time (Figure 8) reveals always a higher value of M_w/M_n for the system MMA/St than for the system MA/St and, in addition, for the system MMA/St a slight increase of the M_w/M_n value with time. At lower temperatures the M_w/M_n values are lower. This is an indication of the fact that the rate constant for side reactions which result in an increase of the polydispersity index through the formation of dead chains are reduced at lower temperatures.

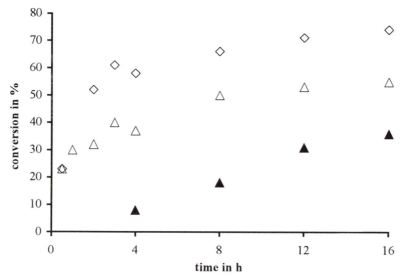

Figure 7. Plot of conversion vs. time for the copolymerization of equimolar amounts of MMA or MA with St with the alkoxyamin initiator HST.
Polymerization conditions: (▲) [MMA +St]$_0$ / [I]$_0$ = 40, T = 110 °C; (△) [MMA +St]$_0$/ [I]$_0$ = 40, T = 130 °C; (◇) [MA +St]$_0$/ [I]$_0$ = 40, T = 130 °C.

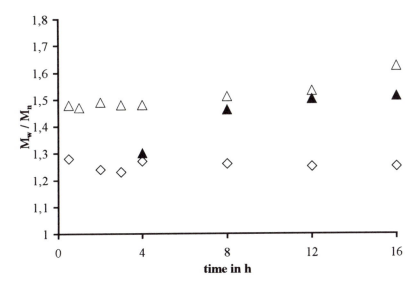

Figure 8. Plot of polydispersity index (M_w/M_n) vs. time for the copolymerization of equimolar amounts of MMA or MA with St with the alkoxyamin initiator HST. Polymerization conditions: (▲) [MMA +St]$_0$ / [I]$_0$ = 40, T = 110 °C; (Δ) [MMA +St]$_0$/ [I]$_0$ = 40, T = 130 °C; (◇) [MA +St]$_0$/ [I]$_0$ = 40, T = 130 °C.

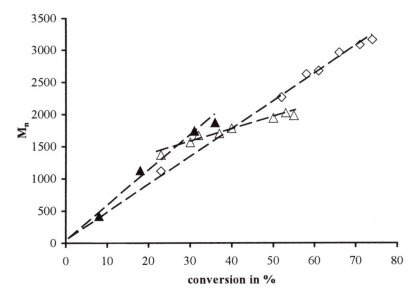

Figure 9. Plot of number average molecular weight M_n vs. conversion for the copolymerization of equimolar amounts of MMA or MA with St with the alkoxyamin initiator HST. Polymerization conditions: (▲) [MMA +St]$_0$/ [I]$_0$ = 40, T = 110 °C; (Δ) [MMA +St]$_0$/ [I]$_0$ = 40, T = 130 °C; (◇) [MA +St]$_0$/ [I]$_0$ = 40, T = 130 °C.

The plot of M_n vs. conversion for the copolymerization of MMA with St (Figure 9) performed at 130 °C reveals a linear dependence, however, the values are not in good agreement with the calculated values indicating that more TEMPO is produced by side reactions than radicals are produced by thermal self initiation. Upon decreasing the temperature to 110 °C the plot of M_n vs. conversion results in a linear dependence close to the calculated one. Obviously, the side reactions leading to dead chain ends and TEMPO in excess show a strong temperature dependence. For the MA/St system a linear plot of M_n vs. conversion is observed even at 130 °C.

Conclusions

The homopolymerization of styrene and methyl acrylate and the copolymerization of methyl acrylate with styrene mediated by TEMPO can be performed in a controlled way up to conversions of 50 % to 90 % depending on the monomer feed. Homopolymerization of methyl methacrylate under the same conditions is an uncontrolled process due to side reactions. The reaction control in the copolymerization of methyl methacrylate with styrene depends on the monomer composition in the feed. At molar fractions of MMA $\leq$ 0.5 a controlled copolymerization is obtained. Parameters like polymerization temperature and conversion are crucial for the reaction control.

References

1. (a) Solomon, D.H.; Rizzardo, E.; Cacioli, P., US patent 4,581,429 (1986) (*Chem. Abstr.* **1986,** *102*, P221335g) (b) Rizzardo, E., *Chem. Aust.* **1987**, *54*, 32. (c) Johnson, C.H.L.; Moad, G.; Solomon, D.H.; Spurling, T.; Uearing, D., *J. Aust. Chem.* **1990**, *43*, 1215. (d) Moad, G.; Rizzardo, E., *Macromolecules* **1995**, *28*, 8722.

2. (a) Georges, M.K.; Veregin, R.P.N.; Kazmaier, P.M.; Hamer, G.K., *Macromolecules* **1993**, *26*, 2987. (b) Georges, M.K.; Veregin, R.P.N.; Kazmaier, P.M.; Hamer, G.K., *Trends Polym. Sci.* **1994**, *2*, 66. (c) Keroshkerian, B.; Georges, M.K.; Boils-Boissier, D., *Macromolecules* **1995**, *28*, 6381.

3. (a) Gaynor, S.; Greszta, D.; Mardare, D.; Teodorescu, M.; Matyjaszewski, K., *J. Macromol. Sci., Pure Appl. Chem.* **1994**, *31*, 1561. (b) Shigemoto, T.; Matyjaszewski, K., *Macromol. Rapid. Commun.* **1996**, *17*, 347. (c) Matyjaszewski, K.; Shigemoto, T.; Frechet, J.M.J.; Leduc, M., *Macromolecules* **1996**, *29*, 4167.

4. (a) Hawker, C.J., *J. Am. Chem. Soc.* **1994**, *116*, 11185. (b) Hawker, C.J.; Hedrick, J.L., *Macromolecules* **1995**, *28*, 2993. (c) Hawker, C.J.; Elce, E.; Dao,

J.; Volksen, W.; Russel, T.P.; Barclay, G.G., *Macromolecules* **1996**, *29*, 2686. (d) Benoit, D.; Harth, E.; Fox, P.; Waymouth, R.M.; Hawker, C.J., *Macromolecules* **2000**, *33*, 363.

5. (a) Fukuda, T.; Terauchi, T.; Goto, A.; Tsuji, Y.; Miyamoto, T., *Macromolecules* **1996**, *29*, 3050. (b) Fukuda, T.; Terauchi, T.; Goto, A.; Ohno, K.; Tsuji, Y.; Miyamoto, T., *Macromolecules* **1996**, *29*, 6393. (c) Goto, A.; Fukuda, T., *Macromolecules* **1999**, *32*, 618.

6. (a) Matyjaszewski, K.; Coca, S.; Gaynor, S.G.; Wie, M.; Woodworth, B.E., *Macromolecules* **1997**, *30*, 7348. (b) Matyjaszewski, K., *Macromol. Symp.* **1996**, *111*, 47. (c) Matyjaszewski, K.; Woodworth, B.E.; Zhang, X.; Gaynor, S.G.; Metzner, Z., *Macromolecules* **1998**, *31*, 5955. (d) Bon, S.A.F. Ph. D. *thesis work*, TU Eindhoven **1998**. (e) Bon, S.A.F.; Chambard, G.; German, A.L., *Macromolecules* **1999**, *32*, 8269.

7. (a) Hawker, C.J.; Barclay, G.G.; Orellana, A.; Dao, J.; Devonport, W., *Macromolecules* **1996**, *29*, 5245. (b) Benoit, D.; Chaplinski, V.; Braslau, R.; Hawker, C.J., *J. Am. Chem. Soc.* **1999**, *121*, 3904. (c) Braslau, R.; Burill, L.C.; Siano, M.; Naik, N.; Howden, R.K.; Mahal, L.K., *Macromolecules* **1997**, *30*, 6445.

8. (a) Bergbreiter, D.E.; Walchuk, B., *Macromolecules* **1998**, *31*, 6380. (b) Miura, Y.; Hirota, K.; Moto, H.; Yamada, B., *Macromolecules* **1992**, *32*, 8356.

9. (a) Hawker, C.J., *Angew. Chem.* **1995**, *13*, 107. (b) Huseman, M.; Malmström, E.E.; McNamara, M.; Mate, M.; Mecerreyes, D.; Benoit, D.G.; Hedrick, J.L.; Mansky, P.; Huang, E.; Russel, T.P.; Hawker, C.J., *Macromolecules* **1999**, *32*, 1424.

10. Kazmaier, P.M.; Daimon, K.; Georges, M.K.; Hamer, G.K.; Veregin, R.P.N., *Macromolecules* **1997**, *30*, 2228.

11. Kazmaier, P.M.; Moffat, K.A.; Georges, M.K.; Veregin, R.P.N.; Hamer, G.K.; *Macromolecules* **1995**, *28*, 1841.

12. Han, C.H.; Drache, M.; Baethge, H.; Schmidt-Naake, G., *Macromol. Chem. Phys.* **1999**, *200*, 1779.

13. Moffat, K.A.; Hamer, G.K.; Georges, M.K., *Macromolecules* **1999**, *32*, 1004.

14. Ohno, K.; Tsuji, J.; Fukuda, T., *Macromolecules* **1997**, *30*, 2503.

15. Li, I.; Howell, B.A.; Matyjaszewski, K.; Shigemoto, T.; Smith, P.B.; Priddy, D.B., *Macromolecules* **1995**, *23*, 6692.

16. (a) Matyjaszewski, K., *Polymer Preprints* **1996**, *37*, 325. (b) Matyjaszewski, K.; Lin, C.H., *Macromol. Chem. Macromol. Symp.* **1991**, *47*, 221.

17. (a) Fischer, H., *J. Am. Chem. Soc.* **1986**, *108*, 3925. (b) Fischer, H., *Macromolecules* **1997**, *30*, 5666. (c) Kothe, T.; Marque, S.; Martschke, R.; Popov, M.; Fischer, H., *J. Chem. Soc. Perkin. Trans.* **1998**, *2*, 1553. (d) Shipp, D.A.; Matyjaszewski, K., *Macromolecules* **1999**, *32*, 2948.

18. Goto, A.; Fukuda, T., *Macromolecules* **1999**, *32*, 618.

19. Veregin, R.P.N.; Odell, G.; Michalak, L.M.; Georges, M.K., *Macromolecules* **1996** *29*, 4161.

20. (a) Greszta, D.; Matyjaszewski, K., *Macromolecules* **1996**, *29*, 7661. (b) Fukuda, T.; Terauchi, T.; Goto, A.; Ohno, K.; Tsuji, Y.; Miyamoto, T., *Macromolecules* **1996**, *29*, 6393.

21. Achten, D., Ph. D. *thesis work*, RWTH Aachen **1999**.

22. (a) Matheson, M.S.; Auer, E.E.; Bevilacqua, E.B.; Hart, E.J., *J. Am. Chem. Soc.* **1951**, *73*, 1700. (b) Matheson, M.S.; Auer, E.E.; Bevilacqua, E.B.; Hart, E.J., *J. Am. Chem. Soc.* **1951**, *73*, 5395.

23. (a) Mayo, F.R., *J. Am. Chem. Soc.* **1953**, *75*, 6133. (b) Mayo, F.R., *J. Am. Chem. Soc.* **1968**, *90*, 1289.

24. Elias H. G. *Makromoleküle*, Vol. 1, Hüthig & Wepf Verlag Basel-Heidelberg-New York **1990**, *5th* edition, p. 523.

25. *Kunststoff Handbuch*, Carl Hansen Verlag München, **1969**; Vol. *5*, p. 599.

26. (a) Kazmaier, P.M.; Daimon, K.; Georges, M.K.; Hamer, G.K.; Veregin, R.P.N., *Macromolecules* **1997**, *30*, 2228. (b) Butz, S.; Baethge, H.; Schmidt-Naake, G., *Macromol. Rapid Commun.* **1997**, *18*, 1049.

27. Roos, S.G.; Müller, A.H.E.; Matyjaszewski, K., *Macromolecules* **1999**, *32*, 8331.

Acknowledgement

Financial support of Bayer AG and Fonds der Chemischen Industrie is acknowledged.

Chapter 30

Potential Impact of Controlled Radical Polymerization on Markets for Polymeric Materials

James Spanswick[1], Elizabeth A. Branstetter[2], and William F. Huber, Jr.[3]

[1]Bridges Technology Innovation Capitalization,
2365 Albright, Wheaton, IL 60187
[2]Creative Marketing Technology, 515A Jefferson, St. Charles, MO 63301
[3]ChemPerspectives, 6S231 Marblehead Court, Naperville, IL 60540

An examination of the patent literature has indicated that industrial research targeting markets for large volume specialty materials is becoming more focused on structures possessing increased levels of functionality.   This   research   effort   reflects   incremental improvements in "standard" commercial polymerization processes, extension of living anionic polymerization process to a broader range of monomers, and evaluation or development, of new controlled polymerization processes.   In this paper we address whether specific products targeted at a number of markets, could be produced through controlled radical polymerization processes (CRP).   The critical commercialization questions that have to be addressed by a product manager seeking to increase market share by introducing materials prepared by CRP, after confirmation of utility, are manufacturing feasibility and cost. These issues will be addressed by examining the cost for production of a model material for graft copolymerization from a commodity polymer based macroinitiator, a self-plasticized PVC.

There has been an explosion of papers on controlled radical polymerization (CRP) in the past six years which have cumulatively presented the material scientist with the tools required to control, or direct the topology and predetermine the functionality of polymeric materials made from a wide range of readily available low cost monomers *(1)*. Functionality is introduced through the use of functional initiators for (co)polymerization of a wide range of monomers, including monomers with functional groups or masked functional groups.   Past work has focused on developing an understanding of the chemistry of the polymerization processes; identification of new catalyst systems in atom transfer radical polymerization (ATRP) *(2-6)*, and extension of the range of suitable stable free radicals for nitroxide mediated polymerization (NMP) *(7-9)*. Reversible addition-fragmentation chain transfer (RAFT) (10-13), another CRP process that shows an ability to polymerize some free radically polymerizable monomers not yet controlled by the other systems has now joined these processes. Both ATRP and NMP now allow an increased range of monomers to be (co)polymerized and other authors at this symposium have defined and report on the conditions required to modify the processes to prepare linear copolymers, gradient, block, and graft copolymers, comb shaped and dendritic materials in organic, aqueous or in other inorganic systems.

We believe these processes are presently at a state that should allow evaluation, from an industrial or commercialization standpoint, of their potential impact in the marketplace for larger volume functional materials. The words function and functionality are used throughout this paper and mean different things when applied to different parts of the polymerization process/product/market commercialization chain. When applied to the preparation of materials it indicates chemical functionality and reactive functional groups on the polymer chain; when applied to products the words refers to the reactivity of the material with added components during fabrication processes; and in evaluation of markets it indicates that the material responds to and fulfills certain service requirements.

We have included the following assumptions in our thinking.

(1) The business units of large chemical corporations do believe their mission statements and are sincerely looking at ways they create value for themselves and their customers.
(2) The customers want products that possess a range of functions, not specific materials.
(3) There is a long term interest in developing sustainable technologies; which means the efficient use of materials, capital and energy.

Discussion

An article that appeared at the time of the meeting on Controlled Polymerization in New Orleans clarified, for one of the authors, one reason behind the changing nature of the materials marketplace. As we collected data on multifunctional products it became apparent that there was an increased level of materials research targeted at a range of markets with well defined service requirements. This increased level of choice initially appeared, at least to a chemist once based in a commodity chemical corporation, to indicate a fragmentation of markets, or over specialization of materials developed to meet the requirements of specific customers in specific applications, that had to result in an unacceptable increase in cost of production, or a decrease in return on investment.

The article(14) discussed W. Michael Cox's solution to the Solow Paradox. It was suggested that the age of mass customization was arriving, and that the impact of information technologies can be seen in a fundamental change in the method of calculation of production costs. Current production methods can result in both low fixed costs and low marginal costs, allowing marketing to target diverse consumers with tailored products. Essentially, technology is allowing close interaction between producer or supplier, and customer or consumer, resulting in an overall increase in choice and a specialization or fragmentation of markets as preferred suppliers uniquely meet specific consumer desires.

This paper will attempt to address the increased intermaterial competition in polymeric materials markets and identify some examples where control over specific material properties can better be achieved by use of the tools developed for controlled radical polymerization; i.e. answering some of the above questions in the positive.

In addition to meeting the specific requirements of large volume applications for functional materials, the markets underlined in Table 1; the expanded level of control over initiation and propagation presented by controlled free radical polymerization afforded by access to an unlimited range of tailored initiators, ranging from small functional molecules through macromolecules to solid substrates, will allow targeting of many high value applications. These high value applications would include those markets not underlined in Table 1 in addition to nonlinear optical materials, materials for microfabrication, nanomaterials, and biocompatible medical implants. Products for these markets will be developed through application of controlled polymerization process and become significant producers of value for the participants, but will not be addressed in this paper.

Table 1 Markets for functional materials

Coating Maretials	Repulpable materials
Elastomers	Alloying Components
Polyurethane Intermediates	Electronics
Epoxies	Drug delivery
Flexible PVC	Composites
Personal Care Products	Implantable material
Bulk Polymers	Separation Materials
Water treatment	Additives
Lubricants	Adhesives

Methodology

There is a choice of CRP processes available, each with some positive features and some perceived drawbacks *(1,15)*. There has also been an expected increase in research in currently commercial technologies in response to this potential technical threat. Specific examples of industrial research directed at producing materials for a number of large markets were taken from the patent literature. If one accepts that these materials, prepared by other technologies, are targeted at valid specific markets, and meet the evolving requirements of the applications, we will argue that CRP can better meet the market demands for performance materials at lower cost. Our arguments will be based on application of technology already demonstrated in CRP processes and will not incorporate any speculative improvements.

CRP processes bring a number of tools to the technologist seeking the development of materials incorporating various functions in one product. Polymerization of a broad range of (co)monomers to prepare molecularly engineered materials exhibiting control over composition, functionality, topology and ultimately rheology and fabricability. It is also possible to prepare homo-, "homo-" and heterotelechelics for subsequent chain extension reactions forming polyurethanes, polyamides, polyesters and epoxies. The term "homo-" is used to describe a polymer with the same functional group at each polymer terminus but that specific group is attached to atoms bearing different substituents. The benefit for the customer is that these "same" functional groups can react at different rates with one curing agent giving controlled response to external stimuli.

Coating Materials

There are over 30 corporation actively patenting in this field in 1999 and the claims in the patents focus on meeting the requirements of specific applications indicating that materials research for the coatings market is been driven by a number of persistent performance targets. These include a desire to reduce VOC's, resulting in a move towards powder or water based coatings. Development of improved multifunctional pigment dispersion aids to reduce the concentration of high priced pigments in the final product. A desire for high gloss clear coats for automobile markets and controlled reactivity to allow reaction under known external stimuli. These targets have been claimed to have been accomplished through end group functionality, or through incorporation of specific functional comonomers leading to in-chain functionality, resulting in improved chain extension, improved adhesion and improved interaction with dispersed solids. There has also been an increase in the number of materials we are describing as core/shell polymers with hermaphrophylicities.

Two costly solutions to the conflicting balance of material properties required for powder coatings are disclosed in *EP 0 773 267 A1 (16)*. This application describes the advantages polymers with known Tg's, and narrow molecular weight distribution possess as film forming polymers when used as a powder coating material. The polymer must possesses functionality suitable for chain extension through use of a curing agent. The controlled composition and viscosity of the coating composition reduces the likelihood for blocking or agglomeration, and produces a high gloss film.

The materials described in the application were produced by group transfer polymerization, or by fractionation of a free radically polymerized copolymer, to produce the desired polymer with narrow molecular weight distribution. We would contend that the more robust polymerization capabilities of CRP processes allow a broader range of materials to be prepared than with GTP. In addition direct synthesis must be able to improve on the cost of large scale polymer fractionation. The products of CRP are much more homogeneous that standard free radical copolymerization with compositional variation seen along the polymer chain as opposed to among the polymer chains. The ability to directly prepare materials with polar substituents, in contrast to ionic polymerization processes, also increases the choice of (co)monomers available to the producer. A move to CRP *(1-13)* should allow production of a lower cost product with improved response to the requirements of the application and materials with better film forming properties will be appearing in the marketplace.

The reputed additional cost for the production of controlled rheology, functional, coating polymers can be offset by the use of less materials, better compounding characteristics, and the tailored functionality can provide improved interaction with color agents, curing compounds and substrates.

Other examples of a desire for controlled macrofunctionality in this market are described in US 5,859,113 (17); WO 99/11681(18); US 5,869,590 (19); and US 5,872,189 (20).

Coatings are not limited to paints but are desired as surface property modifiers in fibers, films and molded articles. An example from *WO 99/05345 (21)*, employs a high temperature-stable flurochemical block copolymers as hydrophobic and oleophobic additives to synthetic organic polymers. In such an application, partial solubility of the block copolymer in the bulk material is desired while migration of the non-compatible block to the surface for surface property modification is to be encouraged. Control over the components of the copolymer and interface is desired.

US 5,789,516 (22) utilizes a copper based redox initiated free radical polymerization process with a silicone based macroinitiator. The question, that we would answer in the positive, is whether the properties can be improved if one has complete control over the composition of all blocks, or grafts, as would be obtained by an ATRP polymerization in the second stage *(5)*.

We would contend that materials prepared by CRP processes, where complete control over all aspects of the polymerization process can be exercised, would truly possess the properties desired in these applications. However if CRP processes are to make inroads into the production of materials for any application discussed in this paper then a number of economic issues have to be addressed.

Broad Economic Issues

The first economic issue that would be addressed by a business unit is whether existing equipment can be used or whether a new plant is required. This will be partially addressed in the discussion on the cost of producing a graft PVC, but the individual answers are dependent on the production capabilities and core manufacturing competencies of the corporation.

Other concerns that have been expressed are initiator cost and availability; the cost for catalyst removal/catalyst recycle for ATRP, or in the case of the other CRP processes the cost of end group stabilization and/or functionalization.

The cost of initiators for ATRP was addressed and resolved by calling potential suppliers of bromo-initiators, the most expensive initiators. Albemarle responded and indicated that a range of small molecule initiators could be supplied at any commercial quantity for $3.50-4.00/lb., and if required macroinitiators could be made available.

Catalyst availability should not solely be a catalyst composition driven exercise, since metal salts in the most stable oxidation state can be added to the reaction and the oxidation state of the catalyst adjusted in-situ by addition of metal zero complexes. This both reduces catalyst cost and increases manufacturing flexibility. In addition supported catalysts are feasible with the transition metal complex supported through the ligand *(4, 23)* or counterion *(5)*. Catalyst removal/recycle is practiced in many large scale industrial processes, such as PTA, and for ATRP a number of separation concepts have been demonstrated. Ligand exchange allows solvent extraction; counterion exchange allows removal on an ion exchange resin; counterion replacement allows precipitation and supported catalysts can be used in continuous processes.

432

In the case of NMP the range of suitable alkoxyamines has been expanded to allow the polymerization of monomers other than substituted styrenes *(7-9)*. The chain transfer agents for RAFT are being developed by corporations from readily available feedstocks *(10-13)*.

Some of the economic questions facing commercial implementation of a CRP process can be addressed by examining the cost of producing a PVC graft copolymer. The analysis should be viewed as a model for the production of graft copolymer from a commercially available macroinitiators *(2-6)* and should not solely be considered as a preliminary cost analysis for one specific material.

Self-plasticized Graft PVC

The market for coating materials is being addressed by a number of technical approaches to new material development and we have indicated that use of CRP can offer advantages for the preparation of functional materials. However a different analysis was run for the flexible PVC market. The market is being assailed by a number of competing materials in addition to sustaining a continuing onslaught from groups concerned about plasticizer migration. The attraction in replacing PVC, even in specific segments of the market lies in the size of the market, 7.0 billion-pound in the US alone. Some of the high profile applications are automotive, medical bags, toys, and poultry packaging. Competitive products being introduced are tailored polyolefins based on metallocene catalysts and, multilayer films based on polyethylene and ethylene vinyl acetate. Presently PVC is holding its own based on low cost and known performance. This inertia is based on several property advantages PVC has over the polyolefin competitors including its processability, weatherability, flame retardancy and electrical properties, with cost a prime factor.

The concept evaluated is production of a self-plasticized graft PVC by a two step suspension copolymerization process. This economic evaluation is based on a product initially produced when exploring the capabilities of controlled polymerization *(4)*. The process involves copolymerization of vinyl chloride with vinyl chloroacetate in a standard peroxide initiated VCM suspension polymerization producing a copolymer that can act as a macroinitiator for an ATRP polymerization. Without isolation of the product, a second vinyl monomer, butyl acrylate is added to the suspension polymerization swelling the PVC particle and is grafted from the macroinitiator in a core/shell type of copolymerization using a copper based ATRP catalyst.

This particular study is a model for grafting from commodity polymers and should also shed some light on the cost of using existing polymerization equipment and commercially available macroinitiators for controlled polymerization *(2-6)*.

Process Evaluation for Production of PVC Graft Copolymer

The process evaluated was a two stage process with the first stage run under normal VCM suspension polymerization conditions. There is additional cost due to the addition of the functional comonomer; a resulting penalty for a 10% slower rate of polymerization and the added cost of stripping VCM from the reaction. This latter cost was calculated as a normal stripping operation as the product from the first stage copolymerization was transferred to a second reactor for suspension graft copolymerization of butyl acrylate onto the PVC particles using the copper based catalyst system. This second stage was run to 80+% conversion and the monomer stripped in a full sized stripper prior to separation and drying. A simplified process flow diagram is shown in Figure 1.

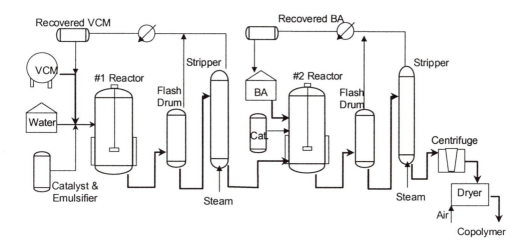

Figure 1. Simplified two stage process flow diagram.

Capital costs were increased over the PVC base case [a 120 mta plant was used for the analysis] by adding an additional flash tank, stripper and a feed tank for the butyl acrylate. The additional $8MM, or 12% of a new plant cost, resulted in an increase in capital costs of 0.7 ¢/lb. Utilities, mainly steam load, increase by 50% and results in a 0.4 ¢/lb. increase in costs. Fixed costs also increase by 0.4 ¢/lb. due to the addition of two operators per shift and additional maintenance expense for the added equipment. Additional initiator and catalyst costs add 0.7 ¢/lb. The catalyst was a copper based catalyst and we used the demonstrated concept *(4)* of mixed ligands to partition the catalyst between the polymerization phase and the suspending media. One of the ligands being removed during the stripping operation pulls the complex into the aqueous phase for separation and recycle.

On a new plant basis this results in a Base Resin cost of 31.9¢/lb. for conventional unplasticized PVC and 49.1 ¢/lb. for Self Plasticized PVC. In the second half of Table 2 we add the costs of converting the Base Resin to Plasticized PVC, and finally, SAR and Transportation costs. Plasticizer is added to the Base Resin in a 1:1 ratio and compounded in an extruder for a 2.5 ¢/lb. processing fee.

The key assumption here is that the cost of plasticizer is the same as the cost of comonomer. In this case our Self Plasticized product cost 55.1 ¢/lb. compared to 50.5 ¢/lb. for the conventional product.

Table 2. Comparative Economics on 120 mta Final Product basis

Product Cost ¢/lb.		
	Conventional PVC	Self Plasticized 50:50
Monomer (VCM)	22	½ lb. @ 22/lb. or 11
Co-monomer		½ lb. @ 52/lb. or 26
Catalyst/additives	.6	1.3
Utilities	0.8	1.2
Plant Fixed	2.4	2.8
Man. Cash Costs	25.9	42.3
Capital Charge	6.1	6.8
Total Cost Base Resin	31.9	49.1
	Plasticized Product	
Base PVC (0.50 x 31.9)	16.0	49.1
Plasticizer (0.50 x 52)	26	0
Compounding Charge	2.5	0
Pkg. & Trans.	3.0	3.0
SAR	3.0	3.0
Delivered Cost of Final Product	50.5	55.1
PVC Plant Investment, $ million	73 (ex compounding)	81

August 1999 Price Comparison

Even with this increased cost penalty for the self-plasticized graft PVC the cost of production is competitive with plasticized PVC if one considers PVC pricing in August of 1999. Our calculated PVC production cash cost of 25.9 ¢/lb. and selling price of 31.9 ¢/lb. compares well with a reported manufacturing cost of 25-30 ¢/lb. and selling price of 35-38 ¢/lb. Compounded plasticized PVC selling price was 50-60 ¢/lb. bracketing our projected selling price for the graft copolymer of 55.1 ¢/lb. [assuming the same cost for plasticizer and butyl acrylate at equivalent volumes.] If one uses the reported cost of butyl acrylate in August 1999 then the selling price is 62 ¢/lb., still below the reported selling price for tailored polyolefins and for specialty PVC grades.

If one envisions that the producers of PVC plasticizers may want to participate in a movement to a self plasticized PVC then a second concept graft PVC could be considered. This would be the use the longer chain alkyl acrylates as graft comonomer, i.e. use of the same long chain alcohols currently employed in the plasticizer to esterify acrylic acid instead of phthalic acid. This would produce a lower Tg product at equivalent acrylate loading. It would also produce a material that could be employed as a surfactant for the present product to reduce migration of plasticizer. Long chain alkyl acrylates have been polymerized by ATRP. This concept product has a projected selling price, within the variability of the cost parameters of the analysis, comparable to commercially available plasticized PVC.

The above analysis of the property targets for new materials in markets for coating materials and the scoping economic evaluation on the cost of producing a graft copolymer using a ATRP catalyst in a suspension process indicates that CRP is a viable approach to meeting the evolving material requirements of a range of applications. The remaining section of the paper will return to an abreviated material property target analysis and allow the reader to decide how the additional control over the functionality and topology of polymers attainable through CRP can impact the development of products for a number of high volume specialty material markets.

Polyurethane Intermediates

The materials in application *WO 98/53013 (24)* were developed as coating materials, but what would be the value of controlled hydrophylicity in slab polyurethane for seating and bedding markets?

An indication of one effort to introduce controlled functionality into a polyurethane precursor is given in *US 5,863,959 (25)*. This patent teaches that incorporation of hydroxy functionality, or amine functionality, into a low viscosity graft polymer dispersion improves the performance of the product. Introduction of such functionality into telechelic polymers has been demonstrated for CRP *(2,3)*.

Epoxies

Similar controlled functional materials are desired for use in epoxy resin manufacture as described in *WO 98/50477 (26)*. The application emphasizes the desire for low viscosity in the components and utilizes telechelics that can readily be prepared by CRP. In *WO 99/16838 (27)*, glycidyl containing acrylic polymers are cured by reaction with acid functional cross linking agents. The questions to be addressed by a business unit is whether the improved functionality, controlled reactivity and improved rheology of materials prepared by CRP that allow unperturbed reactive compounding can provide materials to penetrate a new application, increase market presence, or are required to retain an existing market share.

Personal Care Products

The patent activity of the personal care market demonstrates strong interest in multifunctional materials. There are many products being developed for this market that require control over functionality, polymer composition, and topology to reconcile the different functions the material has to meet. Materials considered in this market are directed towards meeting the requirements of contact lenses, surface modifiers for hair care, surfactants, cleansers and pigment dispersants for cosmetics.

Examples include *WO 98/51261 (28)* and *WO 98/51722(29)* which describe hydrophilic/hydrophobic graft copolymers made by controlled polymerization.
WO 98/50443(30) and *US 5,891,932(31)* describe materials for contact lenses and an ionic hydrophilic copolymer is described in *WO 99/16812(32)*. *WO 99/14295(33)* describes a laundry detergent composition with cellulose based polymers. Cellulose, or indeed any multihydroxy functional material, can readily be modified to form a macroinitiator for CRP and the grafts or arms can be tailored to meet the additional requirements of the final material. *(4,23)*

Lubricants

Lubricants are an extremely large market for functional materials. An example of a functional material tailored for this market is given in *US 5,874,389(34)* where grafting of the desired polar nitrogen containing monomers is attempted through use of an unsaturated "macromonomer" in a standard free radical polymerization in an oil solution. When a macromonomer is employed in a controlled radical copolymerization there is a significant improvement in the homogeneity of the product (35). Any compositional gradient is seen as a change in composition along the polymer chain as opposed to among the polymer chains in a standard polymerization process. This is very important at the beginning of the polymerization process where there can be almost instantaneous incompatibility between the polymer being produced in a standard free radical polymerization and the macromonomer, essentially precluding further incorporation of the macromonomer into the copolymer. This is avoided in a controlled polymerization since the low molecular weight copolymer acts as a surfactant to compatibilize the macromonomer and graft copolymer.

Additives

Another large volume application with well defined quantifiable product requirements is pour point depressants. In *US 5,834,408(36)* the materials are prepared by via anionic polymerization of (meth)acrylic monomers. Can a CRP process produce such a product at a lower cost than a living anionic process and will the additional control over comonomer incorporation into the polymer seen from a radical process can yield an improved product? The approach taught in *US 5,744,523(37)* is use of polyacrylate esters as dispersants. The polyacrylate backbone was prepared by a standard free radical polymerization process and transesterified in a second step with long chain aliphatic alcohols and dialkylaminoalkanols. The product was claimed to be useful as a dispersant for finely divided solids such as fillers and pigments in organic media. Similar materials can be prepared directly by CRP and such products would have improved functionality, controlled rheological properties and improved cocrystalization characteristics*(38)*.

Adhesives

Materials for the adhesives market are continuously being tailored for various specific applications and *US 5,880,217(39)* describes the use of functional polymers prepared by anionic polymerization. Other applications targeted by anionically polymerized materials are polar block copolymers and telefunctional stars. The increased capabilities of CRP processes and the ease with which "homo" and hetero-telechelic materials, polar blocks and functional stars of many topologies are prepared indicate that materials for this type of application, (materials based on tailored curing characteristic), can readily be prepared.

Repulpable Materials

Application *WO 99/11670 (40)* discloses sugar based vinyl monomers and copolymers useful in repulpable adhesives and other applications. CRP products will offer advantages in compounding the materials with additives required for sizing agents and toners, and controlled rheology would offer improved performance in paper board and wood gluing applications over materials produced by standard copolymerization methods. We propose that materials with similar behavior can be prepared by use of sugar based macroinitiators *(15)* and that the resulting star or graft copolymers can be tailored to meet the performance requirement of the application.

Other applications using cellulosic and starch based components are discussed in US *5,854,321 (41); WO 99/14295 (33) and WO 99/25756 (42)* . The later application discusses biodegradable thermoplastics and we believe that with CRP processes one can tailor the susceptibility of the "organic" core to a range of degradation reactions.

Hyperbranched Polymers

Many processes have been developed for the controlled preparation of hyperbranched materials but there is a drive to prepare similar materials by one step processes. One such approach is described in *WO 99/07754 (43)*. The process employed is the free radical copolymerization of mono and multi-unsaturated monomers. Greater control over the polymerization would be gained by controlled (co)polymerization whereby the product would exhibit controlled rheology and where desired, contain functional end groups.

Alloying Components

Acrylonitrile/Styrene/Acrylate blends seek to provide materials that have weatherability and impact resistance. Application *WO 99/15589 (44)* claims to provide a physical blend of a styrene/acrylonitrile copolymer matrix with a dispersed crosslinked acrylate/(meth)acrylate copolymer by forming a substantially uniform blend of the copolymers. Another approach is taught in *US 5,891,962 (45)*.

CRP can offer improved properties in each component in blends and improve the overall properties attainable by the blend. The improvements result from full control of the composition of the copolymers *along* the polymer chain as opposed to *among* the polymer chains, allowing for a wider window for matched refractive indices between the components yielding clarity over a greater temperature range.

438

Summary Comparison of Controlled Polymerization Processes

The cost of low molecular weight bromo-initiators for ATRP is presently $3.50-4.00/lb in commercial quantities with the cost of chloro-initiators significantly lower. The initiators can be easily chosen/tailored for the application and can be used to (co)polymerize a wide range of monomers. While progress is being made in NMP polymerization the projected lowest attainable cost of alkoxyamines are $7.00 – 10.00/lb and are presently limited to polymerization of one type of monomer.

With ATRP functionality can be greater than 95% and there is almost unlimited choice of functionality for every polymer chain end. NMP overall functionality is lower, and the active end group remains thermally labile and presently there is less freedom to modify end groups to introduce desired alternate functionality.

The questions to be answered by a producer who contemplates using ATRP is defining the optimum conditions for the polymerization considering the multiple of adjustments available within the components of the catalyst and polymerization media; indeed many "failed" ATRP polymerizations reported in the literature "fail" due to a limited understanding of the role of all components of the system and the need to tailor the catalyst to the specific polymerization. Removal of the catalyst remains an expressed concern. The best route to convert the halide into another functional end group will depend on the specific end use.

For NMP the question is whether an alkoxyamine is available to carry out the desired (co)polymerization, whether the level and type of functionality allowed by use of that specific alkoxyamine is adequate for the application and whether the higher initiator cost overcomes the concerns residing in recovery/recycle of an ATRP catalyst.

For RAFT and other chain transfer methods the questions are similar to those of NMP with a reputed additional concern of odor and end group transformability.

One additional critical question that has to be addressed before a complete answer to the potential impact of CRP on the market for functional materials is whether the technology is available to corporations interested in developing materials for specific applications. At the present time the fundamental patents for ATRP and some additional patents for specific aspects of NMP are available from University sources.

Conclusions

The majority of the technologies products above, indicate an increased level of responsiveness of the producer to market demands for polymer based materials with a greater range of functionality in a single material. We believe that markets will continue to demand improved products with a greater ability to perform multiple tasks in a single material. The inertia factor, or resistance to incorporate a new technology into material production is evident to some degree in the effort corporations are expending on incremental improvements to existing processes and products. Improvements include the preparation of statistical copolymers by anionic polymerization processes and a desire for increased topological control for specific applications. The increased use of metallocene catalysts in olefin polymerization is allowing the preparation of polyolefins with limited backbone functionality. Standard free radical polymerization processes are being modified to introduce limited control over copolymer composition. These incremental improvements in functional control in materials are being driven by property deficiencies in existing materials, the processing characteristics of the materials, and a desire expressed by customers for lower cost products with a choice of functionality in the building blocks for down stream material applications. The performance characteristics inherent in the products of CRP can more efficiently exploit these deficiencies in current materials.

The continued expansion of understanding of the CRP processes will allow industry to produce materials by controlled polymerization that will compete in the large volume markets discussed in this paper. Control over macrofunctionality will allow the preparation of materials with any combination of phylicities. Control over polymer end group functionality will produce materials that can respond to a variety of external stimuli. Many examples have already been prepared. Building blocks for step growth condensation polymers or crosslinkable systems can be constructed with a range of telechelic and in chain functional groups allowing permanent or degradable linking groups to be used in development of the final product. A degradable linking group can allow recovery of the telechelic building block in recycle operations. Polymers produced by any other polymerization process can be incorporated into CRP processes as initiators or macromonomers, this includes the use of natural products in material synthesis. The multiple functionality capable of being incorporated into CRP products allows for an active interaction of the matrix with fillers, substrates and control of interfaces. The gradual build up of molecular weight in a CRP allows an increased ability to prepare graft copolymers by grafting through macromonomers resulting in the production of more homogeneous graft copolymers.

We do not believe that CRP is likely to produce the next commodity polymer nor do we do believe that it is going to be restricted to the preparation of specialized high value products. CRP can and will compete in the preparation of improved functional materials for large volume applications and will provide the foundation for new specialty chemical businesses. Which CRP process is employed by a specific business unit will depend on valid corporate core strengths and whether a narrow range of monomers will be polymerized or whether polymerization of a broad range of comonomers is required for multiple functionality in the material.

Literature Cited

1. *"Controlled Radical Polymerization"* Matyjaszewski, K., Ed.; ACS Symp. Series #685; ACS Publishing: Washington, D.C., 1998.
2. Matyjaszewski, K.; Coca, S.; Gaynor, S. G.; Gretszta, D.; Patten, T. E.; Wang, J-S.; Xia, J.; U.S. 5,807,937 1998
3. Matyjaszewski, K.; Miller, P.J.; Fossum, G.; Nakagawa, Y.; *Appl. Organometal. Chem.* 1999, *12*, 667.
4. Matyjaszewski, K.; Gaynor, S. G.; Coca, S.; WO 98/40414
5. Matyjaszewski, K.; Coca, S.; Gaynor, S. G.; Nakagawa, Y.; Seong M. J.; U.S. 5,789,487 1998
6. Matyjaszewski, K.; Gaynor, S. G.; Paik, H-J.; Pintauer, T.; Qiu, J. Teodorescu, M.; Xai. J.; Zahn, X.; patent application 1999
7. Tordo, P. et al.; Polymer Preprints 1999, 40(2), 313 and 317
8. Hawker, C. et al.; Polymer Preprints 1999, 40(2), 315

440

9. Puts, R. et al.; Polymer Preprints 1999, 40(2), 323
10. Le, T.P.; Moad, G.; Rizzardo E.; Thang, S.H.; WO 98/01478
11. Moad, G.; Rizzardo E.; Thang, S.H.; WO 99/05099
12. Corpart, P.; Charmot, D.; Zard, S.; Franck, X.; Bouhadir, G.; WO 99/36177
13. Bouhadir, G.; Corpart, P.; Charmot, D.; Zard, S.; WO 99/35178
14. *"The economics of panty hose"* Peter Brimelow; Forbes; August 23, 1999, p 70
15. Matyjaszewski, K.; *PMSE Preprints.* 1999, *40 (2),* 309.
16. Ohkoshi, T.; et al.; EP 0 773 267 A 1.
17. McIntyre, P. F.; King, J. G.; Spinelli H. J.; Jakubauskas, H. G.; US 5,859,113 1999
18. Drujon, X.; et al.; WO 99/11681;
19. Clark, M. D.; Collins, M. J.; Lopez, P.; Taylor, J. W.; US 5,869,590; 1999
20. Bett, Bill; Richard, J.; US 5,872,189 1999.
21. Klun, T. P.; Gasper, A. J.; Temperante, J. A.; WO 99/05345
22. Graiver, D.; Khieu, A. Q.; Nguyen, B. T.; US 5,789,516, 1999
23. Haddleton, D. M.; et al.; PMSE Preprints. 1999, 40 (2), 147.
24. Vandevoorde, P. M.; et al. WO 98/53013
25. Heyman, D. A.; Gallagher, J. A.; US 5,863,959, 1999.
26. Leibelt, U.; Bohler, P.; WO 98/50477.
27. Dumain, E.; WO 99/16838
28. Midha, S.; Nijakowski, T. R.; WO 98/51261
29. Midha, S.; Nijakowski, T. R.; WO 98/51722
30. Henry, D.; WO 98/50443
31. Benz, P. H.; Ors, J. A.; US 5,891,932, 1999
32. Gentilcore, G.; Bekkaoui, A.; Lancaster, I. M.; WO 99/16812
33. Leupin, J. A.; Gosselink, E. P.; WO 99/14295
34. Boden, F. J.; Sauer, R. P.; Goldblatt, I. L.; McHenry, M. E.; US 5,874,389, 1999
35. Mueller, A. H. E; Roos, S. G.; Schmidt, B; Polymer Preprints 1999, 40(2), 140
36. Mishra, M. K.; Saxton, R. G.; US 5,834,408, 1999
37. Barkowski, M.; Fock, J.; Schaefer, D.; US 5,744,523, 1999
38. Fetters L. J.; Polymer Preprints 1999, 40 (2), 972
39. St. Clair, D. J.; Erickson, J. R.; *US 5,880,217*
40. Bloembergen, S.; McLennan I. J.; Narayan, R.; *WO 99/11670*
41. Krause, F.; Klimmek, H.; US *5,854,321, 1999*
42. Berger, W.; Jeromin, L.; Mierau, U.; Opitz, G.; *WO 99/25756* .
43. Campbell, D. J.; Teymour, F.; *WO 99/07754*
44. Hughes, R. E.; *WO 99/15589*
45. Otsuzuki, S.; et.al *US 5,891,962 1999*

Indexes

Author Index

Subject Index

A

N-(2-Acetoxyethyl)maleimide (AEMI)
charge transfer complex (CTC) formation with donor monomers, 386, 387f
copolymerization with isobutyl vinyl ether and vinyl benzoate, 391–392
copolymerization with styrene, 386–390
See also Alternating copolymers of N-substituted maleimides with styrene
Acrylamides
controlled polymerization by atom transfer radical polymerization (ATRP), 350–351
reversible addition–fragmentation chain transfer (RAFT), 290, 291f
Acrylate graft copolymers
conventional radical copolymerization, 362
See also Copolymerization of n-butyl acrylate with methyl methacrylate and PMMA macromonomers
Acrylates
controlled polymerization by ATRP, 349
controlled polymerization with alkoxyamines, 120
reversible addition–fragmentation chain transfer (RAFT), 288–290
well-defined random copolymers with isoprene, 131
See also Polystyrene/polyacrylate block copolymers
Acrylic acid

controlled polymerization by ATRP, 349–350
IR spectrum of poly(acrylic acid), 350
reversible addition–fragmentation chain transfer (RAFT), 288–290
Acrylic graft copolymers
conventional radical copolymerization, 362
See also Copolymerization of n-butyl acrylate with methyl methacrylate and PMMA macromonomers
Acrylonitrile
controlled polymerization by ATRP, 351
frozen state electron paramagnetic resonance (EPR) spectra of ATRP systems, 79
See also Styrene–acrylonitrile (SAN) copolymers
Activation mechanisms
activation rate constants, 30t
Arrhenius parameters for activation rate constant of alkoxyamines, 32t
examples of living radical polymerizations, 31
living radical polymerization, 28–29
Active species, formation, 4
Acylating agents, accelerating additives of TEMPO mediated polymerization of styrene, 119
Addition–fragmentation chemistry
C–C bond, 10
mechanism for equilibration, 9–10
redundant or degenerate loops, 5–6
Additives
comparison with controlled radical polymerization methods, 23t
patent activity, 436

450

476